The Weather Book

Jack Williams

The Weather Book

Revised and Updated

VINTAGE BOOKS
A DIVISION OF RANDOM HOUSE, INC.
NEW YORK

A Vintage Original, Second Edition
May 1997
Copyright © 1992 by USA TODAY
Copyright © 1997 by Gannett New Media

Library of Congress Cataloging-in-Publication Data
Williams, Jack, 1936 -
The weather book / by Jack Williams. — 2nd ed.
p.   cm.
"USA TODAY."
"A Vintage original" — T.p. verso
Includes index.
ISBN 0-679-77665-6 (pbk).
1. United States — Climate.   2. Weather.
I. USA TODAY (Arlington, Va.) II. Title.
[QC983.W55 1992]
551.6973 — dc20   91-50697
CIP

Random House Web address: http://www.randomhouse.com/

Printed in the United States of America
10 9 8 7 6 5 4 3 2 1

CONTENTS

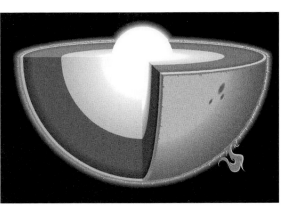

The sun starts our weather. It heats some parts of our world more, some less. And with the Earth's rotation, unequal heating drives our weather. Page 14-15

Know the difference between heavy drizzle and light rain? The size and distance between the drops determine the difference. There are three kinds of drizzle and three kinds of rain. Page 69

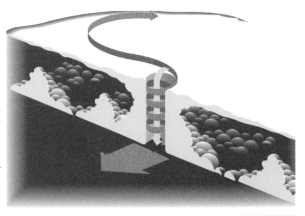

Hurricanes get their energy from water vapor turning to liquid or ice. In one day, a hurricane can release enough energy to power the USA for six months. Page 148-149

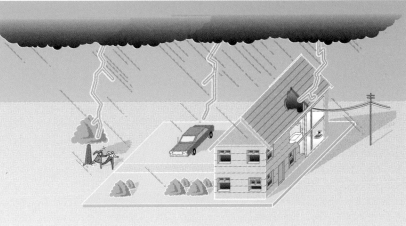

Start counting when you see a lightning flash. If you hear thunder five seconds later, the lightning is a mile away; 10 seconds later, it's two miles away. Page 128

FOREWORD

This book is intended as a guide, an introduction to today's science of weather and climate. It will help you understand the flood of weather information that washes across the USA each day. You can dip into the book at any point or you can start with some science basics in Chapters 1 through 5 before moving on to the chapters that look in more detail at various kinds of weather.

Meteorology is based on mathematics and physics, but you won't find a lot of equations and laws of physics in this book. This doesn't mean that the science of weather is like a simple jigsaw puzzle. The science of weather is an intricate, sometimes exasperating, puzzle with countless pieces that interact and change among themselves.

Meteorology also is a community of men and women, working on different parts of this great puzzle. In this book you will meet some of these people, including many leaders in the field.

In a sense, this book is a report on the atmospheric sciences community as it moves into the 21st century. As the 20th century ends, these scientists are adding a new focus: how man is affecting the very climate of the Earth. The time and patience of many members of the atmospheric sciences community made this book possible. Those who contributed most directly are listed on Pages 223-224.

Working with the atmospheric sciences community and its reception of this book's first edition have been among my life's greatest pleasures. I felt especially honored in 1994 when the American Meteorological Society awarded me its Louis J. Battan Author's Award. The citation describes the book as "a scientifically authentic, easy-to-understand, text with novel visuals."

Special mention should go to Stanley D. Gedzelman of the City College of New York who acted as an adviser for the first edition, suggesting improvements both in the science and in the writing. Much of the credit for scientific accuracy should go to Gedzelman and the other scientists who helped. Any errors are mine.

Ira Geer, director of education programs for the American Meteorological Society, also deserves special mention.

When I began part-time math and science studies at the State University of New York College at Brockport in 1979, Geer's course released a latent interest in meteorology. In 1981 when I began working with prototypes of the USA TODAY Weather Page, Geer had many useful suggestions. He encouraged me to write this book and used it in conjunction with the Society's educational efforts. The AMS Education Office sells color viewgraphs of many of the book's graphics for teachers to use.

Many people at USA TODAY made the book possible, beginning with Peter Prichard, who was the newspaper's editor when we produced the first edition. He gave a talented staff of editors and graphic artists the time and encouragement needed to produce a book rich in scientific detail and graphic imagery. Robert Dubill, the newspaper's executive editor, cleared the way for production of the second edition.

I want to thank Lorraine Cichowski, vice president and general manager of the USA TODAY Information Network, for putting me in charge of weather for USA TODAY's online service. This gave me a chance to become adept at using the World Wide Web for finding weather and climate information, which made updating this book easier.

The book is an outgrowth of USA TODAY's Weather Page and its graphics, which have focused on aspects of the weather each day since the paper's first edition on Sept. 15, 1982.

The book's graphics were conceived by me and all were drawn by USA TODAY's talented artists under the direction of Richard Curtis, USA TODAY's managing editor for graphics and photography. As the overall designer of USA TODAY, he helped conceptualize the newspaper's Weather Page. As the overall manager and designer of this book, he kept all involved directed toward producing a high quality, graphically exciting product. He was also manager and designer for the second edition.

As the book's editor for both editions, Carol Knopes of USA TODAY helped clarify the original, sometimes vague, ideas about the nature of the book and about how topics should be approached. She has a talent for encouraging a writer's best efforts.

Jacqueline Blais of USA TODAY was copy editor for both the first and second editions. I thoroughly appreciate her skills in catching errors both of fact and grammar, making the writing sharper, and asking the questions intelligent readers would have asked. Walt Rykiel and Theresa Barry of USA TODAY assisted with copy editing the first edition.

Jeff Dionise, the book's primary artist, and other USA TODAY artists did a magnificent job of translating the author's sketches and verbal descriptions into illustrations that give a rare and forceful vision to this book.

Alexandra Korab of USA TODAY coordinated the first edition's photos. Researcher Cathy Hainer contributed much of the book's historical information.

Chris Cappella of the USA TODAY Information Network made invaluable contributions to the second edition. He researched and wrote most of the extensively revised chapter on thunderstorms, updated the "USA record breakers, weather makers" section and used his meteorological, reporting and writing skills to help improve other parts of the book.

Thanks also go to our editors at Vintage Books, particularly Marty Asher, senior vice president and editor-in-chief, whose personal interest in weather sparked this book, and to Linda Rosenberg, managing editor, whose fine editing skills improved the work.

The source of information on U.S. climate, such as the wettest and driest cities, is the National Climatic Data Center in Asheville, N.C., which is staffed by knowledgeable meteorologists who went out of their way to supply information.

The public relations staffs at the National Oceanic and Atmospheric Administration, including the National Weather Service, and the National Center for Atmospheric Research continued aiding me as they have since I first started writing about weather in 1980. I have found these public affairs officers to be total professionals who are always able to help find the right scientists to talk with about any aspect of atmospheric research.

No book such as this one can be written without access to first-rate libraries. USA TODAY's library was our starting point. The helpful staff could always be relied upon to come up with facts or to obtain books through interlibrary loan. We also made extensive use of the National Oceanic and Atmospheric Administration library, where the staff always made us feel welcome.

Finally, the book would not have been possible without the constant, loving support of my beloved wife, Darlene. Her sunny disposition always evaporates any storm clouds that peek above the horizon.

Jack Williams

INTRODUCTION

A hurricane barreling up the coast. Thunderstorms spawning tornadoes on the Great Plains. Ice storms glazing roads across the Midwest. Most people see these as frightening, freak acts of nature that exist to inflict death and destruction.

In reality, it's just the opposite. Storms perform a much needed function. They make our Earth livable.

We need to remember there are tremendous differences between heat and cold on the Earth. Because of the way the sun shines on the globe, the tropics are always hot and the polar latitudes are cold. To make the planet tolerable to live on, the warm and cold air must be mixed.

That's what storms are all about.

When a winter storm forms, it's like a giant eggbeater sweeping across the planet mixing cold polar air with warm tropical air.

This process works very well most of the year. But during the summer, the cold air retreats northward toward the North Pole, leaving a huge buildup of heat in the tropics. This heat must be dissipated and nature has a very effective way of doing this. It's called a hurricane.

When a hurricane forms, it means the tropical atmosphere is boiling. Heat is removed from the tropics when the spinning bubble of hot air moves toward the poles.

If human beings would learn to live in harmony with the forces of nature, we could minimize our problems. Building a house on an earthquake fault line in California is not very wise. Likewise, we should limit development along river valleys subject to flooding or on small, sandy coastal islands exposed to devastating hurricanes. But we have decided not to do that in this country, so we are extremely vulnerable to storms.

To live in harmony with nature, we must understand atmospheric forces. We need to understand storms — and all our weather systems. Then we can adjust to them and keep out of harm's way.

This book gives a clear explanation of those forces. From fronts and storms to rainbows and sunsets; from what happens on the molecular level in clouds to big

questions about the climate's future. Jack Williams, USA TODAY's longtime weather editor, has amassed the basic knowledge of the great atmospheric machines. Difficult concepts have been illustrated with the kind of exciting, colorful graphics that have become the trademark of USA TODAY.

Scientific discoveries and weather forecasts are the result of a lot of hard work by a lot of people. Along with explaining how the weather works, this book introduces

you to a few of the people responsible for our knowledge of the atmosphere.

So, fasten your seat belts and prepare to blast-off on an exciting trip into the world of atmospheric phenomena. I trust you will enjoy the trip!

Neil Frank is former director of the National Hurricane Center.

The Weather Book

The USA has the world's wildest weather. A normal year brings hurricanes, blizzards, +100°F heat, -20°F cold, jungle humidity, desert dryness, and the world's most numerous and strongest tornadoes. Dangerous weather hits other countries, of course, but no other nation copes with all the kinds of bad weather in nature's arsenal.

Farmers, hikers, sailors and pilots need to know how to read the weather. Everyone can benefit from being weatherwise. If you know some basics, keep your eye on the sky, and stay tuned to forecasts, you can be out of the way if the weather turns deadly.

People who know that a "Category 3" hurricane can bring a surge of water 12 feet above the highest tides will quickly flee an approaching storm. Someone who knows the "tornado watch" just announced on the radio means conditions are ripe for tornadoes and then sees a thunderstorm with a rotating "wall cloud" will be well-sheltered when the twister hits. If you know how heat and humidity combine to put deadly strain on your body, you'll pass up the tennis match on a steamy day.

In terms of numbers of people affected and the costs, weather and weather forecasts are more important today than they were 100 years ago. In the 1800s a Midwestern blizzard could be deadly for those caught outside. Farmers and ranchers sometimes got lost between the house and barn and froze to death. Those dangers still exist. In addition, a blizzard today could strand more than 150,000 at Chicago's O'Hare Airport for hours, not to mention the thousands of people in cars — from interstates to side roads — blocked by snow.

When cars and airplanes replaced trains as our main way of getting around, weather gained more potential to snarl lives. Just to melt ice on roads, the U.S. uses an estimated 10 million tons of salt each year.

Weather is a factor in about 30 percent of all avia-

A tornado rips through Harvey County, Kansas, March 13, 1990, doing more than $20 million damage and killing one person. The clash of cold and hot air on the Great Plains accounts for a majority of the about 800 tornadoes that hit the United States in an average year. Not only does the United States have more tornadoes than any other nation, it also has the world's strongest.

WILD WEATHER IN THE USA

Photographs from space, especially weather satellite views, teach that our day-to-day, local weather is determined by huge systems moving across the Earth. This imaginary satellite view shows many of the major, global-scale events that shape the USA's weather. All aren't likely to be happening at the same instant, as shown here, but all do occur during the course of a year, giving the United States the world's wildest, most varied weather. From space the Earth is a blue and white ball floating in a black sky. The blue comes from the oceans and the blue sky we see when we look up from the ground. The white areas are clouds. Many of the clouds create swirling patterns that grow or shrink as they move across the oceans and continents. Some are circles with a clear area — the eye — at the center. These are hurricanes, which are born over tropical oceans, but which often move northward. Others look like a comma with a line of clouds trailing off toward the equator. Such patterns are storms known as extratropical cyclones. At times the clouds form lines or circles smaller than those of the big storms. These are thunder-storms, which can produce torna-does and other dangerous winds as well as lightning and heavy rain.

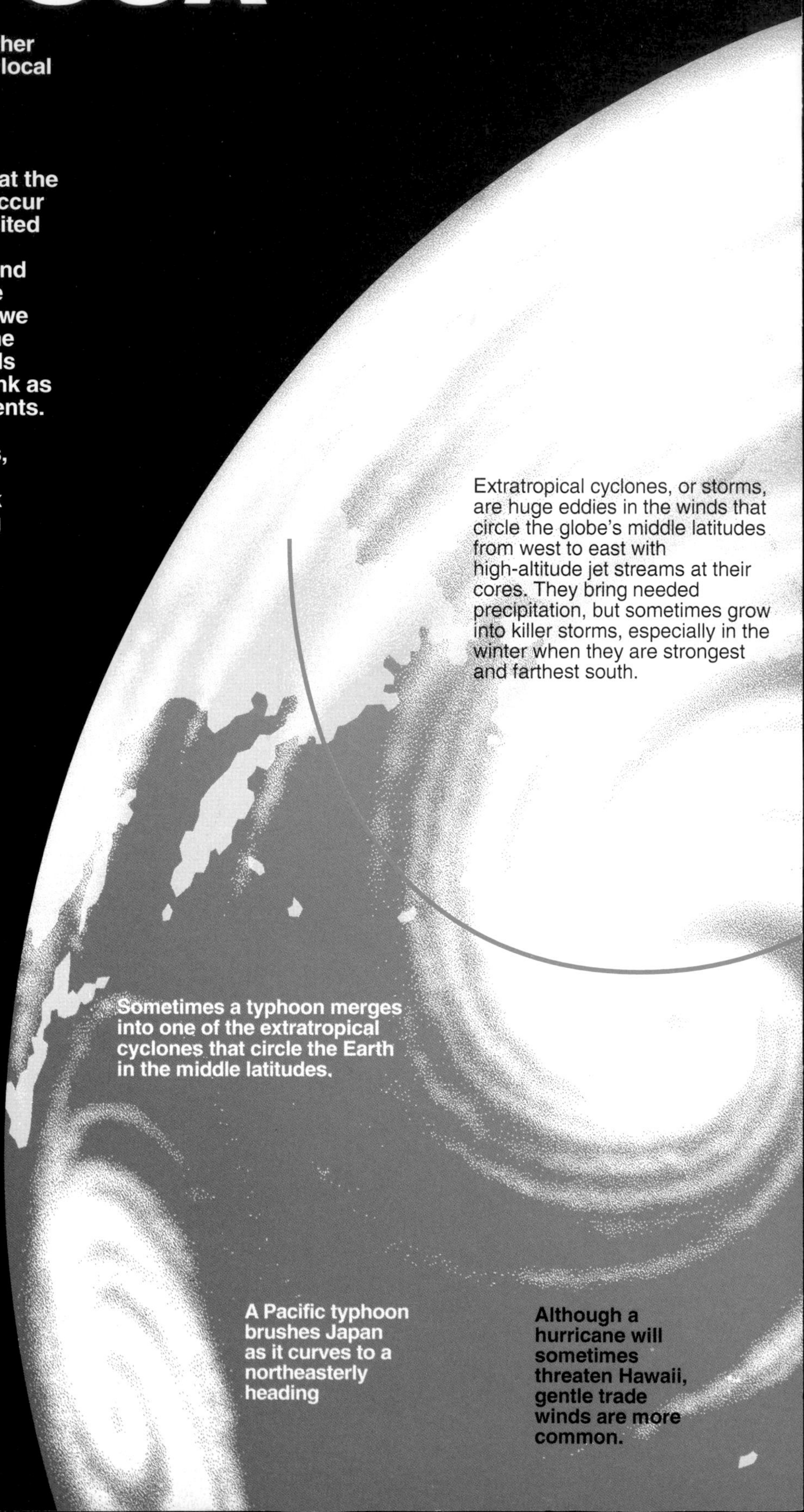

Extratropical cyclones, or storms, are huge eddies in the winds that circle the globe's middle latitudes from west to east with high-altitude jet streams at their cores. They bring needed precipitation, but sometimes grow into killer storms, especially in the winter when they are strongest and farthest south.

Sometimes a typhoon merges into one of the extratropical cyclones that circle the Earth in the middle latitudes.

A Pacific typhoon brushes Japan as it curves to a northeasterly heading

Although a hurricane will sometimes threaten Hawaii, gentle trade winds are more common.

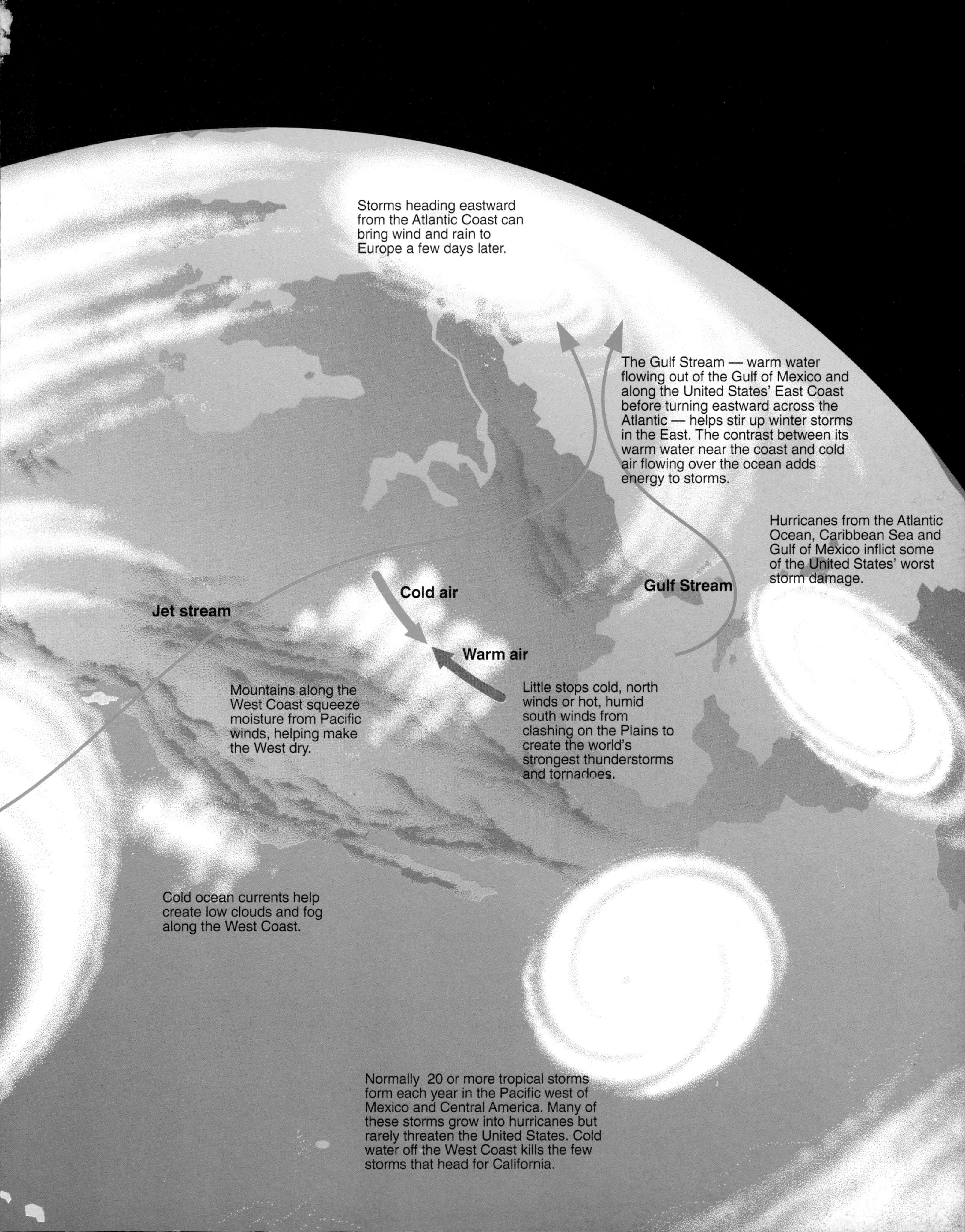

Storms heading eastward from the Atlantic Coast can bring wind and rain to Europe a few days later.

The Gulf Stream — warm water flowing out of the Gulf of Mexico and along the United States' East Coast before turning eastward across the Atlantic — helps stir up winter storms in the East. The contrast between its warm water near the coast and cold air flowing over the ocean adds energy to storms.

Hurricanes from the Atlantic Ocean, Caribbean Sea and Gulf of Mexico inflict some of the United States' worst storm damage.

Cold air

Jet stream

Gulf Stream

Warm air

Mountains along the West Coast squeeze moisture from Pacific winds, helping make the West dry.

Little stops cold, north winds or hot, humid south winds from clashing on the Plains to create the world's strongest thunderstorms and tornadoes.

Cold ocean currents help create low clouds and fog along the West Coast.

Normally 20 or more tropical storms form each year in the Pacific west of Mexico and Central America. Many of these storms grow into hurricanes but rarely threaten the United States. Cold water off the West Coast kills the few storms that head for California.

Galveston's killer hurricane took the low-lying island by surprise

America's greatest weather disaster was a hurricane that killed more than 7,200 people in Galveston, Texas, Sept. 8, 1900. Galveston — at the time the wealthiest city in Texas — was a victim of geography and the inability of forecasters to know, until it was too late, that a hurricane was on the way.

The city sprawls across a low-lying, barrier island. In 1900 the highest point was only nine feet above sea level. One side of Galveston Island fronts on the shallow Gulf of Mexico where hurricanes can push up large storm surges. The ocean can rise in walls of water 20 feet high. Large surges can push water from the bay over the island. In 1900, the backside of the island facing Galveston Bay was a major port accommodating more than 1,000 ships a year.

On the morning of Tuesday, Sept. 6, 1900, the U.S. Weather Bureau telegraphed its Galveston office about a "tropical storm disturbance moving northward over Cuba." For most of the 38,000 Galvestonians this was nothing to worry about; the city had

On the south side, the storm destroyed every building within three blocks of the Gulf.

weathered other storms with minor flooding and wind damage.

By the afternoon of September 7 the wind strengthened, long ocean swells began rolling in from the southeast, and the tide rose higher than usual. Such signs were the only way of knowing a big storm might be stirring up the ocean somewhere over the

tion accidents. The airline industry has estimated that all delays cost airlines more than $3 billion a year and that about 65 percent of those delays are weather related. Airlines say that improved forecasts of winds at high altitudes, where jetliners fly, could save about $90 million a year in fuel costs.

According to Richard Hallgren, the thriving, private weather-forecasting industry in the United States reflects not only our interest in weather but also the wide variety of weather across America. Hallgren was the head of the National Weather Service from February 1979 to May 1988 before becoming executive director of the American Meteorological Society.

In the U.S., the National Weather Service gathers weather data and issues general forecasts and warnings of dangerous weather. But in addition, the U.S. has a $100-million-a-year private forecasting business. Private firms concentrate on specialized forecasts for subscribers including news media, state and local governments, utilities, and farmers. Private forecasting firms depend on National Weather Service observations and generalized forecasts to make their specialized forecasts.

In the rest of the world, weather forecasting is a government monopoly. For example, in the United Kingdom, France, Australia and New Zealand, government-run weather services sell specialized weather information via telephone, fax, and the Internet. In the USA, most of this information would be free.

Weather watchers

Even if you aren't trying to change airplanes at O'Hare the day a blizzard hits, it is likely to be news no horizon in the days before weather satellites and storm-penetrating airplanes.

On the morning of Sept. 8, Isaac Cline, who headed the Galveston Weather Bureau office, was alarmed by the winds, tides and rapidly falling barometer. He rode his horse along the beachfront, urging residents to flee. But most just sought shelter in the brick houses at higher elevations. Few left the island.

Around 4 p.m. tides surging from both the ocean and bay met, covering the highest part of the island with a foot of water and cutting off the bridges to the mainland. At 6:15 p.m. the Weather Bureau anemometer was ripped from its supports after hitting a steady rate of 84 mph with gusts to 100 mph. Less than an hour later, a four-foot wave surged through the city, followed by a 20-foot wave. For the rest of the evening storm waves on top of the high tide — along with winds estimated to be above 120 mph — ripped the city apart, toppling even the substantial brick houses of the city's wealthy.

Cline, his family and others fled to his home. He later wrote about what happened that night: Waves pushing on wreckage of buildings "acted as a battering ram against which it was impossible for any building to stand for any length of time. At 8:30 p.m. my residence went down with about fifty persons who had sought it for safety, and all but eighteen were hurled into eternity. Among the lost was my wife, who never rose above the water after the wreck of the building."

For the next three hours, Cline, his three children, his brother and another woman and her child, drifted on wreckage. From the feel of the waves, he assumed they drifted out to sea and then back to within 300 yards of where his house had been. "While we were drifting," Cline later wrote, "we had to protect ourselves from the flying timbers by holding planks between us and the wind."

Until his house was smashed, Cline had been making careful notes of the water's height in his house and this information helped establish how high the water rose. Afterward, the government honored Cline for "Heroic devotion to duty" during the storm.

The brick Galveston Orphan's Home stood up to the storm, saving the 50 children sheltered there. But the wooden Catholic orphanage nearer to the beach collapsed, killing 90 children.

When the eye of the storm passed Galveston at around 9 p.m., the barometric pressure had dropped to 27.91 inches. The storm poured 10 inches of rain on the city, and the tide reached a record 15.2 feet above normal. The National Weather Service lists the storm in Category 4 on its current 1 to 5 hurricane strength scale.

Original estimates listed more than 6,000 people dead or missing, but more recent estimates have put the toll at 7,200 dead. The exact number will never be known. Galveston was in ruins. More than 3,600 houses, along with businesses and ships, were smashed.

Instead of abandoning the city, residents decided to prepare for future storms by building a sea wall 3 miles long, 17 feet high, with a base 16 feet wide facing the Gulf of Mexico. Behind the sea wall, houses and other buildings were jacked up and sand was pumped in to raise the city's level as much as 17 feet in places.

When another Category 4 hurricane hit Galveston in August 1915, damage was minimal. Since then, the city has grown with many areas unprotected by the sea wall.

Residents count on today's satellites and other technology to help forecasters give warnings in time to flee. Still, hurricane forecasters and experts worry about a storm that could blow up too quickly to allow time for residents or visitors to flee heavily populated barrier islands, such as Galveston, before a big hurricane hits.

matter where you live. Blizzards, tornadoes, hurricanes, and heat and cold waves are always news. Americans' fascination with weather began with the first explorers and hasn't diminished since. Encounters with raging storms made for rip-roaring tales in the early days just as they make interesting news stories now.

"We discovered the winde and waters so much increased with thunder, lightning, and raine, that our mast and sayle blew overboard and such mighty waves overracked us ... that with great labour we kept her from sinking by freeing out the water," Captain John Smith wrote of his 1606 voyage to Chesapeake Bay. "Two dayes we were inforced to inhabite these uninhabited Isles which for the extremitie of gusts, thunder, raine, stormes, and ill weather we called Limbo."

Today we don't have to wait for an explorer to return home and write a book to hear about "extremitie of gusts, thunder, raine, stormes, and ill weather." Thanks to video recorders, any tornado that approaches a town with more than a few hundred people is likely to be captured on tape, and maybe end up on television. Today a regular viewer of the evening news anywhere in the United States is likely to see more tornadoes in a couple of years than a 19th-century centenarian would have seen in a lifetime in Kansas.

The USA's temperate climate

While America's weather makes for rousing stories, the climate — the long-range average of the day-to-day weather — is no less important to people.

Since Cape Cod is as far south as the Mediterranean Coast of France and northern Italy, the first explorers of

'Modern' New York no match for the legendary Blizzard of 1888

The Blizzard of '88 — which struck the East Coast from the Chesapeake Bay to New England in March, 1888 — has become a part of American folklore. Snowfall records set then still stand in parts of the East. But probably even more important, the storm was a preview of how weather can render the most up-to-date technology useless.

When the blizzard hit, New York was the most technologically advanced city in the United States. Miles of new electric, telephone, and telegraph wires laced the streets. The country's best elevated train and streetcar system served the city's 1.5 million people. Yet within hours after snow began falling, wires were being pulled down, cutting electrical, telephone and telegraph services. Trains and streetcars were stuck in snow, halting all transportation except for a few horse-drawn carriages. Ice-covered cables and 80 mph winds prompted police to close the great Brooklyn Bridge.

In addition to the damage on land, the storm was a maritime disaster; sailors called it "the Great White Hurricane." Winds above

By March 12, snow — some 20 feet high — stopped the city in its tracks.

60 mph whipped the seas, sinking, grounding or wrecking more than 200 vessels from the Chesapeake Bay up the New England coast.

On Sunday, March 11, the National Weather Service's predecessor, the U.S.

Signal Service, issued "indications" — a forecast — for New York City that predicted "fresh to brisk easterly winds with rain" Sunday. Monday's forecast was for brisk westerly winds and cooler weather.

Rain that was falling Sunday night had

what is now the United States expected the weather to be milder than England's. But the geography of weather is more complicated than distance from the poles or equator.

Since it stretches from the tropics (Hawaii) to well north of the Arctic Circle (northern Alaska), the United States is guaranteed to see any kind of weather found anywhere outside the Antarctic.

Geography grants small favors: America never turns as cold as the South Pole. Even without the southernmost and northernmost states, however, the United States has a variety of weather unmatched by even larger nations such as Russia and China.

The quip often heard as bitter winds howl across the Plains sums up one reason why we have such variety: "There's nothing between us and the North Pole but a few barbed wire fences." You can say almost the same

thing, substituting "Gulf of Mexico" for "North Pole," on steamy Plains days.

Invasions of extremely cold and extremely hot air give the central United States the world's most violent thunderstorms and tornadoes. Europe and Asia avoid such violent storms because mountain ranges across both continents keep polar and tropical air from colliding as often and as violently as in North America.

Westerly winds from the relatively mild Atlantic Ocean keep western Europe warmer than places much farther south in North America. And the Alps hinder tropical air from moving northward. All of western Europe doesn't see annual temperature swings to match those of North Dakota.

turned to snow by Monday morning with 10 inches on the ground and the temperature falling toward 20°F by the time most people awoke. A few early morning commuters caught elevated trains, but most businesses were closed. The New York Stock Exchange closed at noon when only 33 of the 1,100 employees showed up — the first time since its opening in 1790 that it had been closed by the weather.

By that afternoon, live electrical wires hung from ice-covered poles; metal signs with jagged edges dangled from shop fronts; broken telephone poles blocked the streets; dead horses were buried in giant drifts. Trains on all four elevated lines stopped, stranding an estimated 15,000 people in the unheated cars.

The stranded took shelter in hotels, bars, private residences, even jails. Among those seeking refuge in Fifth Avenue hotels were Civil War hero William Tecumseh Sherman, and author Mark Twain, in from Hartford, Conn., who sent word to his wife that he was "Crusoeing on a desert hotel."

The next day, *The New York Times* reported that the city looked like a "wreck-strewn battlefield," and reported one observer say-

Miles of electric, telephone and telegraph wires fell under the weight of snow and ice.

ing, "Great piles of [snow] rose up like gigantic arctic graves in all directions." In the city, officials reported 400 dead.

While snow fell from Virginia northward, the record amounts were in New York and New England. In New York City, the Central Park weather office had recorded 21

inches of snow by 5:30 p.m., March 12. Elsewhere in the state, Albany recorded 46.7 inches; Troy, 55 inches; and Saratoga Springs, more than 50 inches. In New England, New Haven, Conn., had 42 inches of snow; Worcester, Mass., 32 inches; Concord, N.H., 27; and Boston, 12.

A deadly merger

Sometimes tropical and mid-latitude weather systems gang up on the United States. When Hurricane Agnes hit Florida's Gulf Coast June 19, 1972, it was a weak hurricane with winds a little above 74 mph. After doing relatively minor damage to Florida, it tracked across Georgia, the Carolinas and Virginia into the Atlantic. Agnes then turned back to the northwest, across Long Island into central New York where it merged with a mid-latitude storm.

The merger was deadly. By the time the rain ended, some of the highest floods in years across the Northeast had sent the storm's death toll soaring to 122 and the damage to more than $4 billion.

This isn't to complain about the United States' overall weather. Far from it. In fact, the climate has to be counted among the natural resources that made the

nation's settlement and growth possible.

Early America's weather explorers

Francis Higginson, one of the first settlers in Massachusetts, put it this way:

"The temper of the air of New England is one special thing that commends this place. Experience does manifest that there is hardly more that agrees better with our English bodies. Many that have been weak and sickly in old England, by coming hither, have been thoroughly healed and grown healthful and strong. For here is an extraordinary clear and dry air that is of a most healing nature to all such as are of a cold, melancholy, phlegmatic, rheumatic temper of body.

"In the summertime, in the midst of July and August, it is a good deal hotter than in old England; and in winter, January and February are much colder, as they say; but the spring and autumn are of middle temper."

When President Thomas Jefferson sent Meriwether Lewis and William Clark westward to explore from the Mississippi River to the Pacific Northwest, one of their tasks was to report on the weather and climate.

On Dec. 11, 1804, when the explorers were near what is now Bismarck, N.D., Lewis wrote in his journal: "The weather became so intensely cold that we sent for all the hunters who had remained out with Captain Clark's party, and they returned in the evening several of them frostbitten. The wind was from the north and the thermometer at sunrise stood at twenty-one below 0, the ice in the atmosphere being so thick as to render the weather hazy and give the appearance of two suns reflecting each other."

On April 1, 1805, Lewis wrote about something that strikes any first-time visitor to the West: "The air is remarkably dry and pure in this open country; very little rain or snow, either winter or summer. The atmosphere is more transparent than I ever observed it in any country through which I have passed."

Early explorers in the West also encountered tornadoes on the Plains. While tornadoes, or whirlwinds, are seen in Europe and the eastern United States, those on the Plains are more awesome. First, they're larger and more violent. Also, the clear sky and generally flat land allow you to see more of the thunderstorm and the tornado than you usually see in the East.

"Once as the storm was raging near us, we witnessed a sublime sight," Father Pierre Jean de Smet, a Belgian priest traveling with a party of settlers from Indiana to California, wrote in 1841. He was describing an experience on the Plains before reaching the Rockies. "A spiral abyss seemed to suddenly be formed in the air. The clouds followed each other into it with such velocity, that they attracted all objects around them. The noise we heard in the air was like that of a tempest …

"The column appeared to measure a mile in height; and such was the violence of the winds which came down in a perpendicular direction, that in the twinkling of an eye the trees were torn and uprooted, and their boughs scattered in every direction. But what is violent does not last. After a few minutes, the frightful visitation ceased. The column, not being able to sustain the weight at its base, was dissolved almost as quickly as it had been formed. Soon after the sun re-appeared: all was calm and we pursued our journey."

Folk forecasting

The explorers and settlers who learned about North America's weather and climate were continuing a tradition that goes back to the dawn of the human race. People always have been concerned with the weather and those most directly affected, such as farmers and sailors, developed rules of thumb for forecasting what the weather would do in the future. One of the most famous — and sometimes true — is:

Red sky in morning,
sailor take warning.
Red sky at night,
sailor's delight.

The rhyme works in most of the United States where storms generally move from west to east. The "red" refers to the sky overhead, not at the horizon. The red is caused by sunlight reflecting off clouds.

To have a red sky in the morning, the eastern horizon has to be clear while clouds are moving in from the west. Since that's where storms come from, a storm could be heading your way.

To have a red sky at sunset, clouds have to have moved away from the western horizon — heading east. Since the storm is moving east, clear skies are coming your way.

Red sky in the morning only works when a storm is on the way. You can have a red sky — morning or night — caused by non-storm clouds. Also this rhyme can't predict thunderstorms that develop in an afternoon when the morning has been clear.

Generations have developed folk-forecasting techniques and handed down myths about weather. These often made the things happening in the sky manifestations of deities. It is certainly natural enough to think of lightning as a display of an angry god.

Science takes hold

As most people were going about their lives using traditional, folk-forecasting techniques, a few men and women were developing what we now call science, including the scientific study of the atmosphere. The invention of the barometer in 1643 was the beginning of true scientific study of weather. Progress was slow, however. The modern understanding of storms didn't start until the 1920s. Even now, scientists continue breaking ground. Meteorology still has many unanswered questions.

The invention of the telegraph in 1837 made the first weather-forecasting services possible. Finally people could know instantly what was going on more than a few miles away. Observations showed that storms usually move more or less from west to east. Reports by telegraph from places to the west gave some idea of what was coming. These were the first forecasts.

Telegraph networks, and the widespread observations they encouraged, boosted more scientific forecasting. Still, until well into the 20th century, weather forecasting was mostly what scientists call "empirical" or weak in theory. Meteorology was mainly a "descriptive" science. Study of observations had shown that rain or other kinds of weather were often associated with certain patterns of atmospheric pressure or changes in pressure. These observations were linked to certain kinds of weather, but little theory and mathematical calculation were involved.

Without theory, most scientific effort went into describing weather and apparent connections among meteorological events.

Weather forecasting didn't become "scientific" until the period around World War I. At that time a group of scientists, led by the Norwegian physicist Vilhelm Bjerknes, developed both a picture of the atmosphere and forecasting methods based on the laws of physics. Some of Bjerknes' students brought these techniques to American universities and the U.S. Weather Bureau in the 1930s. Though some of the scientific laws they used had been worked out by Isaac Newton around 1700, they never had been applied to weather forecasting.

You could argue that scientific forecasting didn't become possible until the invention of the computer in the 1950s. Computers quickly solve the mathematical equations of atmospheric movements that scientists had developed by the 1930s.

People like to joke about forecasts that have been totally wrong, and most people can give some notable examples. But those who pay close attention know that weather forecasting has improved over the last few years.

In general, forecasters are better at saying what large-scale weather patterns will be like in three or four days than they were a decade ago. But predicting what small, often fierce, events such as thunderstorms will do in the next two or three hours is still difficult. Snowstorms can be especially troublesome, as Hallgren was reminded one holiday when he was still head of the National Weather Service.

On Veterans Day, Nov. 11, 1987, weather forecasters were predicting that a winter storm would bring about an inch of snow to Washington, D.C. Hallgren, an avid woodworker, decided to help his son install cabinets at his office.

"Before the day was out we were in the bull's-eye of the storm. There were 16 inches on the ground." The director of the National Weather Service had believed his office's forecast and was snowbound. "I thought afterward, 'Good thing that didn't get out.' "

As it turned out, the snow was 16 inches deep only over a small area. A few miles away — at Hallgren's house — the snow was only three inches deep and some places had only an inch or two. A forecast for 16 inches of snow would have been way too high for most of the area. "I went over those charts closely and I wound up siding with the forecaster," Hallgren says. "This shows how small scale the significant [weather events] are."

New technology and the resulting new forecasting techniques promise to improve predictions of small-scale events. In the future forecasters may well say: "Expect 15 inches of snow on the city's north side by evening; only six inches on the south side." But forecasting will never be perfect and no one expects to be able to give detailed forecasts for thunderstorms or local variations in snowstorms more than a few hours in advance. But with some knowledge of how our atmosphere works, anyone can learn what realistically to expect from a forecast.

Even in Hawaii where surface temperaturse are always above 50°F, there is snow. Between 1 and 2 feet of snow fall each year in the mountains. But there is no snowfall in Hawaii at elevations under 5,000 feet.

Fairbanks, Alaska, reached a record 100°F on June 27, 1915. While temperatures in the 90s are rare, on average, every year Fairbanks has seven days with temperatures over 80°F.

A singular career discovering the ways of clouds

At the end of World War II the effort to understand and predict weather was the activity of a few men using simple instruments and pencil-and-paper or slide-rule calculations. Today, it's a high-tech enterprise involving men and women working with satellites, sensitive radars, four-engine research airplanes and some of the world's biggest computers.

Forecasting day-to-day weather is still a major focus, but the concerns have expanded to include long-range climate problems such as the potential "greenhouse" warming of the Earth.

Joanne Simpson has been at the center of these changes. Her career mirrors the growth of meteorology since the late 1940s.

She went from a graduate student who wasn't always welcome in meteorology's male world to become chief scientist for meteorology at NASA's Goddard Space Flight Center Laboratory for Atmospheres. One of her most satisfying duties has been coordinating the work of more than 100 scientists and engineers for the launch and data gathering of the U.S.-Japanese Tropical Rainfall Measuring Mission satellite.

"This is the most exciting project I've ever been involved in," she says of the satellite project. "When I started in meteorology, the best you could do to make measurements of clouds was to hire a small aircraft and go up with a camera and maybe put a thermometer on the plane. You could have someone else drive around on the streets, trying to get another camera to look at the same cloud."

She recalls wearing a football helmet while flying 50 feet above the ocean in a single engine airplane in the 1950s to investigate clouds around Nantucket Island, Mass. "If anybody had told me at that time that before the end of your career you'll be involved in something where we're measuring the crucial properties of these cumulus clouds from space, I would have told them, 'You're in the Buck Rogers funny papers.' "

The changes Joanne Simpson has seen go far beyond weather technology.

When World War II ended, women such as Simpson, who had been teaching meteorology to military officers and aviation cadets, were expected to settle down as housewives. "I didn't want to go home and mop the floor," she says. Simpson wanted to be a scientist with a Ph.D. degree. She says she was told, "No woman has ever earned a Ph.D. in meteorology and even if one did, no one will hire her."

To make her case even more unusual, Simpson wanted to conduct research on cumulus clouds. At the time, top meteorologists were focusing on large-scale weather patterns.

Cloud research was a side issue. Professor Carl-Gustav Rossby, a giant in international meteorology, offered Simpson some advice: Cloud research would be "an excellent problem for a little girl to work on because it is not very important and few people are interested in it." He said, "You should be able to stand out if you work hard."

She did. After earning her University of Chicago Ph.D. in 1949, Simpson continued studying clouds – from the ground, from boats, and from airplanes – always using physics and mathematics to make sense of what she was seeing and measuring.

Today, women scientists are no longer rare and clouds are a central topic in atmospheric science.

Cumulus clouds range from popcorn-like puffs dotting a blue sky to the great thunderstorm clouds that reach more than 50,000 feet in the air to produce downpours, damaging hail and tornadoes. Cumulus clouds also are a key part of hurricanes.

Thanks to work by Simpson and others, scientists now realize that in addition to producing more than three-quarters of the world's rain and some of its most dangerous weather, cumulus clouds are, in Simpson's words, "an important part of the driving force of the atmosphere." They help regulate how much energy reaches the Earth from the sun and how much energy escapes back into space. They also provide most of the fuel to drive the atmospheric wind system by releasing heat as water vapor is turned into liquid cloud droplets.

By the 1940s scientists had long thought of the atmosphere as a "heat engine" that draws its power from hot-cold temperature contrasts.

To explain some of her most important work, Simpson uses the analogy of the atmosphere being like a car's engine. Work she and Herbert Riehl did in the 1950s showed "large clouds in the tropics are, in fact, the cylinders of the automobile that is the atmospheric heat engine. You have to have those big, buoyant, hot towers [cumulus clouds] to 50,000 feet to convert and transport heat energy to high altitudes, where it can be shipped [toward the poles] to help drive high-latitude circulation.

"The atmosphere has variable cylinders. It doesn't run at the same speed all the time. It has a terrible gas gauge and we don't know how much fuel is there and how much is being released. Until we can understand these things better, we aren't going to understand how this car operates or predict how it will operate."

Simpson expects the Tropical Rainfall Measuring Mission to answer questions about the Earth's tropical "cylinders." It should supply some of the information needed to figure out whether the greenhouse effect is warming the Earth. It also could produce data for computer programs that predict day-to-day weather.

Perhaps the biggest changes Simpson has seen in weather science are in the use of satellites and computers. In 1955, while at Imperial College in London on a Guggenheim Fellowship, Simpson and her colleagues had developed a mathematical model of cloud growth and "had started working it out on what they had for hand calculators in those days -- the ones that go grrrr, grrrr. We were getting nowhere; it took six months to do one time step, fractions of a minute."

Professor Rossby happened to visit, looked over Simpson's work and said, "You're crazy. You're going to be 90 before you get that even up to the first two minutes. Come over to Stockholm and I'll let you use my computer." It was available only from 3 a.m. to 6 a.m. and no one was around the first night when Simpson showed up to start work. She hired a locksmith to break in. She used the computer to complete the first scientific paper ever written on a mathematical model of clouds. It was dedicated to Rossby.

Simpson calls that mid-1950s cloud model "very simplified," but it illustrates why today's atmospheric scientists use some of the world's biggest, fastest computers. Writing mathematical equations to describe the atmosphere is hard because "everything affects everything else." In her simple cloud model, for instance, the motion would change the temperature and the temperature would change the motion.

To become an atmospheric scientist today, "you've got to be good at math; you've got to be interested in it," she says. "You've got to be good at physics and you've got to be interested in it. It's a real science that requires physics and math. It's more and more requiring chem-

Joanne Simpson's new project will use a satellite to measure tropical rainfall.

istry and biology."

"Science isn't just cold, hard facts," she says. "The crux is fitting unfitted things together and making them hang together. It's similar to composing music or art or poetry."

In science, you do the subconscious, intuitive work and then test your ideas rigorously. And that, she says, requires work. "You've got to be willing to work your shirt off. It's not only sitting there having intuition, [it's] making numerical or observational tests.

"A theory may be crazy, or right, or partly right, which is usually the case. I learned from Rossby, if you're a person full of ideas, you'll have 10 or more wrong ones for every right one. Often the wrong ones do more good than the right ones because you get up and give a speech or write a paper and everyone will dump all over you. In the process of arguing out the criticism something good will often come out of it."

Simpson's scientific skill and willingness to advance new ideas have been recognized in many ways over the years. One that meant the most to her came in 1983 when the American Meteorological Society presented her its highest honor, the Carl-Gustav Rossby Research Medal. She was later elected as the Society's president.

"I figure that I'm getting paid to do what I want to do. I don't have to go into some law office. Ugh! I think that anybody who has a job they enjoy and makes enough money to send their kids to college -- what more could you want?"

What makes weather?

The sun.

Through the seasons, it heats our world, some parts more, some less. The Earth's yearlong trip around the sun creates a giant engine that drives the weather.

Anyone who ever wilted on a sweltering summer day can easily understand that the sun creates hot weather. But the connections are not so clear between the sun and hurricanes, or tornadoes, or blizzards, or the falling rain, or even long sieges of arctic cold.

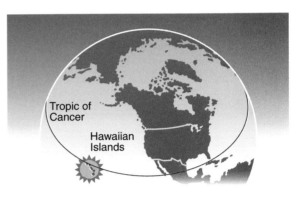

How the sun creates energy

For centuries people thought of the sun as a giant object on fire, burning the way campfires burn. But imagine how much fuel would be needed to keep that fire going. If something the size of the sun were made of gasoline and had enough oxygen, it would burn about 10,000 years. That's a long time but far short of the four and a half billion years the sun and Earth have existed.

Early in the 1900s, as scientists started learning about radioactivity, they conjectured that nuclear energy might explain the sun's power.

In 1938, Hans Bethe of Cornell University and Charles L. Critchfield of George Washington University produced a theory explaining how nuclear fusion fuels the sun and other stars. A fusion reaction fuses atoms together, creating other kinds of atoms and giving off energy. (Fission reactions, like those in today's nuclear power plants, split atoms to produce energy.)

Only a small share of the sun's tremendous energy reaches the Earth, but that's enough to maintain life and drive the weather.

How the sun powers weather

The sun's electromagnetic energy, including light, ultraviolet energy and infrared energy, flows into the

On a broiling summer's day in southern Arizona, you might feel as if the sun is beating down directly on your head. But it isn't. Hawaii, the southernmost state, is the only state where the sun is ever exactly overhead. At the summer solstice, around June 21, the sun is directly over the Tropic of Cancer that runs north of Hawaii. The sun is never overhead any farther north than that. If you're in Hawaii in May as the sun is moving northward toward the summer solstice or again in August as it heads south, you'll see it shining directly down at noon.

INSIDE THE SUN:
THE ENGINE THAT POWERS (

Ancient people worshiped the sun. They recognized its importance to life on Earth even though they didn't understand how it makes green plants grow and drives the weather. Knowing many of the details of how the sun works, if anything, should make it even more awesome to us. Here's a quick look at the highlights of what we know today about what is happening inside the sun.

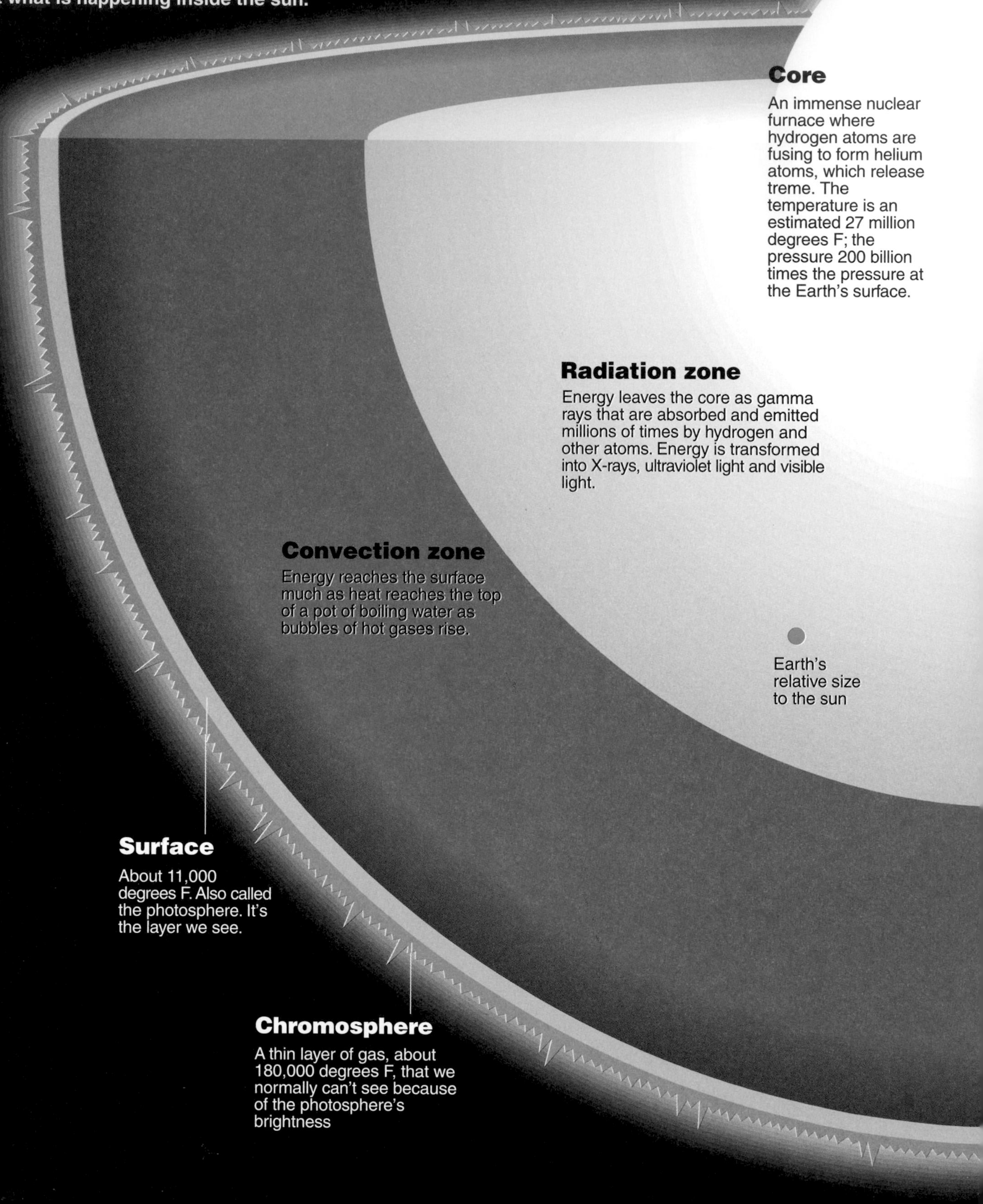

Core

An immense nuclear furnace where hydrogen atoms are fusing to form helium atoms, which release treme. The temperature is an estimated 27 million degrees F; the pressure 200 billion times the pressure at the Earth's surface.

Radiation zone

Energy leaves the core as gamma rays that are absorbed and emitted millions of times by hydrogen and other atoms. Energy is transformed into X-rays, ultraviolet light and visible light.

Convection zone

Energy reaches the surface much as heat reaches the top of a pot of boiling water as bubbles of hot gases rise.

Earth's relative size to the sun

Surface

About 11,000 degrees F. Also called the photosphere. It's the layer we see.

Chromosphere

A thin layer of gas, about 180,000 degrees F, that we normally can't see because of the photosphere's brightness

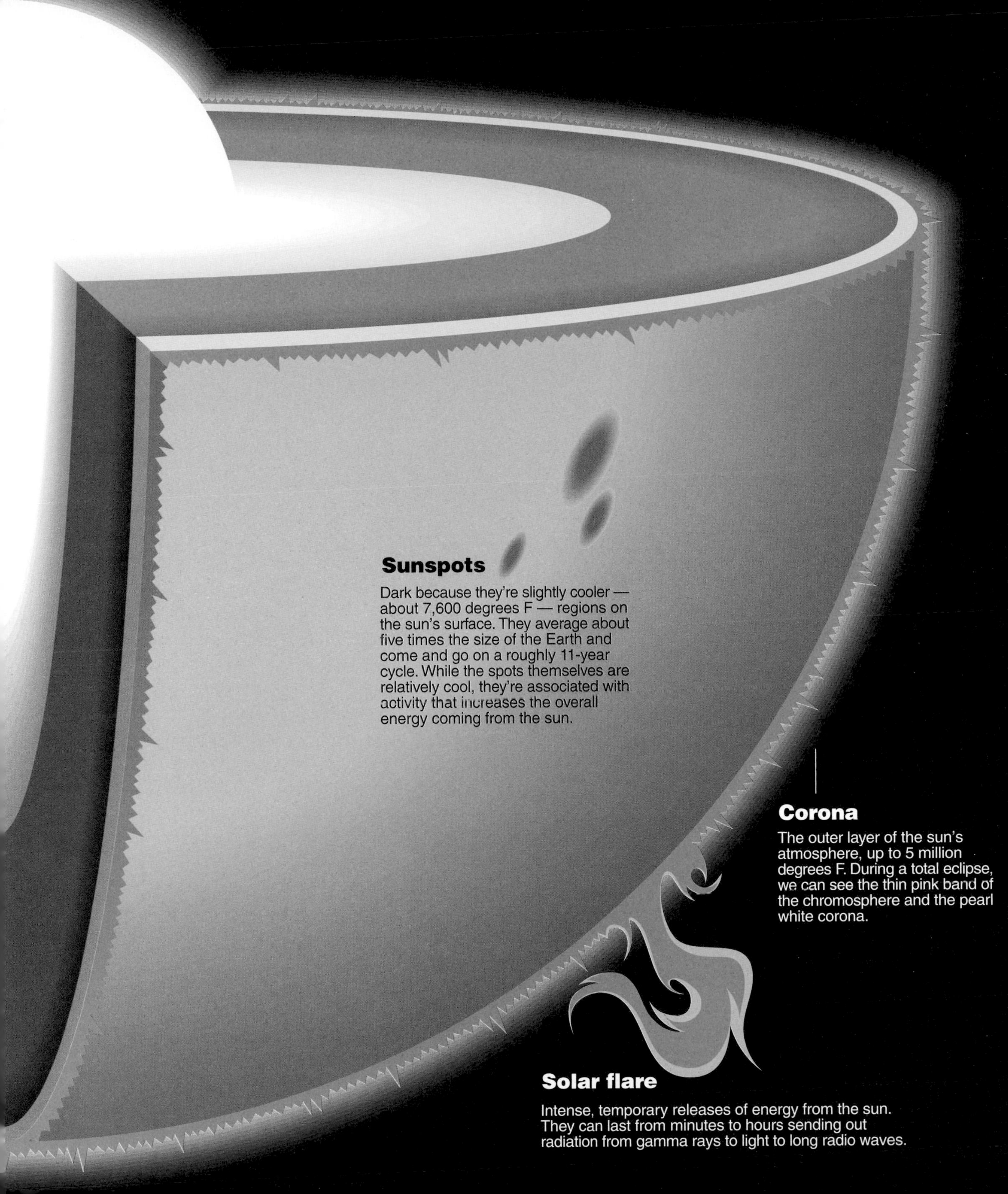

Sunspots

Dark because they're slightly cooler — about 7,600 degrees F — regions on the sun's surface. They average about five times the size of the Earth and come and go on a roughly 11-year cycle. While the spots themselves are relatively cool, they're associated with activity that increases the overall energy coming from the sun.

Corona

The outer layer of the sun's atmosphere, up to 5 million degrees F. During a total eclipse, we can see the thin pink band of the chromosphere and the pearl white corona.

Solar flare

Intense, temporary releases of energy from the sun. They can last from minutes to hours sending out radiation from gamma rays to light to long radio waves.

The Earth's year around the sun

March 21

June 21

Dec. 21

Sept. 21

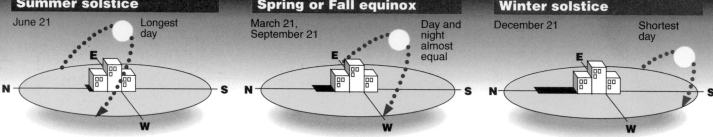

Summer solstice
June 21 — Longest day
N — E — S — W

Spring or Fall equinox
March 21, September 21 — Day and night almost equal
N — E — S — W

Winter solstice
December 21 — Shortest day
N — E — S — W

atmosphere to warm the air, oceans, and land.

Besides the sun, the only source of heat on Earth is its molten core and radioactive rocks. While this heat supplies the energy to slowly push the continents around and build mountains, it hardly affects the weather.

As the seasons change, the sun bathes different parts of the Earth in widely varying amounts of energy. The result is a planet sharing frigid polar cold, sweltering tropical heat and all the temperatures in between, all at the same time.

This sets up a huge "heat engine" in the atmosphere that powers the weather.

A heat engine depends on hot-cold contrasts to produce power. Your car is a good example. Power comes from the heat created by burning a mixture of gasoline and air in a cylinder. As the engine runs, its cylinders alternate between being hot and cooler. Power is produced.

In 1824, French scientist Sadi Carnot published a treatise showing that the bigger the contrast between hot and cold in a heat engine, the more power it produces. That's why automotive researchers are looking for new materials that can withstand more heat than today's engine. An engine that could stand more heat

Light is most concentrated from overhead source.

Light hitting at an angle is less concentrated.

would have bigger contrasts and produce the same power with less gasoline.

The atmosphere is similar. The stronger the hot-cold contrast, usually the stronger the storm.

Water holds the heat better

In addition to differences caused by seasons, the Earth heats unequally because land and water absorb and give up heat. But land warms up and cools off more readily than water.

Do your own science experiment. Go to the beach barefoot on a hot day at about 3 o'clock in the afternoon. The sand is almost too hot to walk on; the water is refreshingly cool. Now go back at midnight, especially on a clear night. The sand is chilly while the water has remained comfortably warm.

On a large scale, tropical oceans hold tremendous amounts of heat while polar water can be frigid year-round. Land areas, on the other hand, quickly warm up or cool off with the seasons.

Earth's tilt makes the seasons

The seasons bring the biggest swings between hot and cold. They are caused by the Earth's yearlong trip around the sun.

People often wonder why the year's hottest days

Space weather

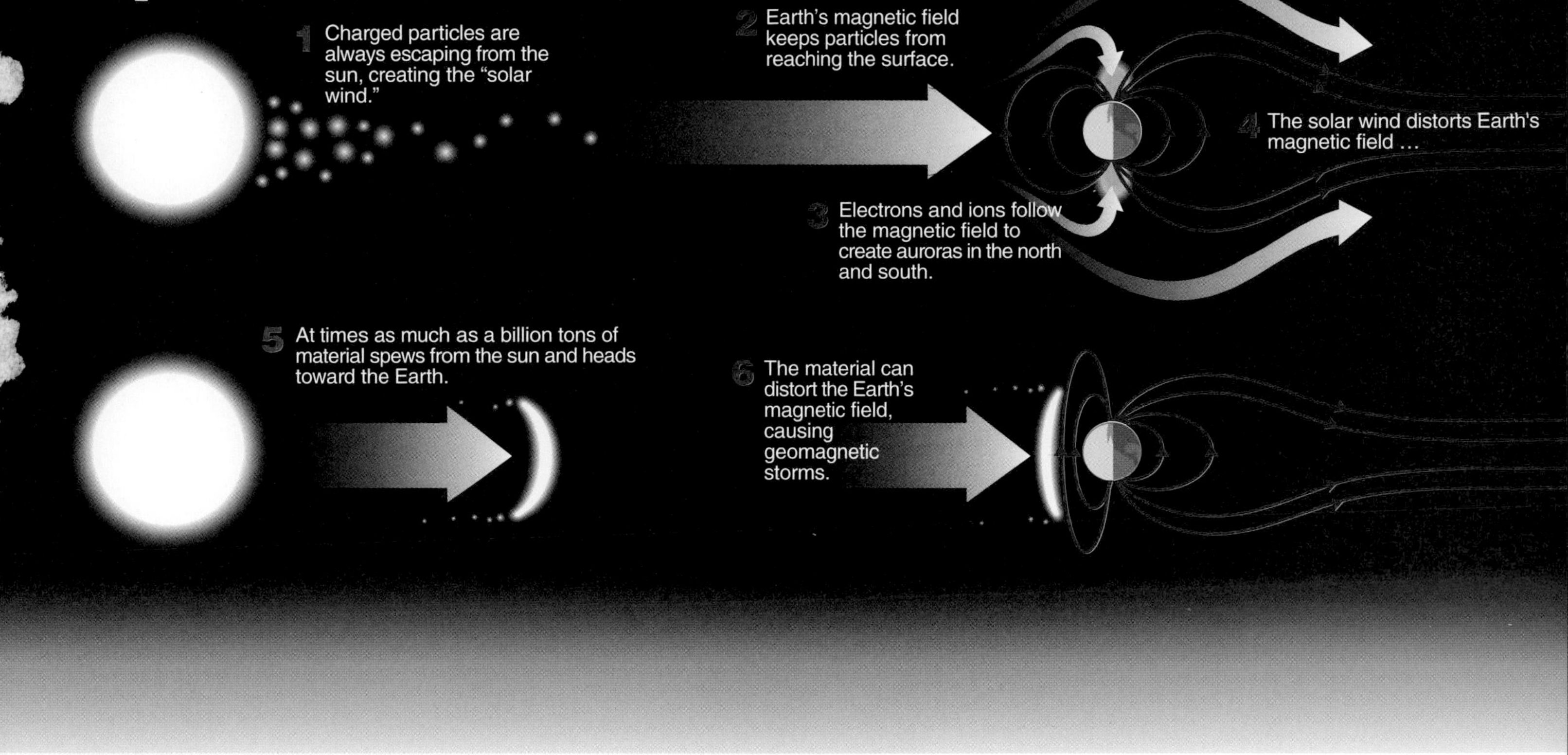

1 Charged particles are always escaping from the sun, creating the "solar wind."

2 Earth's magnetic field keeps particles from reaching the surface.

3 Electrons and ions follow the magnetic field to create auroras in the north and south.

4 The solar wind distorts Earth's magnetic field ...

5 At times as much as a billion tons of material spews from the sun and heads toward the Earth.

6 The material can distort the Earth's magnetic field, causing geomagnetic storms.

come a month or more after the summer solstice when days are longest.

The air at any one location is like a bank account. If money is added, the account grows. If heat is added to the air, it grows warmer. Earth is always losing heat, like a bank account with some money being regularly withdrawn.

As days grow longer after the spring equinox, eventually more heat is arriving than leaving. This is like adding money to the account faster than withdrawing it. The air grows warmer. On the longest days, the amount of heat arriving is greatest. But even after the days begin growing shorter, the amount of heat arriving is more than the amount leaving. It's like continuing to add money to the bank account, even though you're adding less than before.

Sometime in the late summer (usually mid-August in the Northern Hemisphere), the heat "account" is in balance. From then on, more heat is leaving than arriving and the days grow cooler.

In December, when days are shortest, the "withdrawals" from the heat account are greatest. But even after the days start growing longer, more heat is leaving than arriving. The heat account is growing smaller, even though less heat is leaving. This is why the year's coldest Northern Hemisphere days are normally in January or even early February, not on winter solstice Dec. 21.

Sunlight is only the beginning

Earth's orbit determines how much solar energy hits the top of the atmosphere, but not all of this energy reaches the ground. The amount reaching the surface to be absorbed is a key to determining the Earth's temperature. How much energy reaches the ground or oceans depends on whether the sky is clear or cloudy and whether the surface is dark or light.

Thick clouds, for instance, can reflect as much as 95 percent of the solar energy hitting them. If the skies are perfectly clear, more solar energy will reach the ground. But if the ground is covered with sparkling, white snow, it will reflect as much as 95 percent of the sun's energy. A forest or ocean, on the other hand, will reflect less than 10 percent of the sunlight hitting them.

Most of the sun's radiation is visible light and what's known as the "near infrared." The sun also radiates ultraviolet waves and even tiny amounts of microwaves and radio waves. When they reach the Earth, these forms of radiation help start the complex chain of events that give our planet its pattern of air and water temperatures.

Just as important, the Earth and its atmosphere aren't only absorbing solar energy, they're also giving off energy. Everything, including you, emits energy. The kinds and amounts depend on temperature.

Coals in a barbecue are a good example. The flame

People like to call the solstices and equinoxes the "official" beginnings of seasons. But meteorologists in the U.S. consider winter to run from December through February; spring is March through May; summer is June through August; and fall is September through November.

17

Solving 'The Case of the Missing Sunspots'

At first glance, looking for a link between sunspots and the weather makes sense. The sun drives the weather. Sunspots — large black spots on the sun — appear when the sun's energy output increases, so shouldn't these predictable sun changes affect temperature?

Since the 1870s people have been finding apparent links. Water levels on Africa's Lake Victoria seemed to go up and down with the number of sunspots. Rainfall here and there seemed to vary with them. So did some droughts.

"Sunspots have an almost hypnotic fascination for people who work with statistics," says solar astronomer Jack Eddy. "Sunspots seem so predictable and regular. We want something so badly that will give us a handle for predictions. We can tell when the next sunspot minimum and maximum are going to be. How nice if we could use them to predict monsoons in India."

The result? "A trail of broken hearts," says Eddy. Weather events that seem to be marching in step with sunspots for a few cycles stumble sooner or later.

"From everything we know about the inputs of the sun, everything we know about the Earth's atmosphere, any effects of solar activity on the short-term weather, shorter than a decade, will be minor," Eddy says.

This is where the critical difference between climate and weather comes in. Generally, climate is the average weather in a location over a long period of time. Most climate averages are based on 30 years of records. And yes, sunspots can affect climate. Jack Eddy showed that.

He was the "detective" who solved *The Case of the Missing Sunspots*. That was the title of his well-known 1977 *Scientific American* article. As an astronomer at the High Altitude Observatory in Boulder, Colo., Eddy unearthed convincing evidence that a prolonged cold spell in northern Europe — called the Little Ice Age — was connected to an almost complete lack of sunspots between about 1645 and 1715. He also showed that over the years the sun has gone through other periods of decreased energy output that affected climate.

Solar astronomer Jack Eddy: Sunspots can cause changes in the climate but are not reliable guides to the weather.

The sun's energy output increases when the most spots appear. Few spots mean the sun is producing slightly less energy. The difference is tiny. The decrease from the peak to the valley of the 11-year sunspot cycle is less than two-tenths of one percent.

Eddy offers a simple analogy to explain why a slight solar energy decrease could cause the "Little Ice Age," but have little effect on day-to-day weather. If you flip a room's thermostat that had been set at 70°F from 67°F to 73°F every few minutes, the temperature would stay near 70°F. The temperature of a room can't change by even a couple of degrees in a couple of minutes. But, if you set the thermostat at 67°F for an hour or so, the room cools.

Attempts to use sunspots to forecast weather fail because weather is too complicated. It's extremely complex, even if you consider only what the atmosphere is doing. When you toss in the oceans and the Earth's living things, along with the sun — all of which affect the weather and climate — basing predictions on sunspots alone is just too simplistic.

Even though he's shown how solar changes affect climate, Eddy says such changes are, at best, minor players in possible future climate shifts. "My own prejudice is that in the short term the sun is a very benign and reliable star." At the worst, solar changes could heat or cool the earth by about two degrees. In contrast, he says, atmospheric changes being caused by people — the "greenhouse effect" — could warm the Earth by seven degrees in the next half century.

Sunspots were reported as far back as the first century B.C. and astronomers in the Orient kept records of spots seen with the unaided eye, mostly at sunrise and sunset. But sunspots were mostly ignored in the West until 1611, when Galileo observed them with his telescope. In 1843, Heinrich Schwabe, a German amateur astronomer, was the first person to note that sunspots come and go in a more or less regular pattern. During the down part of the roughly 11-year cycle, there can be as few as a dozen sunspots in a year. During a peak in the cycle, they are seen by the thousands.

Boiling water shows how heat travels

Conduction
The transfer of heat by the collisions of molecules. Metal conducts heat very well. Air and other insulators are poor heat conductors.

Convection
The transfer of heat by actual movement of the heated material. Meteorologists usually use "convection" to refer to up and down motions of air. They use "advection" to describe the horizontal movement.

Radiation
The transfer of heat by wave motion. Heat, light and other kinds of waves radiate through the near vacuum of outer space from the sun to the Earth.

The stove's heat causes molecules in the pan's bottom to vibrate faster making it hotter. These vibrating molecules collide with their neighboring molecules, making them vibrate faster too. After a while, molecules in the pan's handle are vibrating so fast it's too hot to touch.

As water in the bottom of the pan heats up, some of it vaporizes making bubbles. The bubbles rise because they're lighter than the surrounding water. Water sinks from the top to replace the rising bubbles. This up and down movement eventually heats all the water.

Even though air conducts very little heat, you can feel the heat coming from the hot stove and the pan. It's radiating through the air.

These drawings look at the various ways heat travels, including how solar energy gets from the sun to the Earth. While we're most familiar with the sun's light, it is also sending energy in the form of X-rays, ultraviolet rays, infrared rays and even microwaves and radio waves. All of these different kinds of energy are electromagnetic radiation. This energy travels through space as waves, but unlike ocean waves that move the water they travel in, electromagnetic radiation consists of magnetic fields and electric fields that do not move the air. We can't see these fields, but they are real and can travel through vacuums.

starts out bright orange or red but turns to a dull red as it dies. Even after all the glow is gone and no more visible light energy is being emitted, you can still feel the infrared energy, the heat, from the coals.

The mostly infrared energy being emitted by the Earth, by the atmosphere and by clouds combines with incoming solar energy in a beautiful balancing act that determines the temperatures of the air and the Earth's surface.

How the balancing act works

About 30 percent of the solar energy reaching the Earth is reflected; it never helps heat the Earth. About 20 percent is absorbed at various heights in the atmosphere. The remaining 50 percent is absorbed by the ground and oceans.

Much ultraviolet energy is absorbed high in the atmosphere, while the sun's infrared energy is absorbed at lower altitudes. Most of the visible light makes it through the atmosphere without warming the atmosphere and is finally either reflected or absorbed by the ground.

Anything that absorbs energy warms up. Absorbed ultraviolet energy warms the upper atmosphere. Infrared energy, which is always being radiated by the Earth, is absorbed by the tiny water drops in clouds and by water

How much solar energy is reflected back into space

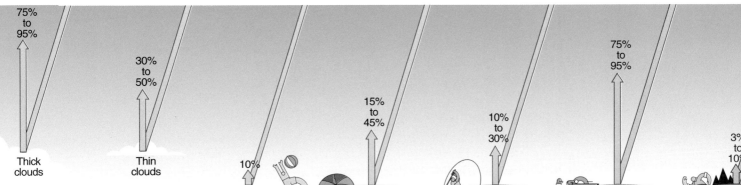

75% to 95% — Thick clouds

30% to 50% — Thin clouds

10% — Water

15% to 45% — Sand

10% to 30% — Grassy field

75% to 95% — Fresh snow

3% to 10% — Forest

vapor and other gases in the atmosphere. This warms the air. Clouds that absorb infrared energy radiate some back to space, some back toward Earth. This is why, everything else being equal, cloudy nights tend to be warmer than clear nights. Clouds act like an infrared blanket.

Heat and cold balance out

If we look at the entire Earth over long periods of time, the heat account seems to be in balance. While many scientists think the Earth is slowly warming up — it has cooled and warmed many times in the past — the evidence isn't clear-cut.

This is what the debate over the "greenhouse effect" is all about. Many scientists fear that pollution is changing the atmosphere in ways that will reduce the amount of infrared energy being sent into space. If that happens, the Earth will warm up before a new balance between incoming and outgoing energy is struck.

While the Earth's heat budget is in balance overall, there are individual places on Earth that are out of balance. Winter cold and summer heat are examples of local imbalances that change with the seasons.

Tropical oceans or forests under clear skies will absorb a huge share of the solar energy hitting the Earth during 12 hours of daylight. The air warms up.

In Alaska snow-covered ground will absorb hardly any solar energy during the three hours of winter daylight. But the cold ground will continue to radiate away infrared energy, growing even colder. The coldest air on Earth comes from polar regions where winter nights are cloudless. Clear skies allow even more infrared energy to be lost to space. The ground grows colder because it's losing more energy than it's receiving.

Scientists are just beginning to develop ways to measure energy deficits and surpluses over wide areas. At the National Aeronautics and Space Administration's (NASA's) Langley Research Center in Hampton, Va., scientists are using satellites to get worldwide measurements of the amounts of energy coming and going from the ground.

The maps on Page 21 are from NASA's Langley scientists. These maps, made in January and July, show which areas of the Earth are losing radiative energy and which ones are gaining. They do not show energy — heat and cold — being moved by the air and oceans.

Air picks up heat and cold

Water and land heat or cool the air above them, creating masses of hot, warm, cool or cold air.

These air masses are huge, at least the size of several states. They form when the same air stays over an area for days, perhaps even a week or two, long enough for the air to warm up or cool down as it tries to match the temperature of the ground or ocean below. In addition to becoming warm or cool, air masses also grow dry or humid to match the area where they form.

Air masses form mostly in the polar or tropical regions; the air in the middle latitudes, which include most of the U.S., doesn't stay still long enough. When air masses move from their home regions they bring us heat waves and cold spells.

The air masses that control the USA's weather are:

• Continental polar air masses, which often originate over Alaska and northwestern Canada. The illustration on Page 22 shows a typical early winter air mass that might grow up around Fairbanks, Alaska, and then move south into the Rockies or Plains.

This cold air is warmed as it moves south over warmer ground. Even in the dead of winter, such heating can be significant. An air mass that leaves Alaska or southern Canada in January with daily high-low ranges from -20°F to -40°F will warm to a range of perhaps 25°F to 50°F by the time it reaches the Gulf Coast, Florida or Southern California.

Bitterly cold air, the kind that drops temperatures to around 20°F in Orlando, isn't likely to reach Florida at all unless the ground over much of the Midwest and South is covered by snow.

• Maritime polar air masses, which build up over the northern oceans. These aren't as cold as continental air masses — they rarely cool below 32°F — but they are more humid. The chilly, damp weather that moves into the Pacific Northwest is caused by maritime polar air masses from the northern Pacific Ocean.

• Maritime tropical air masses, which form over the warm water of the Gulf of Mexico and the Caribbean Sea. Warm air masses that grow over the Pacific Ocean south and west of California normally have little effect on the United States. But the Pacific Ocean water off the California Coast is chilly, cooling tropical Pacific air heading for California. Sometimes chilly, polar maritime air masses move into California in the winter. California's coldest winter weather, however, comes from continental polar air masses that move

Where the Earth gains/loses radiative energy in July ...

These computer-generated pictures from a half dozen satellites show how the Earth uses the sun's energy in summer and winter. The images combine measurements of solar energy arriving at the ground; solar energy reflected away; infrared energy emitted by the Earth, and infrared energy the Earth absorbs from the atmosphere to show total energy being gained or lost through radiation by the ground and oceans.

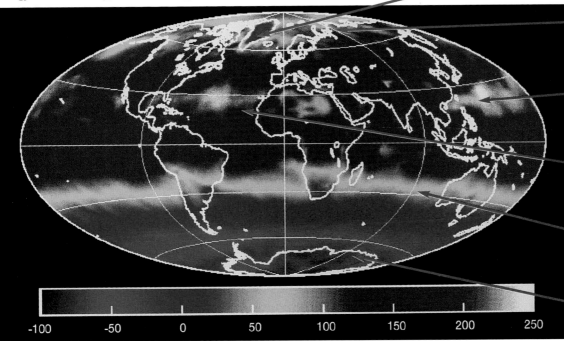

Greenland's ice cap reflects sunlight; it isn't warming up.

Snow and ice have melted in other parts of the Arctic; ground is warming.

Greatest warming is in ocean areas with few clouds overhead.

The sun is more directly overhead here, but clouds block some of its energy.

Days are too short, sun too low in the sky to do much warming south of this area.

With little or no sunlight, Antarctica grows colder and colder.

The color keys on both pictures show the average number of watts per square meter of energy being absorbed or lost by the land and sea. Minus numbers show where the Earth's surface is losing heat. Zero shows where the energy coming in and going out is in balance. Plus numbers show energy gain.

... and in a typical January

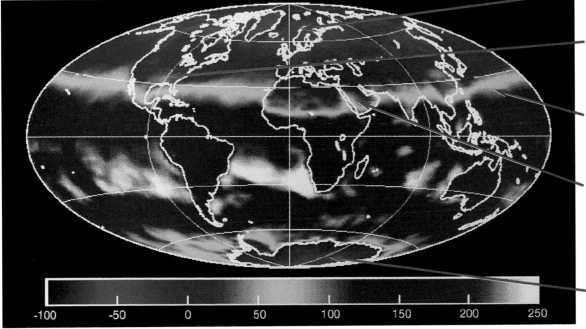

Polar regions are still cooling off even as days grow longer.

Heat gains and losses are close to being in balance for most of the United States.

Yellow band shows where sun's energy is strong enough to begin heating the surface.

Deserts of North Africa and Saudi Arabia reflect more solar energy than nearby oceans; clear skies allow more infrared energy to escape. Result: cold desert nights.

Antarctic snow reflects most sunlight. Nearly 24 hours of sunlight warms the ground very little.

south, then across the Sierras into California.

• Continental tropical air masses, which form over the deserts of the Southwest and Mexico. These air masses give the Southwest its warm, dry climate, but they don't move into other parts of the United States as often as the maritime tropical air masses or polar air masses.

Water and air move heat

If the polar and tropical air masses didn't move south and north, the Earth's poles would grow colder because the polar summer is too short to make up for winter's heat deficit. The tropics would grow warmer since they have no real winter to counteract the heat surplus. But the poles aren't getting colder and the tropics aren't getting hotter.

The Earth balances its heat budget by moving cold water and air from the poles toward the tropics and warm water and air from the tropics toward the poles.

Ocean currents, such as the Gulf Stream in the Atlantic and the Alaska Current in the Pacific, account for 40 to 50 percent of Earth's heat budget balancing. Warm and cold air masses moving north or south do the rest.

The hot-cold balancing act is always going on. This is where our weather begins, bringing not just hot or cold temperatures, but also — as we'll see — stirring up other kinds of weather.

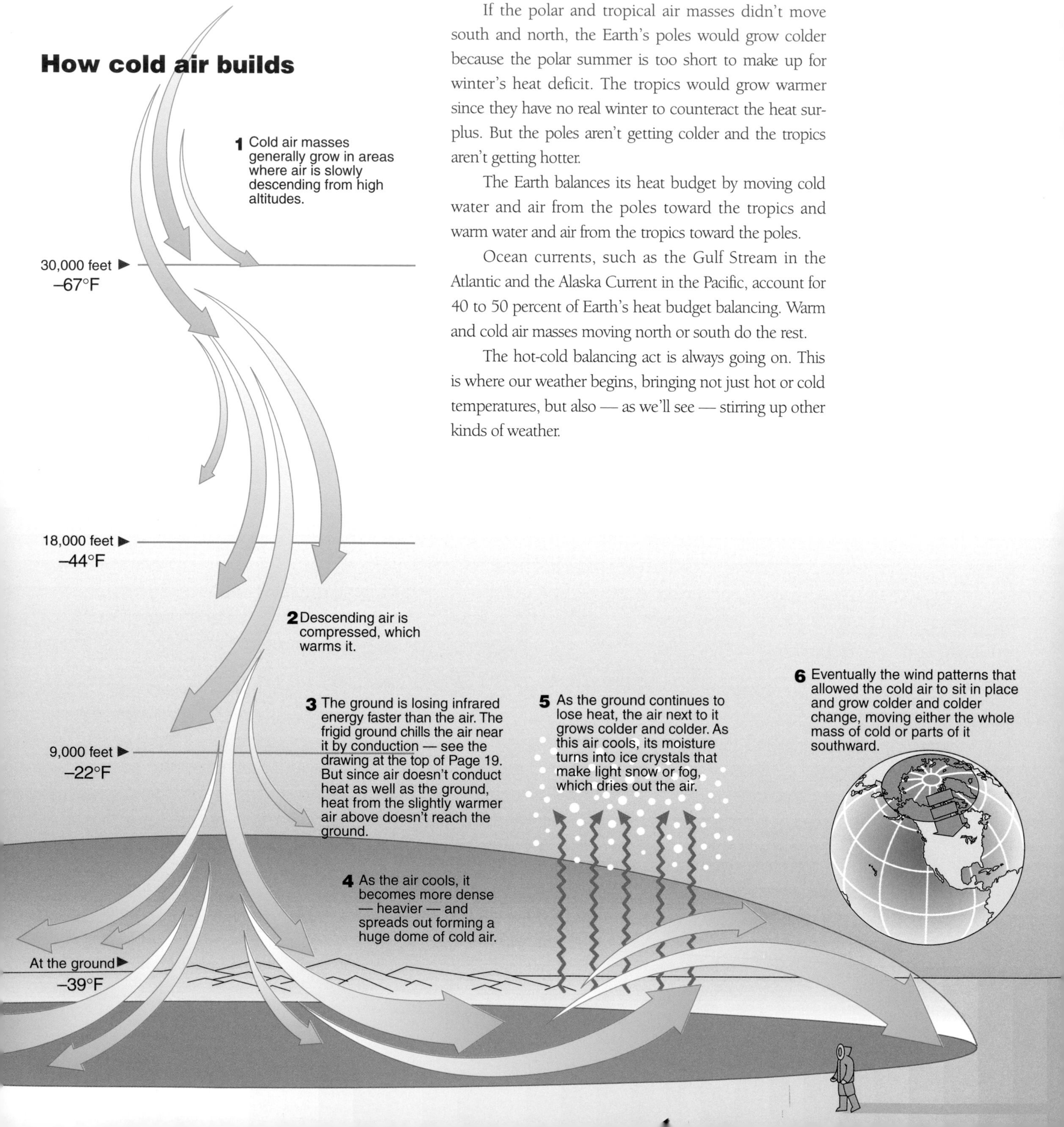

How cold air builds

1 Cold air masses generally grow in areas where air is slowly descending from high altitudes.

30,000 feet ▶
−67°F

18,000 feet ▶
−44°F

2 Descending air is compressed, which warms it.

3 The ground is losing infrared energy faster than the air. The frigid ground chills the air near it by conduction — see the drawing at the top of Page 19. But since air doesn't conduct heat as well as the ground, heat from the slightly warmer air above doesn't reach the ground.

5 As the ground continues to lose heat, the air next to it grows colder and colder. As this air cools, its moisture turns into ice crystals that make light snow or fog, which dries out the air.

6 Eventually the wind patterns that allowed the cold air to sit in place and grow colder and colder change, moving either the whole mass of cold or parts of it southward.

9,000 feet ▶
−22°F

4 As the air cools, it becomes more dense — heavier — and spreads out forming a huge dome of cold air.

At the ground ▶
−39°F

How snow enhances cold

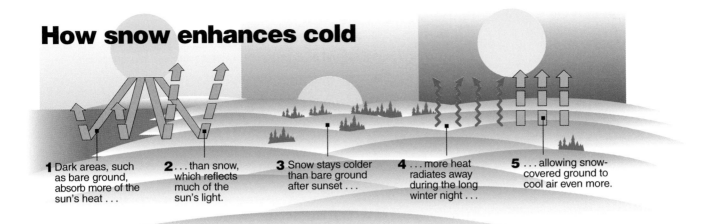

1 Dark areas, such as bare ground, absorb more of the sun's heat . . .

2 . . . than snow, which reflects much of the sun's light.

3 Snow stays colder than bare ground after sunset . . .

4 . . . more heat radiates away during the long winter night . . .

5 . . . allowing snow-covered ground to cool air even more.

More effects of the sun

Long before they began writing down history, our ancestors realized the sun makes the Earth livable. They worshipped sun gods. They celebrated the winter solstice, when the sun begins to spend more time in the sky each day until midsummer.

The sun was a benign god that brought heat and light but no violence.

In the far north, shimmering lights in the night sky inspired awe, but no one associated these auroras with the sun.

In reality, the sun's atmosphere is an incredibly violent place that sends a "solar wind" of charged particles at speeds faster than 600,000 mph far into space, around and past the Earth. (See graphic, Page 17.)

Earth's magnetic field shields us from the solar wind's particle bombardment. As we're being shielded, the solar wind molds our magnetic field into a cometlike shape with a long "magnetotail" stretching 4 million miles out from the Earth's dark side.

Harmless auroras in the northern and southern skies are the only visible signs of the solar wind's bombardment. The solar wind energizes electrons and other charged particles in the Earth's magnetic field. These particles follow the field's lines to the northern and southern magnetic poles. Here, more than 50 miles above the ground, they collide with molecules of nitrogen and oxygen, making them glow as auroras. The more vigorous the solar wind, the bigger the auroras.

Scientific understanding of all this has grown with the space age as researchers collect data from satellites and probes of the solar system.

This knowledge of "space weather" is becoming more important because what goes on inside and around the sun affects satellites, radio transmissions and even the electricity coming into our homes.

On the night of March 12-13, 1989, National Weather Service stations as far south as Brunswick, Ga., reported auroras. These are rarely seen very far south of the Canadian border.

At 2:44 a.m. on the 13th, a voltage regulator on a power line running from hydropower dams in northern Quebec shut down. Less than 90 seconds later, lights flickered out in Montreal and most of the rest of Canada's Quebec Province. Some places were without power

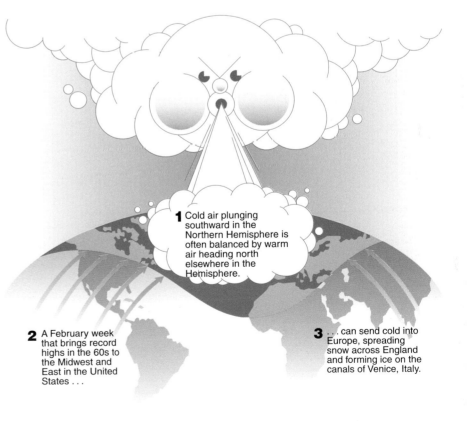

1 Cold air plunging southward in the Northern Hemisphere is often balanced by warm air heading north elsewhere in the Hemisphere.

2 A February week that brings record highs in the 60s to the Midwest and East in the United States . . .

3 . . . can send cold into Europe, spreading snow across England and forming ice on the canals of Venice, Italy.

How hot air builds up

Most of the hot, humid air masses that bring sweltering weather to the United States build up over the Gulf of Mexico and the Caribbean Sea. Hot, humid air also reaches the United States from the Atlantic and Pacific oceans, but not as often. Sometimes summer heat waves form as hot, dry air over the deserts of the Southwestern United States and northern Mexico.

Maritime tropical air masses

2 The air above the ocean warms to similar temperatures.

3 Warm water evaporates into the air, making it extremely humid.

4 Rising warm, humid air often condenses into clouds, showers or thunderstorms.

5 As hot, humid air moves inland in summer the warm ground can heat it even more.

1 Tropical oceans warm into the 80s in the summer.

Continental tropical air masses

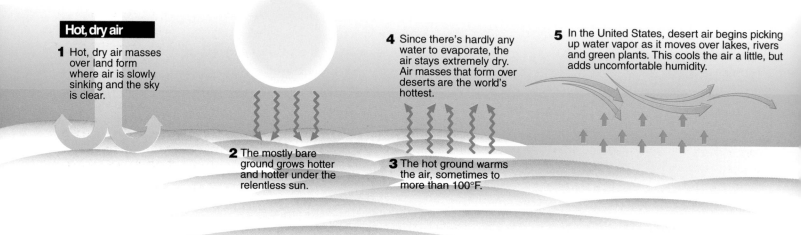

Hot, dry air

1 Hot, dry air masses over land form where air is slowly sinking and the sky is clear.

2 The mostly bare ground grows hotter and hotter under the relentless sun.

3 The hot ground warms the air, sometimes to more than 100°F.

4 Since there's hardly any water to evaporate, the air stays extremely dry. Air masses that form over deserts are the world's hottest.

5 In the United States, desert air begins picking up water vapor as it moves over lakes, rivers and green plants. This cools the air a little, but adds uncomfortable humidity.

for more than a week.

The villain was a "geomagnetic storm," triggered when a huge blob of charged particles smashed into the Earth's magnetic field.

The storm caused the magnetic field to bounce back and forth, creating or inducing currents in long electrical conductors such as power lines, pipelines and railroad tracks. Moving an electrical conductor, such as a wire, through a magnetic field induces an electric current in the conductor. This is how generators and alternators make electricity.

These currents are relatively weak, but they are enough to burn out electrical equipment and shut down power grids. In pipelines, they cause flow meters to give false readings. In railroad tracks, they can interfere with electrical devices that help locate trains.

Satellites, especially the more than 600 geostationary satellites in orbits about 22,500 miles above the equator, are in even more danger. These satellites include those that handle an increasing amount of the globe's communications as well as ones that watch the weather.

Over the years builders have created lighter satellites with smaller electronic systems to handle more complex tasks. But these smaller, lighter satellites are more at risk. Geomagnetic storms can cause computer chips to issue false commands, including commands to turn on thruster rockets, which control satellite movements. This wastes valuable fuel and shortens the satellite's useful life.

Solar activity has little effect on television and commercial radio stations, but it does disrupt ground-to-air, ship-to-shore, and shortwave radio transmissions.

The Global Positioning System, which provides precise navigation data via radio from a system of satellites, is vulnerable to solar disruption.

These problems are likely to get worse at the turn of the century. Solar activity varies in a roughly 11-year cycle that is expected to build to a new peak around the year 2000. The Quebec power outage came near the peak of the last cycle.

Little wonder we are hearing more about the study of "space weather."

For more than 30 years, the National Oceanic and Atmospheric Administration and the Air Force operated the Space Environment Services Center in Boulder, Colo. It is the national and world warning center for solar disturbances.

In 1996, the center became part of the National Weather Service's National Centers for Environmental Prediction, which includes the groups that do the basic weather forecasting for the U.S. as well as specialized operations such as the National Hurricane Center. The move shows the growing importance of space weather forecasting.

But this new science is in its infancy. Space weather scientists say their knowledge today is roughly at the stage ordinary meteorology was in 1945. They need better tools to make proper forecasts.

Over the next few years, experimental satellites will be put in orbits between the sun and Earth. The extra data will help forecasters. But, at best, the new information would let forecasters give a few hours warning that a solar storm is on the way.

Scientists hope one day to understand enough about how the sun works to be able to predict solar storms before they are heading for us.

Heat is the USA's biggest weather killer

In 1936, the hottest summer on record for many parts of the U.S., 4,678 deaths were blamed on heat.

Nearly 60 years later, 1,000 people died in a July 1995 Midwest heat wave.

How could a heat wave have contributed to so many deaths in 1995 in a country with the world's most sophisticated weather forecasting system?

A disaster survey team from the National Oceanic and Atmospheric Administration noted in its report that "the National Weather Service issued warnings of the developing heat wave several days in advance, which were quickly broadcast by the local media. Given this advance warning, many, if not all, of the heat-related deaths associated with this event were preventable."

What went wrong?

People didn't consider the impending heat wave as life-threatening. Many who would have evacuated for a flood or hurricane figured a heat wave couldn't be so bad. Older people didn't move to air-conditioned shelters or take other precautions. The disaster survey team said the heat wave's magnitude was so unusual that it was not immediately recognized as an emergency. "Unfortunately, a heat wave connotes discomfort, not violence; inconvenience, not alarm," the report said.

While heat does kill young, healthy people who push themselves too hard working or exercising outdoors, most victims are poor, elderly people.

The elderly are the most vulnerable because heat puts more stress on weak hearts and on bodies less capable of controlling internal temperatures. Poor people aren't likely to have air conditioning. Heat kills more poor people in the large cities of the Midwest and East than in the South, mainly because of housing differences.

Laurence Kalkstein, a University of Delaware climatologist, studies and even predicts how death rates climb during heat waves. In a city such as Chicago, "If it gets to be 100 outside," Kalkstein says, "and you're on the top floor in a multistory tenement made of red brick, with a flat, black tar roof and there's not much cloud cover, your room could get to 120 or 130 degrees." Such buildings are easier to heat during cold winters.

In a Southern city, in contrast, a poor person is more likely to live "in a one-story building with windows on all four sides. It's likely to be white with a steep roof made of a shiny metal. Those factors add up to a much cooler dwelling."

Fear of crime adds to the danger. Newspaper photos from heat waves early in the 1900s show people in cities escaping their hot houses by sleeping on fire escapes or in city parks, practices that would be considered unsafe today.

The National Weather Service uses the "heat index" or "apparent temperature," based on a chart like the one on Page 26 as a measure of danger. The weather service issues an excessive heat warning when the heat index is forecast to reach more than 105° F for more than three hours a day for two consecutive days or when the index is expected to top 115° F for any length of time.

To understand the deadliness of the 1995 Chicago heat wave, look at high and low temperatures for the peak days, July 12-16. In July, Chicago's average high temperature is about 83°F and the average low is about 63°F. But over those five days, the afternoons were hot and the nights never got below 75°F, a sign of high humidity. The figures here are from Chicago's Midway Airport, which was hotter than Chicago O'Hare International. Heat index figures are for each day's high temperature. An index of 105°F to 130°F is in the dangerous "very hot" range.

Chicago Midway Airport

Date	Daily high	Daily low	Heat index
7/12	97	75	105
7/13	106	81	126
7/14	102	84	122
7/15	98	77	112
7/16	93	76	102

When winter winds howl, you hear warnings about the wind chill factor. Wind chill shows how cold the wind makes exposed flesh feel and is a good indication of how much danger you face of frostbite or hypothermia. The lower the chill, the more you should bundle up.

How to find wind chill

Draw a line down from the temperature on top. Draw another line across from wind speed on the left. Where they meet is how cold it feels.

Wind chill index
Temperature in degrees Farenheit

Wind	30°	25°	20°	15°	10°	5°	0°
15 mph	9°	2°	-5°	-11°	-18°	-25°	-31°
20 mph	4°	-3°	-10°	-17°	-24°	-31°	-39°
25 mph	1°	-7°	-15°	-22°	-29°	-36°	-44°
30 mph	-2°	-10°	-18°	-25°	-33°	-41°	-49°

Example: With 20 mph wind, 10°F temperature, the wind chill is –24°F.

The coldest NFL Championship Game was Dec. 31, 1967, when the temperature was -13°F at Green Bay, Wis. The Green Bay Packers defeated the Dallas Cowboys 21-17 as 50,861 fans watched. The Packers went on to win the Super Bowl.

While the heat index might be a good guide to whether someone should play tennis at 2 p.m. on a hot day, it's not a good measure of the total heat danger.

More than temperature and humidity are involved in the most dangerous heat waves. "People respond to the entire umbrella of air [or air mass] that surrounds them rather than to individual weather elements such as temperature," Kalkstein says.

How to find apparent temperature

The combination of heat and humidity gives the "apparent temperature," which is a measure of how dangerous the combination is. To find the apparent temperature: Find the temperature on the left side. Find the relative humidity across the top. The curved line where they meet is the apparent temperature. Follow the temperature and humidity lines. The color of the area where they meet tells you the danger.

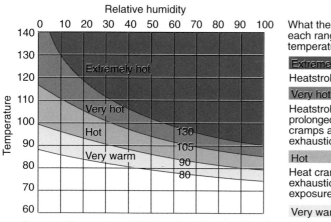

Example: Temperature 100° F and relative humidity of 60 percent. The apparent temperature is 130° on the edge of the "extremely hot" range.

What the dangers are in each range of apparent temperatures

Extremely hot
Heatstroke imminent.

Very hot
Heatstroke possible with prolonged exposure. Heat cramps and heat exhaustion likely.

Hot
Heat cramps and heat exhaustion possible with exposure.

Very warm
Physical activity could be more fatiguing than usual.

In their research, Kalkstein and his colleagues have grouped various combinations of weather factors that commonly go together into air-mass categories. The air temperature, humidity, the amount of sky covered by clouds, the air pressure, the wind speed and the wind direction are all used to determine the type of air mass that's over a city each day. Clouds are important, for instance, because the sun burning down in a clear sky heats the roofs of buildings faster than on cloudy days.

While heat and humidity are most important, these other factors also help determine the heat danger. The number of days a dangerous air mass has been around is also important because as a heat wave continues, the death rate goes up.

During the summer of 1995, Kalkstein's group at the University of Delaware worked with Philadelphia health officials and the National Weather Service office to issue heat warnings based on the kind of air mass over the area, not just how hot and humid a day was.

"The warnings did save lives," Kalkstein says "but it's hard to say how many." The study calculated how many people might die during particular weather conditions based on past Philadelphia heat waves. "One reason for thinking the warnings saved lives is that as the season wore on, our calculations of deaths more and more overestimated the actual number," Kalkstein says. Local health officials and weather services are starting to use similar warning systems in other cities. But issuing more accurate warnings is only the beginning.

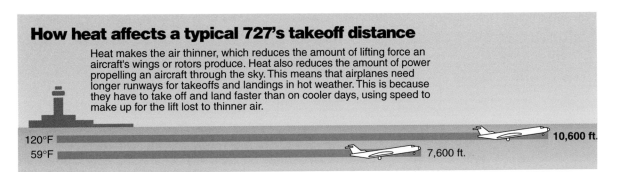

How heat affects a typical 727's takeoff distance

Heat makes the air thinner, which reduces the amount of lifting force an aircraft's wings or rotors produce. Heat also reduces the amount of power propelling an aircraft through the sky. This means that airplanes need longer runways for takeoffs and landings in hot weather. This is because they have to take off and land faster than on cooler days, using speed to make up for the lift lost to thinner air.

120°F — 10,600 ft.
59°F — 7,600 ft.

When heat alerts are issued, more cities are making special efforts to get those at risk, such as the elderly, to go to air-conditioned shelters for at least part of the day. They also try to make sure that someone checks regularly on those at risk.

In the past, cities and charities often gave electric fans to poor people who were endangered by heat. Kalkstein says cities are stopping the practice because in temperatures above the mid-80s, a hot breeze can add to the problem. "Imagine standing in front of the open door of a pizza oven with a fan blowing the heat onto you," he says. Dry breezes can be dangerous because they speed dehydration, one of heat's biggest dangers even to healthy people.

Illnesses caused directly by heat range from annoying but relatively harmless heat cramps to life-threatening heatstroke:

• Heat cramps. Exercising in hot weather can lead to muscle cramps, especially in the legs, because of brief imbalances in body salts.

• Heat syncope or fainting. Those not used to exercising in the heat can have a quick drop in blood pressure that can lead to fainting. Recovery is usually quick. As with heat cramps, the cure is to take it easy until the body is used to the heat.

• Heat exhaustion. Losing fluid and salts through perspiration — or replacing them in an imbalanced way — can lead to dizziness and over all weakness. Body temperature might rise, but usually not above 102°F. Heat exhaustion is more likely after several days of hot weather than when a heat wave is just beginning. In some cases, victims, especially the elderly, must be hospitalized. The best defenses are to take it easy, get to a cool place, and drink plenty of water. Don't take salt tablets without consulting a doctor.

• Heatstroke. In some cases, the body's thermostat can falter in hot weather, causing the temperature to rise to 105°F or more. Other effects are lethargy, confusion, unconsciousness and perhaps death. If you suspect that someone might be suffering from heatstroke, get that person immediate medical aid.

Extreme cold

Kalkstein and his colleagues have found that extreme cold doesn't increase death rates nearly as much as extreme heat. Still, cold weather can be a killer.

Your body strives to keep its temperature close to 98.6°F. When the body's thermostat senses that it's becoming too cold, it tries to warm up by constricting blood vessels near the skin. This reduces the amount of blood flowing near the skin, so the blood loses less heat to the air. Shivering begins and that also warms the body.

The two key dangers of becoming too cold are hypothermia and frostbite.

• Hypothermia is the condition when the core body temperature drops below about 95°F. At first shivering is violent, but as the temperature falls, shivering decreases. Disorientation is often a sign of hypothermia. The result can be unconsciousness and possibly death.

Bitter cold isn't necessary for hypothermia. People who are very old or very young are more vulnerable because the body is less able to regulate its temperature. Older people who have cut their heat to save on fuel bills often become victims of hypothermia.

A key defense outdoors is to wear wool, not cotton, clothing. When cotton gets wet, it draws heat away from the body. Wool continues to insulate even when it's wet.

Temperatures sometimes climb above 120°F in the Southwest deserts, the hottest part of the United States. During a June 1990 heat wave in the Southwest, Phoenix's official temperature reached 122°F and one television station's thermometer "topped out" at 124°F.

In Phoenix, airlines grounded their Boeing 737s during the hottest part of the day because the charts pilots use to calculate takeoff distance didn't go above 120°F.

Energy from solar cells

Solar cells convert only a fraction of the sun's energy that hits them into electricity. Let's look at how much power a 10 percent efficient cell would produce in summer and winter across the Northern United States.

10 kilowatt hours of electricity produced

100 kilowatt hours of incoming solar energy

The average American house uses about 9,000 kilowatt-hours of electricity each year. Below, we see how large an area of 10 percent efficient solar cells would be needed to produce 9,000 kilowatt hours of electricity on June 21 and Dec. 21.

June 21 — Solar cells covering two football fields produce 9,000 kilowatt hours of electricity

Dec. 21 — Solar cells covering three football fields plus another 20 yards produce 9,000 kilowatt hours of electricity

Geomagnetic storms produce mayhem, mystery and marvelous challenge

The solar mysteries that JoAnn Joselyn is helping to solve make a lot of science fiction look unimaginative.

Imagine the sun spitting out energetic particles in blobs as large as the sun itself. Some of these blobs race toward Earth where they create geomagnetic storms that can knock out power grids and radio transmissions, send satellites into crazy motions, and create northern lights seen as far south as Georgia.

Until recently, scientists blamed solar flares, eruptions of energy from the sun, for geomagnetic storms.

But in the late 1970s, Joselyn was among the scientists piecing together facts and theories to show that a different kind of event, called "coronal mass ejections" (CMEs), cause geomagnetic storms on Earth. Joselyn is a senior scientist at the National Oceanic and Atmospheric Administration's Space Environment Center in Boulder, Colo., and one of the top solar scientists in the country.

The early recognition that flares were not the story "is the most satisfying event of my scientific career," Joselyn says. Solar scientists didn't generally accept the role of CMEs until the 1990s.

Plenty of questions remain about CMEs, says Joselyn. "Is this something that's just incidental to the sun, is it just a cute trick? Or is it something needed for the sun to be the sun?"

Joselyn credits being a "Sputnik child" for her scientific career. The Russians launched the Earth's first artificial satellite in October 1957, "on my 14th birthday," she says. Sputnik stirred efforts to get U.S. students interested in science. "It was very patriotic to go into space studies," she says. "We had to catch up with the Russians."

She earned a bachelor's degree in applied mathematics and a Ph.D. in astrogeophysics from the University of Colorado in Boulder.

"I've worked in Boulder my entire career

JoAnn Joselyn helped discover the cause of geomagnetic storms — "coronal mass ejections," which the sun spits out.

except for one little bit in Houston for NASA during the Gemini era."

Gemini was a two-man spacecraft that astronauts flew in 1965 and 1966. Gemini's job was to develop techniques that would be needed later for the Apollo spacecraft that carried astronauts to the moon and back beginning in 1969.

"During Gemini I worked on solar radiation problems. We were trying to think of how to predict giant episodes." These are the solar events, now known to be coronal mass ejections, that could have sent enough radiation into space to endanger astronauts on the moon.

While the Earth's magnetic field protects everyone on Earth from such radiation, astronauts, especially those working outside spacecraft with nothing but their space suits for protection, could be endangered.

But the solar danger to technology on Earth is growing. Learning more about CMEs could help improve predictions of geomagnetic storms, which makes Joselyn's work not only fascinating but also potentially vital to keeping today's technology going.

Joselyn recalls the excitement of an American Geophysical Union conference on CMEs that she helped organize in Bozeman, Mont.,

in the summer of 1996.

"I was so energized by the meeting," she says. "The new discoveries, seeing the new observations. We had 117 scientists from around the world with everyone trying out their explanations of what's going on."

Scientists at the meeting agreed to keep on using the term "coronal mass ejection." "We're still making up the terminology in this field. It's such a threshold science that even the words are still being invented."

Joselyn was also "pleased to see the progress women scientists are making." About 20 percent of the scientists there were women. She and Nancy Crooker of Boston University managed the conference's scientific program.

"There were enough women that you don't feel alone, unlike at Colorado University, where there were four women out of 500 or more in my class" at the engineering school.

When Joselyn was working on her undergraduate degree at Colorado's engineering school, "computers were just coming on as one of the things you could do with applied mathematics. I often wonder what would have happened to my career if I hadn't tracked off into space."

Maybe she would have become one of the millionaires who got into computers at the beginning.

"Working for the government, you won't make a million," she says. "But I wouldn't trade it for anything. It's a wonderful career. I find every day exciting as I try to understand what's happening on the sun. It's a terrific life."

Why 1995 Midwest heat was a killer

During July 1995 searing heat combined with extremely high humidity to kill more than 1,000 people in the Midwest. Researchers have found why the humidity was so high and are looking into past heat waves to see how often similar situations arise.

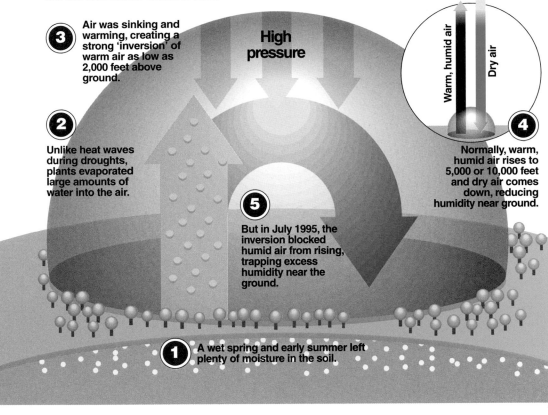

3 Air was sinking and warming, creating a strong 'inversion' of warm air as low as 2,000 feet above ground.

2 Unlike heat waves during droughts, plants evaporated large amounts of water into the air.

High pressure

Warm, humid air

Dry air

4 Normally, warm, humid air rises to 5,000 or 10,000 feet and dry air comes down, reducing humidity near ground.

5 But in July 1995, the inversion blocked humid air from rising, trapping excess humidity near the ground.

1 A wet spring and early summer left plenty of moisture in the soil.

No one really knows how many people heat kills because medical examiners define heat deaths in different ways. In many places, heat is listed as the cause of death only for those obviously and directly killed by heat, such as heat stroke victims. In other places, heat is considered the cause of death if it contributes in any way. The primary cause of death might have been a heart attack or other illness, but if heat is a contributing factor many medical examiners consider that a heat death.

If skin becomes cold enough, it can actually freeze. Ice crystals form and damage tissue. This is frostbite. Ears, nose, hands and feet are the most vulnerable. Mild frostbite usually causes no lasting problems. But more serious cases can lead to extreme sensitivity to cold that can last for years. In the most serious cases amputation may be necessary.

The defense is to be aware of the danger when it's bitterly cold, especially when the wind is blowing. Mittens, hats, a warm covering for the face and warm, dry socks can keep frostbite away.

The wind chill factor

In cold weather the human body works hard to stay warm. But wind can undercut the body's efforts and make it seem far colder than the thermometer reports.

Heat escaping from a body warms the air next to it. If the air is calm or nearly calm, this warmed air stays next to the body. But if the wind is blowing, it carries warm air away from around the body. The faster the wind, the faster the heat is carried away. More heat escapes; it feels colder.

The wind chill index factors in temperature and wind speed to explain not how cold it is, but how cold it feels. The wind chill index is a guide to bundling up.

Wind chill temperatures apply only to people and animals, not to objects like cars. If the temperature is 10°F with 30 mph winds — that's a wind chill of -33°F — a car will still cool only to 10°F.

Once a car cools to the actual air temperature — 10°F in this case — it can't get any colder no matter how hard the wind blows. This is an example of a basic scientific principle called the Second Law of Thermodynamics. Heat cannot "flow" to a warmer object. So on a 10°F night, no matter how hard the wind blows, a car can't get colder than 10°F.

Hot and cold air masses are only the beginning of weather. To create the constant panorama of changing skies, winds and temperatures, something has to set those air masses in motion. That something is air pressure — the force air exerts on everything it touches.

Air pressure is important for three reasons. It creates the wind — from the great seas of wind that blow hundreds of miles across the Earth to the lightest local sea breezes. Up and down movement of air in low- and high-pressure areas creates cloudy or clear skies. Finally, changes in air pressure give vital clues about what the weather will do over the next several hours.

Since the air's pressure at any one place rarely stays the same for long, the winds and weather are constantly changing.

What makes air pressure?

Even though we can't see it, air is real and has weight. It's made up mostly of molecules of nitrogen and oxygen. In the lower part of the atmosphere, air is about 78 percent nitrogen and 21 percent oxygen with other gases accounting for the other one percent. Here we will use the term "air molecules" to refer to all molecules of all the gases in air. As in any gas, air molecules are zipping around at incredible speeds. Near the Earth's surface, they're traveling an average of about 1,600 feet per second or 1,090 mph. Warm up the air and the average speed increases; cool it off, and it slows down. Air molecules are bumping into and bouncing off each other — and anything else that gets in their way.

The impacts of those billions of bouncing molecules cause pressure. Imagine a bathroom scale attached to a wall. Now start throwing baseballs at the scale. The weight indicator jumps every time a ball hits it. It's showing the force of the ball colliding with the scale. If enough people lobbed enough baseballs at the

Regional winds

<u>Chinook:</u> a warm wind down the east slope of the Rocky Mountains, often reaching hundreds of miles into the high plains.

<u>Haboob:</u> extremely severe dust storms that occur mostly north of the Sudan. Walls of dust can rise to several thousand feet.

<u>Santa Ana:</u> a warm wind descending from mountains into Southern California.

<u>Sirocco:</u> a warm wind of the Mediterranean, sweeping northward from the hot Sahara.

scale, the indicator wouldn't have time to move down between hits. The balls would apply such constant pressure, the indicator's tiny movements wouldn't be noticeable. What the indicator reports — the amount of force the balls are applying — would depend on how fast the baseballs are going and how frequently they are hitting the scale.

Right now air molecules are bouncing off you from every direction. You don't feel each tiny impact because the air has so many extremely tiny molecules moving so quickly. Near the ground, a box one inch on each side would hold around 400 sextillion air molecules. (That's 400 followed by 21 zeros.)

Some rules of air pressure

Air pressure depends on the number of air molecules in a given space and how fast they're moving. Heating air — or any other gas — increases its pressure unless you give it more room. Cooling air will decrease its pressure, if it's kept in the same amount of space. Squeezing air into less space increases its pressure. Allowing air to expand into a larger space decreases

pressure. Also, if you heated air and squeezed it into a smaller space at the same time, its pressure would zoom. The science behind all this is shown in the simple experiment below.

Air and gravity

Every molecule of air, even those 200 miles above the Earth where the atmosphere trails away to outer space, is being pulled downward by the force of gravity. So why aren't all of the air molecules pulled down into a huge clump of nitrogen and oxygen here at the surface? Air pressure is the answer.

Air molecules are zipping around creating pressure that pushes in all directions, including upward. That is how air pressure offsets gravity.

But gravity does have an effect. Pressure is greater at lower levels because the air's molecules are squeezed under the weight of the air above. At sea level, the average pressure is 14.7 pounds per square inch. Go up to 1,000 feet and it drops to about 14.1 pounds per square inch. At about 18,000 feet, the pressure is 7.3 pounds per square inch; about half of the atmosphere's

How air reacts to being squeezed, heated, cooled

You can use an ordinary bicycle pump to see how air pressure applies a force and how outside forces, as well as heat and cold change air pressure. First, plug the pump's air hose so that no air can escape.

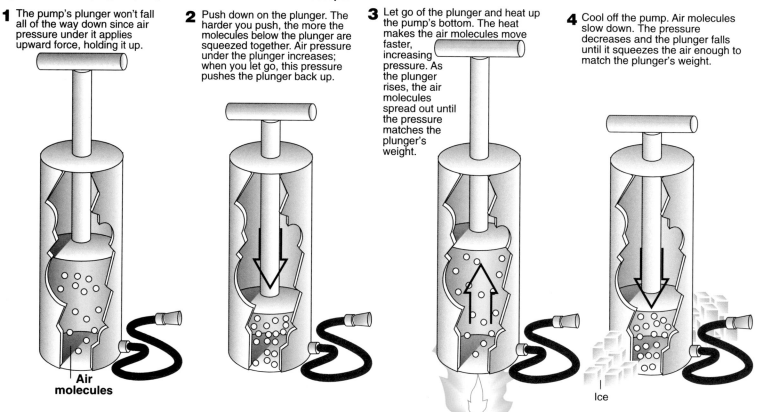

1 The pump's plunger won't fall all of the way down since air pressure under it applies upward force, holding it up.

Air molecules

2 Push down on the plunger. The harder you push, the more the molecules below the plunger are squeezed together. Air pressure under the plunger increases; when you let go, this pressure pushes the plunger back up.

3 Let go of the plunger and heat up the pump's bottom. The heat makes the air molecules move faster, increasing pressure. As the plunger rises, the air molecules spread out until the pressure matches the plunger's weight.

4 Cool off the pump. Air molecules slow down. The pressure decreases and the plunger falls until it squeezes the air enough to match the plunger's weight.

Ice

air molecules are below you; half above.

Air pressure aches

Since human beings evolved in an environment where the air pressure doesn't change rapidly, our bodies have trouble handling rapid decreases or increases in air pressure. That's why airplanes or even fast elevators can make us uncomfortable.

These quick changes in air pressure can cause several uncomfortable consequences:

Why do your ears "pop" as you go up or down in an airplane or a fast elevator? Air pressure inside and outside the middle ear normally is equal. But, as a person goes up, the outside pressure decreases, leaving the pressure inside the middle ear higher. The air inside the ear usually moves through the Eustachian tube to the throat, equalizing the pressure. However, if a person has a cold, the Eustachian tube might be blocked, making it harder for the air to leave the middle ear and causing a sense of fullness or pain.

Ear discomfort isn't the only symptom of pressure imbalance:

Sinuses are cavities in the bones in the front of the skull. If their openings are stopped up, say by swelling caused by an allergy, pressure changes can cause pain as the higher pressure inside the sinuses presses against nerves.

Gases in the stomach, for example from a carbonated drink, need to escape to lower pressure outside the body as you gain altitude.

Dental problems, such as a tooth abscess, can create tiny air pockets. When the air inside an abscess can't escape and presses against a tooth's nerve, the pain can be excruciating.

But there are far more serious side effects of a decrease in air pressure. If airplanes were not pressurized, as an airplane climbed, the decreasing air pressure would mean a decrease in oxygen.

Oxygen is vital to all the body, but oxygen depletion affects brain cells first with the most dangerous consequences, such as the loss of the ability to think clearly. Early symptoms of oxygen starvation — called hypoxia — vary from person to person. But eventually a victim becomes unconscious.

Most people who fly never have to worry about hypoxia because airliners are pressurized; the air pressure is kept at near-ground levels no matter how high the plane flies. For instance, the pressure in the cabin of an airliner flying around 35,000 feet is kept at

How pressure changes affect ears

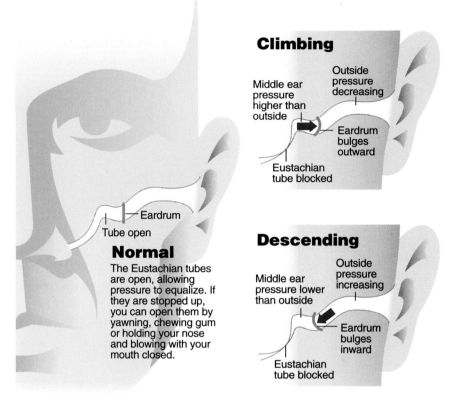

Climbing

Middle ear pressure higher than outside

Outside pressure decreasing

Eardrum bulges outward

Eustachian tube blocked

Eardrum
Tube open

Normal
The Eustachian tubes are open, allowing pressure to equalize. If they are stopped up, you can open them by yawning, chewing gum or holding your nose and blowing with your mouth closed.

Descending

Middle ear pressure lower than outside

Outside pressure increasing

Eardrum bulges inward

Eustachian tube blocked

the normal air pressure for 5,000 to 6,000 feet. But emergency oxygen masks must be available. And, the crews of non-pressurized aircraft must use oxygen masks whenever they are above 14,000 feet.

As an airliner climbs away from the airport, the air pressure inside is gradually lowered. While the plane might be climbing 2,000 feet each minute, the pressure changes as if in a 500 feet-per-minute climb, a more tolerable pressure change for humans. When the plane descends, the pressure normally changes at the rate of about 300 feet each minute, even though the actual descent is faster.

Mapping highs and lows

Stopped-up ears or aching sinuses tell people about air-pressure changes caused by going higher or lower in the atmosphere. Most people's bodies don't tell them about the important air-pressure changes associated with weather. Pressure changes too slowly from one place to another on the Earth's surface for most people to feel. These differences, however, are the ones that cause the winds. They're marked on weather maps by "H"s and "L"s.

The letters denote areas of high and low air

Most people need a barometer to detect air pressure changes, but some can "feel it in their bones." People with old injuries or arthritis often feel pain when bad weather is coming. One important factor is probably that the affected tissue is sensitive to changes in air pressure and normally the pressure drops when stormy weather is on the way.

How barometers work

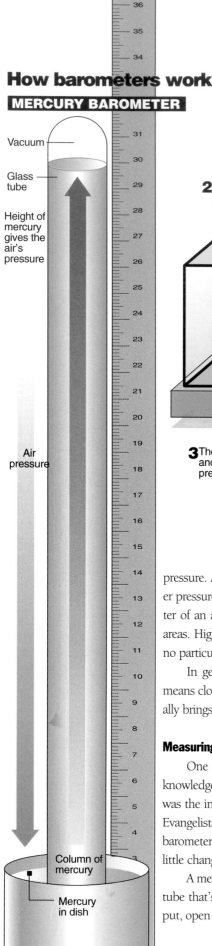

MERCURY BAROMETER

Vacuum

Glass tube

Height of mercury gives the air's pressure

Air pressure

Column of mercury

Mercury in dish

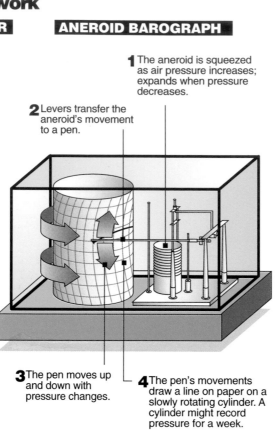

ANEROID BAROGRAPH

1 The aneroid is squeezed as air pressure increases; expands when pressure decreases.

2 Levers transfer the aneroid's movement to a pen.

3 The pen moves up and down with pressure changes.

4 The pen's movements draw a line on paper on a slowly rotating cylinder. A cylinder might record pressure for a week.

---

pressure. An "H" marks the center of an area with higher pressure than its surroundings. An "L" marks the center of an area with lower air pressure than surrounding areas. High and low pressure are always relative. There's no particular pressure number that divides high and low.

In general, low air pressure at the Earth's surface means clouds and precipitation while high pressure usually brings clear skies.

Measuring air pressure

One of the results of the flowering of scientific knowledge that came with the end of the Middle Ages was the invention of the mercury barometer in 1643 by Evangelista Torricelli, one of Galileo's assistants. Mercury barometers used today to make air pressure readings are little changed from Torricelli's original.

A mercury barometer (left) is a simple instrument. A tube that's closed at one end is filled with mercury and put, open end down, into a container of mercury. A ruler

is attached next to the tube. Air pressure on the mercury in the container keeps the mercury in the tube from flowing out. The greater the air pressure, the higher the mercury will rise in the tube. The air's pressure pushes down harder on the mercury in the dish, forcing more of it up into the tube. Because there is an inch-ruler next to the mercury tube, meteorologists measure air pressure in inches of mercury—instead of pounds per square inch. Very high pressure at the surface can push the mercury up approximately 32 inches into the tube.

Since a 32-inch tube of mercury is bulky and often inconvenient, aneroid barometers are more common than mercury barometers. An aneroid is a flexible metal bellows that's been tightly sealed after having some air removed. It might remind you of a tiny accordion. Increased outside air pressure squeezes the flexible metal; decreased pressure allows it to expand. It can be attached to a pointer or a moving pen to keep a pressure record as shown at left.

Millibars and kilopascals

In this chapter, pressure usually will be discussed in pounds per square inch. (The sea level pressure of 14.7 pounds per square inch, for example, means 14.7 pounds of force is pushing against each square inch of any surface exposed to the air. Pressure pushes equally in all directions, up, down and sideways.)

Meteorologists, however, usually use other measurements. In the United States, inches of mercury is the most common measurement used for pressure at the Earth's surface. This comes from the mercury barometer. Most of the rest of the world uses a metric measurement called the millibar. It is a direct pressure measurement like pounds per square inch. In some places these measurements are given in kilopascals, which are millibars divided by 10. In the United States upper-atmosphere pressures are always measured in millibars and weather reports meant for scientists use millibars for surface pressure.

Pressure and altitude

Because air pressure decreases as altitude increases, airplane pilots can determine their plane's altitude with a special kind of aneroid barometer called an altimeter. Altimeters are more precise than many aneroid barometers and can be adjusted to account for weather changes. A change in air pressure of one-tenth of an inch of mercury will make an altimeter read 100 feet higher or lower.

Climbing 2,000 feet up a mountain or in an airplane can create a difference in air pressure of about two inches of mercury. On the ground, to get that kind of pressure change you would have to go from the center of a strong high-pressure system to the center of a strong low-pressure system. You would be in the middle of quite a storm.

The balance between gravity's pull and pressure keeps vertical air pressure changes from causing winds to blow up or down. But, relatively tiny pressure differences between places at the same altitude can cause roaring winds.

A pressure change of 1.7 inches of mercury might not sound like much. In fact that's the change that occurs on an elevator ride up the 1,707 feet to the top of the Sears Tower in Chicago. But, if you traveled across the ocean for about 150 miles and came through that same pressure change, you'd be in a hurricane. In fact, the normal air pressure at the top of the Sears Tower is what you would find a few feet above the ocean in the center of a hurricane with winds above 110 mph.

Starting the winds moving

Winds are the result of the action of various forces, all of which are acting at the same time. In order to understand the winds, we have to look at these forces one at a time; we have to pretend that the wind begins with one force, then another and so on. Just remember, all the forces we talk about are acting together.

The drawings below show how differences in air pressure create one of the forces that causes the winds. The top drawing shows two containers of air connected by a pipe. This is a push-of-war instead of a tug-of-war. Even though the air in both containers is applying a force, the air in the pipe will move in the direction the stronger one is pushing. The greater the difference in strength between the two containers, the faster the air will go.

Winds are stronger if the high- and low-pressure areas are close together. If the tube between the two air containers were longer, as shown in the bottom drawing, the added air in the tube would mean the pressure would have more weight to push. Keeping the same force, but adding more distance — therefore, more weight — means the air wouldn't be pushed as fast.

Wind makers

Three important forces work together to determine how strong the winds will be and where they will blow:

Pressure Gradient Force: This goes back to our push-of-war. To forecast winds, meteorologists note the size of the pressure difference and the distance between the high- and low-pressure areas. This combination is called the Pressure Gradient Force (PGF).

Friction: As wind blows near the surface of the Earth, it rubs against trees, hills, buildings and other

Here are four ways of expressing the same average air pressure at sea level:

- 14.7 pounds per square inch

- 29.92 inches of mercury

- 1013.25 millibars

- 101.325 kilopascals

How pressure differences create wind

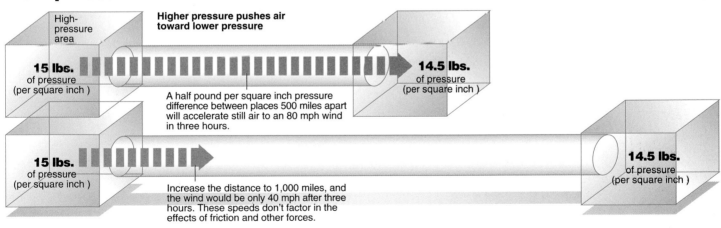

High-pressure area

Higher pressure pushes air toward lower pressure

15 lbs. of pressure (per square inch)

14.5 lbs. of pressure (per square inch)

A half pound per square inch pressure difference between places 500 miles apart will accelerate still air to an 80 mph wind in three hours.

15 lbs. of pressure (per square inch)

14.5 lbs. of pressure (per square inch)

Increase the distance to 1,000 miles, and the wind would be only 40 mph after three hours. These speeds don't factor in the effects of friction and other forces.

objects that slow it down. That is the force of friction. There is far less friction over water than over land.

Coriolis Effect: Pressure Gradient Force and friction would explain everything about winds if the Earth were not rotating. The wind could blow straight from high pressure to low pressure and die down when the pressures equalize. But the Earth is spinning, which is why winds curve as they blow.

Things that move freely across the Earth — rockets, ocean currents, the wind — that might follow a straight path if the Earth wasn't rotating end up following a curved path. The Earth is rotating under them.

In the Northern Hemisphere the wind, ocean currents or rockets act as though they're being pushed to their right. In the Southern Hemisphere the push is to the left. This curving motion is called the Coriolis Effect — named after Gustave-Gaspard Coriolis, a French scientist who first described it mathematically in 1835. An important point: The Coriolis Effect depends on speed. The faster a rocket or the wind is moving, the stronger the Coriolis Effect.

Vertical movement of air

Imagine putting in a wood screw. Turn the screwdriver clockwise and the screw goes down. Turn counterclockwise, the screw comes up. This will help you remember that in the Northern Hemisphere, air goes clockwise and down around high pressure. Air goes counterclockwise and up around low pressure.

Sinking air in a high-pressure area is warmed by the increase of pressure. As it warms, any water in the air evaporates into vapor. Even more important, the warming keeps water vapor from condensing into clouds and precipitation. That makes for generally good weather.

Rising air in low-pressure areas becomes cooler. It condenses water vapor into clouds and precipitation — wet weather. Eventually air rising in low-pressure areas comes down in high-pressure areas at the surface.

Jet stream winds

How gases react to temperature and pressure, how high pressure pushes air, and the three forces on wind are the building blocks of jet streams, a major influence on our weather.

Jet streams are relatively narrow bands of high-speed winds in the upper atmosphere. Their general direction is west to east — but they can swoop northward and southward. They develop when strong tem-

How the Earth curves winds

The wind, like anything moving freely across the Earth, follows a curved path over the ground. You can use a turntable, a round piece of cardboard, a ruler and a marking pen to see why this happens.

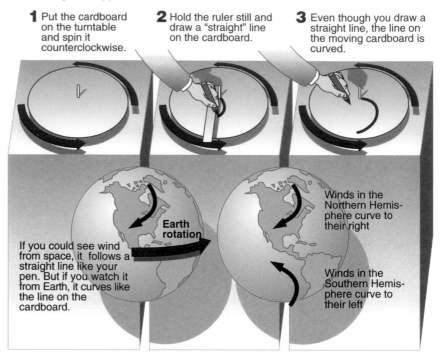

1 Put the cardboard on the turntable and spin it counterclockwise.

2 Hold the ruler still and draw a "straight" line on the cardboard.

3 Even though you draw a straight line, the line on the moving cardboard is curved.

If you could see wind from space, it follows a straight line like your pen. But if you watch it from Earth, it curves like the line on the cardboard.

Earth rotation

Winds in the Northern Hemisphere curve to their right

Winds in the Southern Hemisphere curve to their left

perature differences cause great pressure differences at high altitudes. In general, the twists and turns of jet streams follow the boundaries between warm and cold air.

In winter, because temperature contrasts are greater,

MAKING THE WINDS BLOW

Now it's time to put all the forces to work together to see how they make the winds blow. Since we can't see the wind, we'll use a balloon that doesn't rise or sink — it just goes where the forces push it. We'll start high above the Earth where friction doesn't complicate the picture. Pressure Gradient Force begins pushing the air — including the balloon — from the high pressure toward the low pressure. But Coriolis Effect turns the wind to the right. Eventually, these forces balance and the wind flows parallel to the pressure areas.

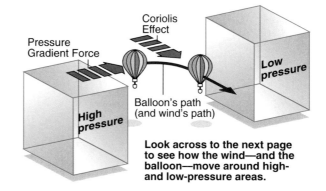

Coriolis Effect

Pressure Gradient Force

Low pressure

High pressure

Balloon's path (and wind's path)

Look across to the next page to see how the wind—and the balloon—move around high- and low-pressure areas.

Why wind directions change as you go higher

The drawings on the previous page and at the bottom of this page show how winds flow around high- and low-pressure areas above the ground in the Northern Hemisphere. Now, let's see why the wind flows into the center of low-pressure areas near the surface, but around lows in the upper atmosphere and also why winds blow from different directions at different altitudes.

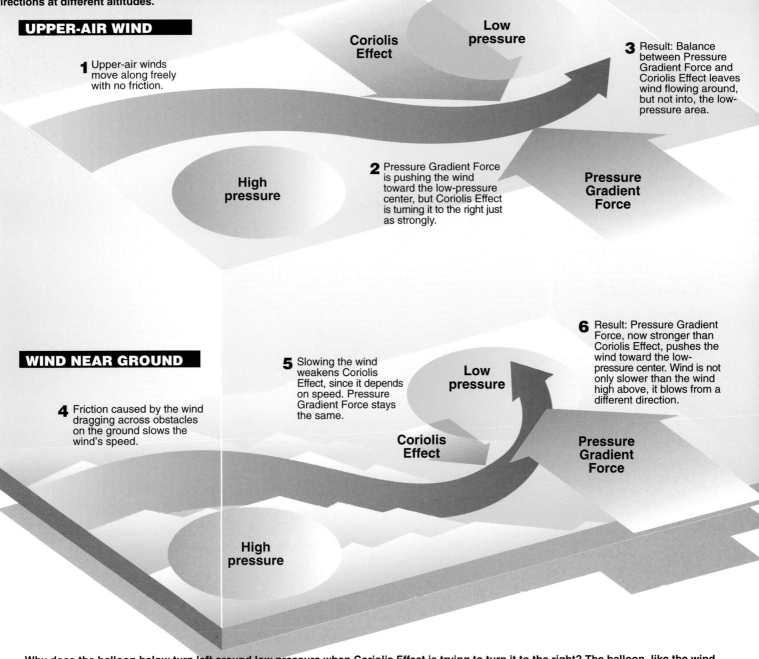

UPPER-AIR WIND

1 Upper-air winds move along freely with no friction.

Coriolis Effect

Low pressure

3 Result: Balance between Pressure Gradient Force and Coriolis Effect leaves wind flowing around, but not into, the low-pressure area.

High pressure

2 Pressure Gradient Force is pushing the wind toward the low-pressure center, but Coriolis Effect is turning it to the right just as strongly.

Pressure Gradient Force

WIND NEAR GROUND

4 Friction caused by the wind dragging across obstacles on the ground slows the wind's speed.

5 Slowing the wind weakens Coriolis Effect, since it depends on speed. Pressure Gradient Force stays the same.

Low pressure

6 Result: Pressure Gradient Force, now stronger than Coriolis Effect, pushes the wind toward the low-pressure center. Wind is not only slower than the wind high above, it blows from a different direction.

Coriolis Effect

Pressure Gradient Force

High pressure

Why does the balloon below turn left around low pressure when Coriolis Effect is trying to turn it to the right? The balloon, like the wind, follows a line of constant pressure that keeps Pressure Gradient and Coriolis forces in balance. If it tries to stray from this line, the forces become unbalanced and the balloon is pushed back in line. To see how this works in the Northern Hemisphere, look at the two different pressure systems below.

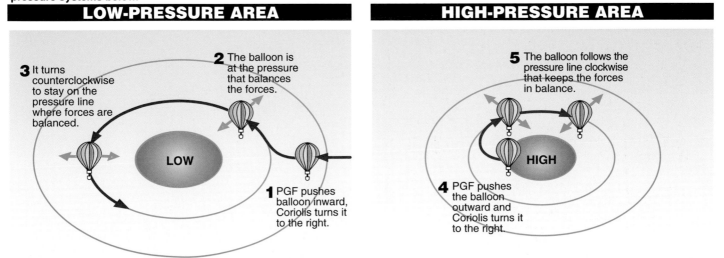

LOW-PRESSURE AREA

3 It turns counterclockwise to stay on the pressure line where forces are balanced.

2 The balloon is at the pressure that balances the forces.

LOW

1 PGF pushes balloon inward, Coriolis turns it to the right.

HIGH-PRESSURE AREA

5 The balloon follows the pressure line clockwise that keeps the forces in balance.

HIGH

4 PGF pushes the balloon outward and Coriolis turns it to the right.

JET STREAMS: THE WORLD'S FASTEST WINDS

There's more to jet streams than you might think from looking at television weather maps, which usually depict them as a ribbon, or flashing arrows snaking across the United States. If you caught a ride in a jet stream's core — the area of fastest winds — you might find yourself going 100 mph; then speeding up to more than 150 mph for a couple of hundred miles. Then you might slow to maybe 90 mph for a few hundred miles before speeding up again.

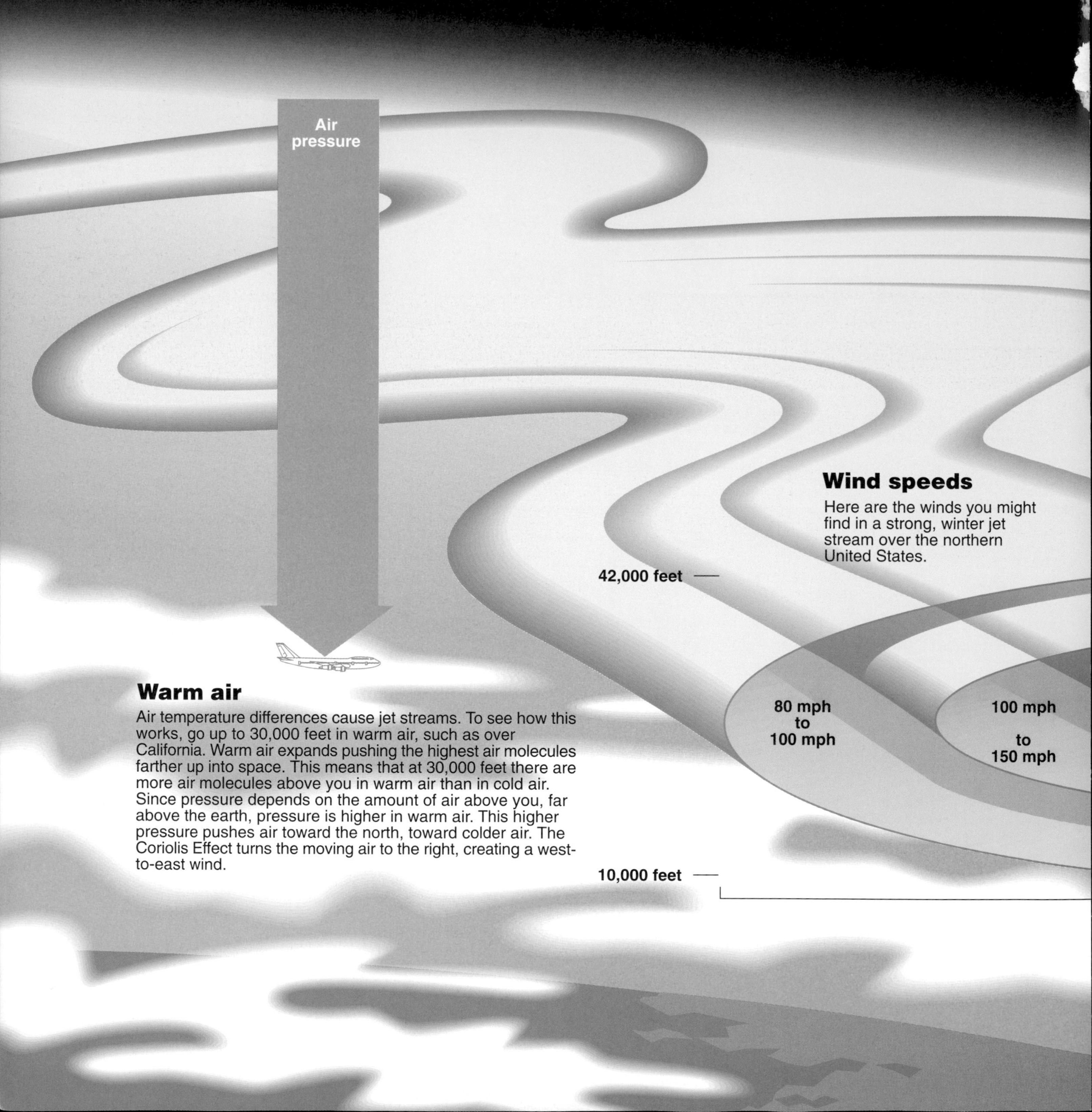

Air pressure

Wind speeds

Here are the winds you might find in a strong, winter jet stream over the northern United States.

42,000 feet —

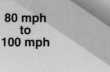

80 mph to 100 mph

100 mph to 150 mph

Warm air

Air temperature differences cause jet streams. To see how this works, go up to 30,000 feet in warm air, such as over California. Warm air expands pushing the highest air molecules farther up into space. This means that at 30,000 feet there are more air molecules above you in warm air than in cold air. Since pressure depends on the amount of air above you, far above the earth, pressure is higher in warm air. This higher pressure pushes air toward the north, toward colder air. The Coriolis Effect turns the moving air to the right, creating a west-to-east wind.

10,000 feet —

Jet stream positions

Summer

In summer, the jet weakens and is farther north.

Winter

In the winter, with its great temperature contrasts, jet streams are stronger and farther south.

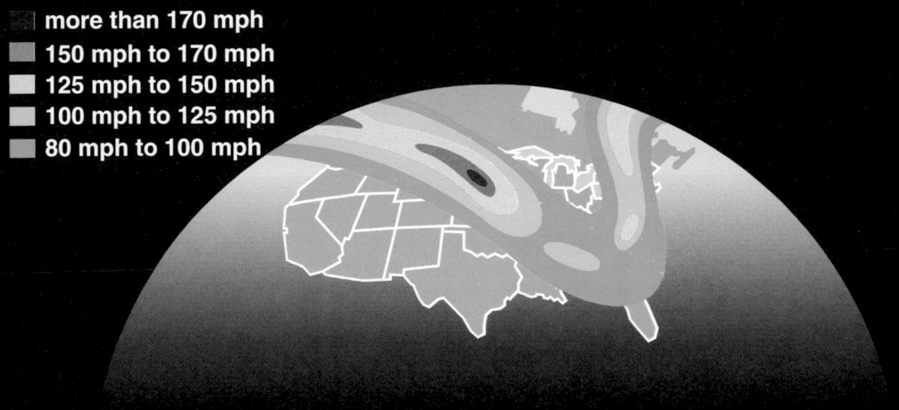

Wind speeds

- more than 170 mph
- 150 mph to 170 mph
- 125 mph to 150 mph
- 100 mph to 125 mph
- 80 mph to 100 mph

The greater the temperature contrasts, the stronger the pressure differences between the warm and cold air. Stronger pressure differences create stronger winds. This is one reason why jet stream winds vary so much in speed. If you looked down on the core of a typical early winter jet stream, its pattern of wind speeds might look like the map above. The large drawing below shows a cross section of such a jet stream in the region of highest winds.

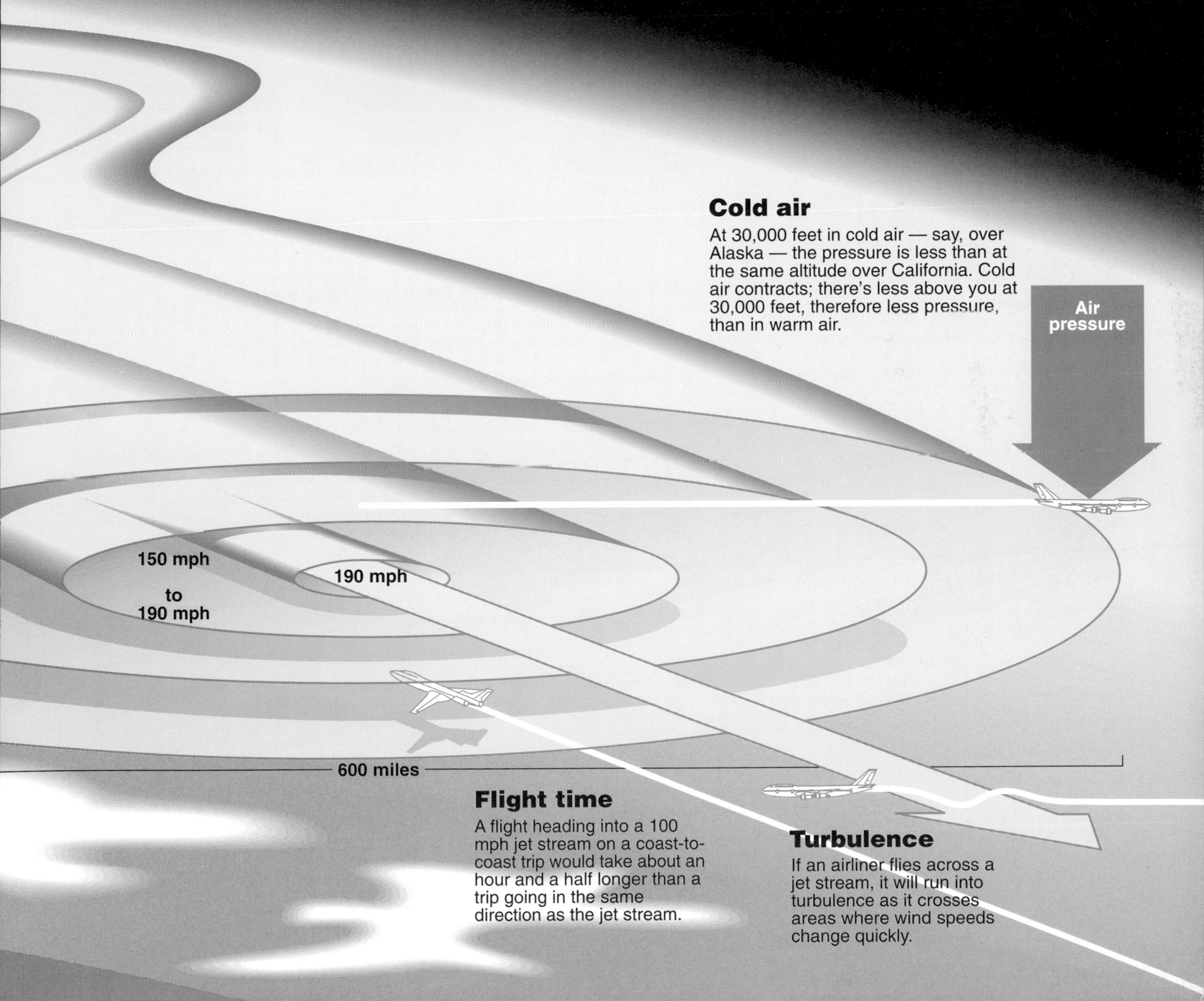

Cold air

At 30,000 feet in cold air — say, over Alaska — the pressure is less than at the same altitude over California. Cold air contracts; there's less above you at 30,000 feet, therefore less pressure, than in warm air.

Air pressure

150 mph to 190 mph

190 mph

600 miles

Flight time

A flight heading into a 100 mph jet stream on a coast-to-coast trip would take about an hour and a half longer than a trip going in the same direction as the jet stream.

Turbulence

If an airliner flies across a jet stream, it will run into turbulence as it crosses areas where wind speeds change quickly.

jet streams tend to blow faster. In the Northern Hemisphere they move south as polar air pushes south, shoving the warm-cold boundary southward. Polar jet streams cross Canada during the summer and slip southward over the United States in the winter. They move back and forth with the advance and retreat of cold air. Subtropical jet streams often flow over the southern United States in the winter.

Jet streams blow from west to east in both the Northern and Southern Hemispheres. The drawings on Pages 38 and 39 show how in the Northern Hemisphere cold, high-altitude temperatures in the north and warm, high-altitude temperatures in the south create pressure differences that push air around 30,000 feet above the ground from south to north — toward the North Pole. Coriolis Effect turns the resulting wind to its right, creating west-to-east jet streams.

In the Southern Hemisphere, the high-altitude temperature differences are similar, but the cold is in the south, the warm in the north. The resulting pressure differences push wind southward, toward the South Pole. But, in the Southern Hemisphere, Coriolis Effect turns the wind to its left. Since the wind is heading south, a left turn means it ends up blowing from west to east.

Jet streams and jets

Jet streams are extremely important to pilots of high-flying airplanes — and to airlines. In fact, the right jet stream winds are money in the bank for an airline. Even on the relatively short Chicago to New York flight, catching a 150 mph, eastbound jet stream can save 400 to 500 gallons of fuel for a Boeing 727. It works the other way, too. A westbound airliner from Europe to Chicago fighting a strong eastbound jet stream might have to land early to take on more fuel.

Jet streams account for many of the bumps you feel in a high-flying airliner. As an airplane flies through any area where the wind's speed or direction changes rapidly, the ride will be bumpy because of wind shear. In recent years the term "wind shear" has come to be associated with a particular kind of wind near the ground called a "downburst," which has caused several airline accidents. But wind shear is any sudden change in wind speed or direction. It can happen near the ground or at 35,000 feet. It can be vertical or horizontal. The bigger the change in speed or direction, the greater the turbulence. The International Civil Aviation Organization defines a wind-speed change of more than

10 mph in a 100-foot distance as "strong" wind shear. A change of more than 14 mph in 100 feet is "severe" wind shear.

Temperate zones

The parts of the Earth from roughly 30 degrees to 60 degrees latitude in both hemispheres are known as the temperate zones. Most of the United States except northern Alaska and Hawaii is in this zone. While the average temperatures in these regions are "temperate" compared with constant polar cold or tropical heat, anyone who lives in the mainland United States knows the weather can be anything but temperate.

War zones might be a better term. These middle latitudes are where polar cold and tropical heat battle, creating storms. Storms, which are low-pressure areas, move across the land and sea, followed by high-pressure areas. As the lows and highs pass, winds from these clashes can swing from all directions. But the ever-present, strong jet streams often guide these storms. That is why most storms generally follow a path from west to east.

In the Northern Hemisphere winds encounter plenty of resistance; they run into the Rockies and other large western mountains and the smaller Appalachians in the eastern United States. The large mountain ranges of Europe and Asia also slow the winds. But in the

Some of Bob Rice's (story on next page) major balloon flight forecasts:

__August 12-17, 1978:__ First trans-Atlantic balloon flight. *Double Eagle II,* from Presque Isle, Maine to near Paris, 3,120 miles.

__May 8 - 12, 1980:__ First non-stop balloon flight across North America. Maxie Anderson and Kris Anderson in the *Kitty Hawk* from Fort Baker, Calif., to Matane, Quebec, 2,417 miles.

__July 2-3, 1987:__ First hot-air balloon Atlantic crossing — others were helium balloons — *Virgin Atlantic Flyer* from Sugar Loaf, Maine to Northern Ireland, 3,075 miles.

__Jan. 15-17, 1991:__ First hot-air balloon Pacific crossing. *Virgin Pacific Flyer* from Miyakonojo, Japan, to Lac la Martre, Canada, 6,761 miles.

Wind shear found at all altitudes

Wind shear refers to a quick change in wind speed or direction. Wind shear is found in jet streams 35,000 feet up, at the ground and all altitudes in between.

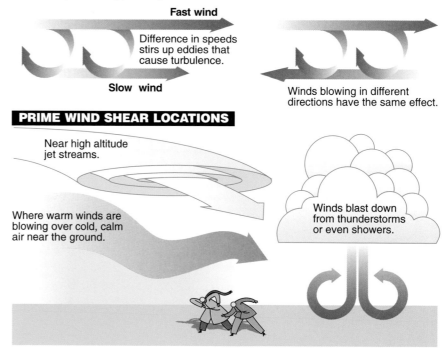

Fast wind

Difference in speeds stirs up eddies that cause turbulence.

Slow wind

Winds blowing in different directions have the same effect.

PRIME WIND SHEAR LOCATIONS

Near high altitude jet streams.

Where warm winds are blowing over cold, calm air near the ground.

Winds blast down from thunderstorms or even showers.

Finding the winds for the record books

For every long-distance balloon flight, someone has to figure out where and how fast the winds will blow. For record-setting flights since the mid-1970s, that someone usually has been meteorologist Bob Rice.

Balloons live and die by the wind. The pilot can make the balloon rise or sink, but to go in a particular di-rection the pilot has to find wind going in that direction. Rice has built a reputation for finding the winds to take the pilot where he wants to go.

Rice doesn't make claims of pinpoint accuracy. Speaking of his first big success, the *Double Eagle II* flight across the Atlantic in 1978, Rice says, "I'd like to tell you we were aiming right at Paris. In fact I will tell you that. But we were really aiming for somewhere north of the Mediterranean and south of the Baltic."

Even though every time a big flight is over he swears he's going to quit, Rice is

Bob Rice does more than forecast for balloonists. He forecasts for yacht races, 'round-the-world sails, or any event that needs wind predictions.

unable to resist when another balloonist calls.

Balloon flights "chew up a lot of time," he says. "They're very exciting, but by the time you get around to flying, these [pilots] are friends. You know the penalties for failure. I worry. You've got one or two or three lives up there. That's not a lot in a context of an airliner crashing or a war, but it's people you know that are in your hands.

"It's not only their safety," he says. "It's their goals; you know what they're after.

It's meant so much to them. They've spent the time and money creating a balloon. It's all now in one ball of wax that's either going to succeed or fail. Each flight ages me by 10 years; there's so much riding on it."

Like all modern-day forecasters, Rice starts with computer projections of world-wide wind and weather patterns. "The art starts to come in when you start asking what are the probable errors or what are the penalties for various errors" in the forecast. "There are always surprises."

Forecasting for balloon flights requires understanding both the weather and balloons. Rice has forecast for both helium and hot-air balloons, which require different techniques. In addition to the winds, Rice worries about storms, which can damage a balloon, and cold clouds, which can coat it with too much ice to fly.

"If you're going to be successful," Rice says, "you've got to ... visualize what the air is doing in three dimensions. I think of myself as weather mechanic rather than a [theoretical] meteorologist. I understand how this fluid is working and moving. That's what makes a forecaster."

Southern Hemisphere only the tip of South America extends far south of 40 degrees south latitude. Winds blow mostly over open ocean. This allows the winds to build up to higher speeds and kick up huge waves that can circle the globe without hitting land. Because of these winds, sailors started calling the region "the roaring 40s."

Global wind patterns

Trade winds blow generally from northeast to southwest in the Northern Hemisphere. For hundreds of years, sailing ship captains — among them Christopher Columbus — sailed far enough south to catch the steady, easterly trade winds that pushed them westward to the Caribbean. To return to Europe, they sailed northward close to the North American coast until they found the westerly winds that predominate in the temperate latitudes.

The Intertropical Convergence Zone, where the trade winds come together — converge — shows up on satellite photos as a band of white clouds near the equator. Sailors know this zone as "the doldrums." Here the winds blow lightly or not at all. As the trade winds approach this area, they die out as air warmed by the tropical oceans begins rising.

During the Northern Hemisphere summer, as shown on Page 43, the Intertropical Convergence Zone creeps northward especially over Asia. The zone moves more directly under the sun where it's hotter. In the fall it moves southward to the equator, or, in some places, even south of the equator.

Air rising from the Intertropical Convergence Zone's thunderstorms flows northward or southward in the upper atmosphere and sinks when it gets about as far north as Florida or as far south as Australia. Then it forms huge high-pressure areas at the surface with light

Why the trade winds blow

1 Air, heated by warm water, rises over the tropical oceans, forming huge thunderstorms.

3 In the Northern Hemisphere, Coriolis effect turns the southward-bound wind to its right, creating northeast winds.

5 The region of thunderstorms, where the trade winds converge, is called the Intertropical Convergence Zone.

2 Air flows in to replace the rising air.

4 In the Southern Hemisphere, Coriolis effect turns the northward-bound wind to its left, creating southeast winds.

winds.

On land these high-pressure areas help create the Earth's large deserts, such as in the southwestern United States and across northern Africa and Saudi Arabia. At sea these high-pressure areas, where air is slowly descending, create calm or light, variable winds and mostly clear skies.

Polar winds: Around the poles, the winds are variable, but blow from the east on average. As cold air piles up, it creates strong high pressure at the surface. As cold air flows away from the North and South Poles, it warms slightly as it reaches about 60 degrees latitude, where the ocean is a little warmer. Warming causes the air to rise, which creates low-pressure areas. Meteorologists call these "semipermanent" lows, since they are in the same places most of the time, but not always. Air flowing away from both the North and South

Poles into these low-pressure areas causes easterly winds around the polar regions in both hemispheres.

The clear skies of the high-pressure areas and cloudy skies of the lows are important global weather makers. The pattern of winds around the semi-permanent areas of high and low pressure also makes west winds dominate at the surface in the temperate zones. The flow of the wind around these big highs and lows is a weather predictor.

The drawings on the next page show how these semipermanent-pressure areas and the winds around them change from January to July.

As you look at the drawings you'll see that low-pressure areas over the oceans near the poles grow in the winter, shrink in the summer. The high-pressure areas over tropical oceans, in contrast, grow in summer and shrink in winter.

Hitching a free ride on mountain winds

Wind hides few of its secrets from glider pilot Tom Knauff.

Knauff, who runs the Ridge Soaring Gliderport near State College, Pa., combined his ability to fly and read the winds with a perfect weather pattern on April 23, 1983, to set a gliding round-trip world record.

A combination of a low-pressure center over New York and

high pressure over Missouri helped give him a free ride 1,023 miles from Williamsport, Pa., to Flat Creek Church near Knoxville, Tenn., and back in 10.5 hours — with no engine.

The counterclockwise wind around the low over New York and the clockwise wind around the high over Missouri funneled the wind at nearly a right angle against the Appalachians. It pushed up creating what meteorologists call "ridge lift" a couple hundred feet above the Appalachians.

The highest and fastest parts of the trip were on "mountain waves," created when winds blowing above 30 mph hit the mountains and moved upward.

The world's major pressure patterns

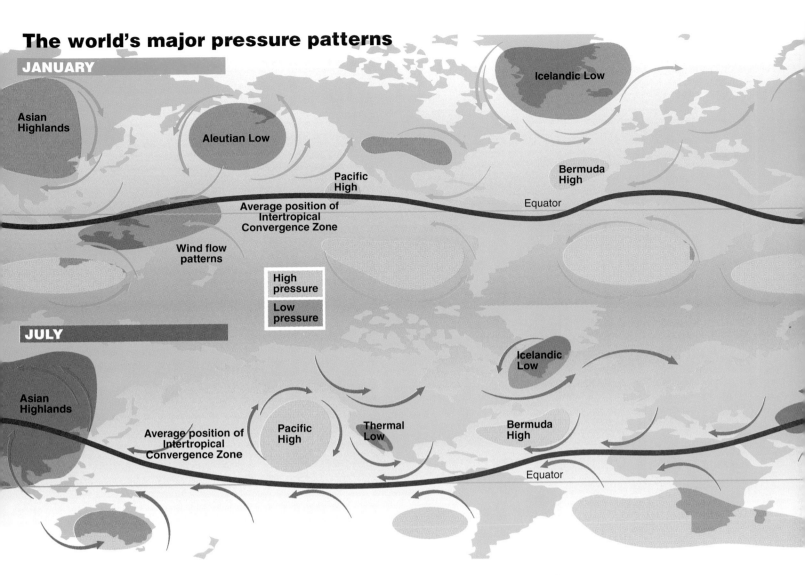

JANUARY

Asian Highlands

Aleutian Low

Icelandic Low

Pacific High

Bermuda High

Equator

Average position of Intertropical Convergence Zone

Wind flow patterns

High pressure

Low pressure

JULY

Asian Highlands

Average position of Intertropical Convergence Zone

Pacific High

Thermal Low

Icelandic Low

Bermuda High

Equator

Low-pressure areas in a particular latitude form over the warmer parts of the globe. In winter, oceans near the poles are warmer than the land. Air over the warm water is warmed more than air over land and begins rising, creating low pressure — and stormy weather. Areas such as the Aleutian Low spawn winter storms. In summer, as the land warms up, the contrast between land and ocean temperatures decreases and the low-pressure areas weaken.

High-pressure areas in a particular latitude form where the surface is cooler. In the summer, the contrast between the hot, tropical land and the cooler oceans is

Large mountains, such as the Rockies or California's Sierra Nevada, can create waves at least 50,000 feet high that are felt for miles downstream. Sometimes if you feel bumps on an airliner over the Plains, you could be running into mountain waves from the Rockies. The smaller Appalachians don't make waves this large. But at one point in his record-breaking trip, Knauff rode a mountain wave to 14,000 feet, his glider skimming through the air at 180 mph.

Knauff also used thermals, warm air rising from the ground, to gain altitude to cross gaps in the mountain ridges.

Since they can't see the winds directly, glider pilots have to develop an eye for the sky's subtle clues. Sometimes birds can lead to rising air. "When most people see birds circling, they think of buzzards circling something dead on the ground," Knauff says. "We like to think they're doing it because they're having fun. Birds get tired of sitting around all the time; their legs aren't built for sitting. Birds can fly higher to cool off; they use thermals to search for food. Some birds are quite good at soaring; we learn to tell the difference between them and other birds." The leaves on the trees on a hillside or the swirls wind makes in a wheat field can help a glider pilot puzzle out what the wind's doing.

"But clouds are our number one indicator of where lift is or where it might be found," Knauff says. A tiny, puffy cloud often caps the top of a thermal. If the air is humid enough lens-shaped clouds can mark the crests of mountain waves. Other times, especially in the East, rolls of clouds show the wind is creating waves.

Since conditions are constantly changing, a long flight "is quite stressful," Knauff says. "You're making decisions every few seconds. It never lets up, you can't pull over to a rest area and have a cup of coffee. The sun is going through the sky; it's going to set and you're going to land."

How mountain waves work

When the wind hits a mountain range at nearly a right angle, it can create "ridge lift" and "mountain waves" that glider pilots use.

4 Glider pilots ride the upward moving part of the wave to 40,000 feet or higher.

2 Wind thousands of feet above mountains can also be pushed up by the air disturbance below.

3 If conditions are right, the wind comes down on other side of mountains, overshoots its original altitude and goes up, starting a series of gradually diminishing waves.

Zone of best lift

5 Sometimes lenticular — lens shaped — clouds mark top of waves.

1 Wind is forced upward when it hits a mountain, creating ridge lift on the side the wind is coming from. It keeps gliders aloft as they fly along the ridge.

6 A turbulent "rotor" is often found under waves.

Pushing the beauty of gliding to new heights

Mark Palmer doesn't care how long he glides — he cares only how high.

As he catches a wave created by winds flowing over the Colorado Rockies, "the only impression you have is the ground is sinking away from you," he says. On a typical flight the glider climbs as fast as 2,000 feet per minute after Palmer releases the rope that connects him to the tow plane. The wind — and the pilot's skill at reading it — makes the glider climb as high as jetliners fly.

"I try to turn the sailplane directly into the wind and match the glider's airspeed to the wind," Palmer says. "You stay in one posi-tion over the ground. You might have to move back and forth as the lift ebbs and flows, but you generally stay in one spot." Gliders using mountain waves in the West regularly climb above 35,000 feet; some have climbed above 49,000 feet.

"We kind of take pride in soaring," Palmer says. "It's pretty much a solar-pow-ered, natural kind of sport. We don't use too much fuel; we work with nature rather than bulldoze our way through."

As his glider carries him upward, Palmer sees the Rockies spread out below through the sailplane's canopy. "You get to really expe-rience nature in one of its most powerful aspects. You can see Cheyenne to the north and Albuquerque to the south and into Utah to the west. To see the Rockies like that is gor-geous. You can take all the time you want and just look at it."

greater than in winter. This is why the high-pressure areas over tropical oceans grow in the summer.

Local winds

While global-scale winds, such as the trades, determine the weather over large areas, smaller winds create local weather.

Sea breezes: One of the most common local winds is a sea breeze along the ocean or near large lakes. During the day the land heats up more than the water. Warm air rises over the land and cooler air flows in from the water to replace it. If the water is warm enough, such as off the coast of Florida, a land breeze can begin at night as the land cools.

Mountain winds: Mountains also create their own winds. During the day the sun can heat mountain slopes, especially bare rock. The warmed air above rises and more air flows upward from valleys to replace it. This creates a wind blowing up the mountain during the afternoon. As the mountain slopes cool at night, cold air, which is heavier than warm air, begins flowing down into the valleys.

Winds, whether they're the jet streams racing overhead, the gales in the roaring 40s, or the first warm breezes of spring, move masses of air around the Earth. The clash of these moving air masses creates the storms that dominate our ever-changing weather.

The highest wind ever recorded at the Earth's surface, 231 mph, was atop Mount Washington, N.H., at an elevation of 6,262 feet. Average wind speed there is 35 mph.

Alaska is the windiest state with the most local records for highest mean wind speed.

Place	Mean wind speed (in mph)
St. Paul Island, Alaska	18.3
Cold Bay, Alaska	16.9
Amarillo, Texas	13.7
Boston	12.4
Wichita, Kan.	12.4
Buffalo, N.Y.	12.1
Honolulu	11.6
Minneapolis-St. Paul	10.5
Chicago	10.3
New York	9.4
Miami	9.2
Atlanta	9.1
Anchorage	6.8
Los Angeles	6.2

Different ways of measuring wind

Beaufort number	Wind speed (mph)	Seaman's term	Effects at sea	Effects on land
0	Under 1	Calm	Sea like mirror.	Calm; smoke rises vertically.
1	1-3	Light air	Ripples with appearance of "fish scales"; no foam crests.	Smoke drift indicates wind direction; vanes do not move.
2	4-7	Light breeze	Small wavelets; crests of glassy appearance not breaking.	Wind felt on face; leaves rustle; vanes begin to move.
3	8-12	Gentle breeze	Large wavelets; crests begin to break; scattered whitecaps.	Leaves, small twigs in constant motion; light flags extended.
4	13-18	Moderate breeze	Small waves; becoming longer; numerous whitecaps.	Dust, leaves and loose paper raised up; small branches move.
5	19-24	Fresh breeze	Moderate waves; becoming longer; many whitecaps; some spray.	Small trees in leaf begin to sway.
6	25-31	Strong breeze	Larger waves forming; whitecaps everywhere; more spray.	Large branches of trees in motion; whistling heard in wires.
7	32-38	Moderate gale	Sea heaps up; white foam from breaking waves begins to be blown in streaks.	Whole trees in motion; resistance felt in walking against wind.
8	39-46	Fresh gale	Moderately high waves of greater length; foam is blown in well-marked streaks.	Twigs and small branches broken off trees.
9	47-54	Strong gale	High waves, sea begins to roll; dense streaks of foam; spray may reduce visibility.	Slight structural damage occurs; slate blown from roofs.
10	55-63	Whole gale	Very high waves with overhanging crests; sea takes white appearance; visibility reduced.	Seldom experienced on land; trees broken; structural damage occurs.
11	64-72	Storm	Exceptionally high waves; sea covered with white foam patches.	Very rarely experienced on land; usually with widespread damage.
12	73 or higher	Hurricane force	Air filled with foam; sea completely white with driving spray; visibility greatly reduced.	Violence and destruction.

STORMS AND FRONTS

When winds begin moving great seas of cold or warm air away from the regions where they form, air masses come into conflict. The arrival of a new air mass means the old one has to be shoved out of the way. There's a war on.

The place where air masses clash is called a "front," a concept based on battle fronts, developed by Norwegian meteorologists after World War I. Fronts are often the scene of the most dramatic weather changes and sometimes the most violent weather. Weather fronts can stay in place or move. When they move, one air mass is advancing, another retreating.

When the warmer air is advancing, the boundary is a warm front. When the colder air is taking over, it's a cold front. When neither side is winning, it's a stationary front.

Fronts normally are parts of larger weather systems centered on areas of low atmospheric pressure. They are parts of weather systems often called storms. The word "storm" is used in so many ways, however, that it's often confusing. The large-scale weather systems can be the cause of snowstorms. They can cause small-scale thunderstorms or dust storms. In this chapter we'll use "storm" to refer to these large-scale systems that include fronts.

The scientific name for such a storm is "extratropical cyclone." "Extratropical" means the storm forms outside the tropics.

Storms that form in the tropics are different in some fundamental ways — notably, they don't have fronts. "Cyclone" refers to any weather system with winds around a low-pressure area. But that term, too, can be confusing. People in the Midwest once called tornadoes cyclones — that use is dying out — and today tropical cyclones in the Indian Ocean are usually called cyclones. Both tornadoes and Indian Ocean tropical

A roll cloud, formed by cool air advancing into warmer air, frames Cleveland's Lake Erie shoreline. Such clashes of warm and cool air and the resulting clouds and precipitation are among the many localized results of huge disturbances that regularly sweep across North America. Spring storms are usually the most violent — often with wind-driven snow, powerful thunderstorms, and wide temperature swings.

cyclones are particular kinds of cyclones.

While big storms can close down several states with wind-driven snow, whip up waves that sink ships, or flood hundreds of square miles, most storms do little damage. They move cold air southward and warm air northward, helping the Earth balance its heat budget. They bring needed rain and snow.

How a storm picks up energy

Extratropical cyclones draw much of their energy from temperature contrasts created by the sun's unequal heating of the Earth. While such contrasts are strongest along fronts, storms don't need fronts to form.

When conditions are right, gravity pulls cold, heavier air under warmer, lighter air. Contrasts between warm and cold air have potential energy, just as a skier at the top of a slope has potential energy. When the skier pushes off and starts down the slope, her potential energy becomes kinetic energy — energy of motion. When something happens to start warm and cold air masses moving over and under each other, their potential energy becomes a storm's kinetic energy.

Energy is added to storms when air rises and the water vapor in it begins condensing into clouds and precipitation. When water vapor condenses or turns into ice, it releases heat called "latent heat." This heat adds to a storm's energy. Also, storms draw some energy from the high-speed winds of the upper atmosphere.

After a storm starts, the winds can pull in more cold air and warm air, enhancing the temperature con-

trasts and keeping the storm going.

A storm's wind speeds depend on the differences between air pressures around the storm and in its low-pressure center. The greater the difference, the stronger the winds. Another factor also comes into play. Think how a spinning ice skater can speed up or slow down by pulling in her arms or holding them out. When the skater pulls in her arms, she spins faster. When she holds them out, she spins slower. Scientists call this the "conservation of angular momentum" and it works for storms too. As winds spiral into a low-pressure area they make a smaller and smaller circle. Like the ice skater who pulls her arms in, the winds spin faster.

Getting a picture of a storm

Until the first weather satellite started sending photos to Earth on April 1, 1960, no one had ever seen an entire extratropical cyclone. Even airplanes couldn't get above the storm to get a full picture. But before satellites, scientists had been piecing together the giant puzzle of how storms form and move.

Benjamin Franklin took some of the first steps toward creating the picture of storms we have now. In the 18th century, people thought their local storms were isolated events — the winds would bring clouds and then rain or snow; or maybe the storm would spring up in their area. People didn't see storms as parts of large, moving, organized systems of wind, clouds and precipitation. Ben Franklin began questioning this view in 1743. Franklin, in Philadelphia, and his brother James, in Boston, were set to watch a lunar eclipse. But the night of the eclipse, clouds combined with winds from the northeast to block the view in Philadelphia.

Franklin was amazed later to find out from his brother that the storm didn't arrive in Boston until after the eclipse. If the northeast winds brought clouds, the clouds should have covered Boston first.

Franklin wrote about it to the Rev. Jared Eliot in Connecticut: "… tho' the Coarse [sic] of the Wind is from N.E. to S.W., yet the Coarse of the Storm is from S.W. to NE. i.e. the Air is in violent motion in Virginia before it moves in Connecticut, and in Connecticut before it moves at Cape Sable, &c."

This is a good description of the movement of most storms along the East Coast. But instead of seeing the storm as a system of counterclockwise winds moving up the coast, Franklin thought it began with warm air rising over the Southeast and Gulf of Mexico to be replaced by

One of the worst storm-caused naval disasters was the sinking of the steamer *Home* Oct. 9, 1837, off Ocracoke, N.C. Ninety people died. As a direct result of this tragedy, Congress passed legislation requiring all seagoing vessels to carry a life preserver for each passenger on board.

Why fronts can give birth to storms

To demonstrate why fronts can give birth to storms, try an experiment that requires a small glass container. Divide the container into two parts with an easily removable partition.

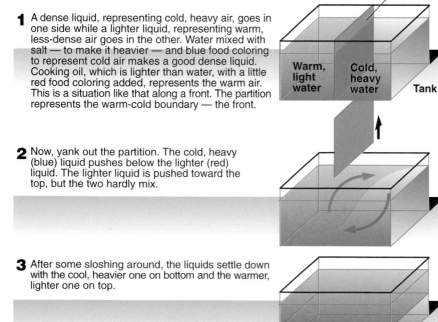

1 A dense liquid, representing cold, heavy air, goes in one side while a lighter liquid, representing warm, less-dense air goes in the other. Water mixed with salt — to make it heavier — and blue food coloring to represent cold air makes a good dense liquid. Cooking oil, which is lighter than water, with a little red food coloring added, represents the warm air. This is a situation like that along a front. The partition represents the warm-cold boundary — the front.

2 Now, yank out the partition. The cold, heavy (blue) liquid pushes below the lighter (red) liquid. The lighter liquid is pushed toward the top, but the two hardly mix.

3 After some sloshing around, the liquids settle down with the cool, heavier one on bottom and the warmer, lighter one on top.

Partition

Warm, light water

Cold, heavy water

Tank

Understanding fronts

Here is what happens along fronts and the weather map symbol for each. The symbols are lines showing where the cold-warm boundary touches the ground.

Cold front

Cold air is displacing warm air. The heavier, cold air is shoving under the warm air, pushing it upward. Unless the air is extremely dry, clouds form. Often the clouds grow into thunderstorms. The slope of the front is fairly steep, especially if it's moving fast, around 25 mph for example. About 30 miles back from the front the warm-cold boundary would be at around 3,000 feet above the ground.

Warm front

Warm air is replacing colder air. The lighter warm air slides over the heavier cold air, creating a boundary with a gentle slope. You could be around 100 miles ahead of the warm front — still in the cold air — and the warm-cold boundary could be only 3,000 feet above you. If you're 600 or 700 miles ahead of the front, in the cold air, wispy clouds at about 30,000 feet overhead could be the first sign of the approaching warm air. As the front moves toward you — or if you move toward the front — the clouds overhead grow thicker and lower and eventually rain, snow, sleet or freezing rain begin falling.

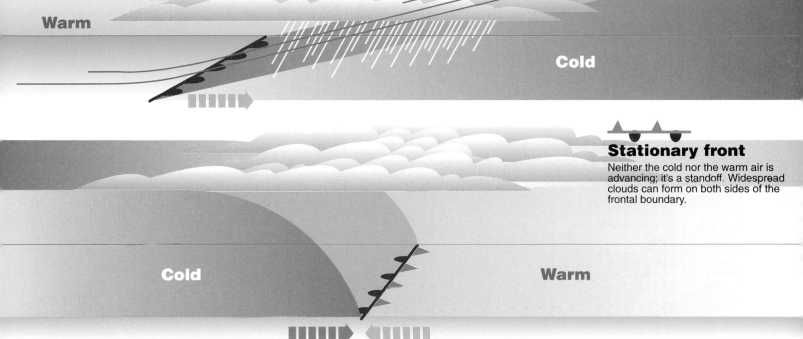

Stationary front

Neither the cold nor the warm air is advancing; it's a standoff. Widespread clouds can form on both sides of the frontal boundary.

Occluded fronts

Occluded fronts, or occlusions, are created when cold, warm and cool air come in conflict, forming boundaries above the ground as well as at the surface. They are often described as being caused by a cold front catching up with a warm front, but that seldom happens. Their clouds and precipitation are a mix of typical cold-front and warm-front clouds.

COLD OCCLUSION

Cold air is shoving under cool air at the Earth's surface — much like the cold front shown above. Thus the name 'cold occlusion.' The cold-warm air boundary aloft is often west of the surface front.

WARM OCCLUSION

Cool air rises over cold air at the surface — much like the warm front shown above. Thus the name 'warm occlusion.' The warm-cold air boundary aloft is often east of the surface front.

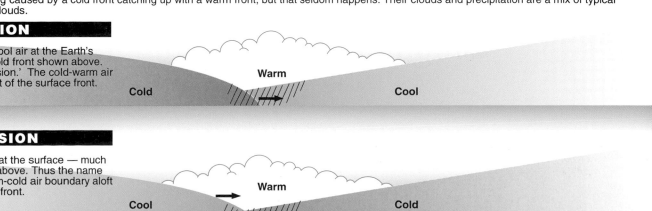

How storms form

Fronts are often parts of larger storms. Storms form in zones where warm and cold air are close together, sometimes along fronts.

Warm front
Cold front

1 Both the warm and cold masses of air are high-pressure areas with clockwise winds. The boundary does not have to be sharp enough to be called a front.

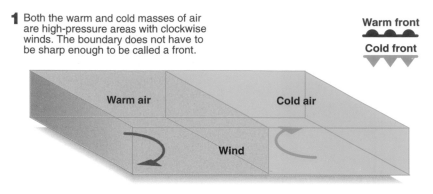

2 A low-pressure area forms on the boundary and counterclockwise winds around it begin moving the air. Warm air begins advancing on the east side — creating a warm front. Cold air begins advancing on the west side — creating a cold front. The fronts and low-pressure area begin stirring up clouds and precipitation.

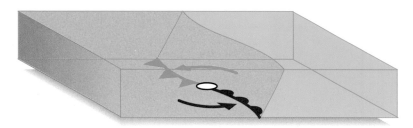

3 The low-pressure area grows stronger. Its pressure decreases. Winds increase in speed, and clouds and precipitation spread.

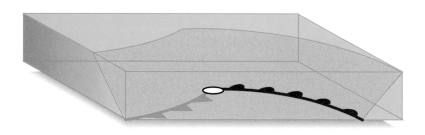

4 Sometimes the cold front catches up with the warm front, forming an occluded front, but this is rare. Scientists are still working out the details of the structure of occluded fronts and how they form. Often, the formation of an occluded front is the beginning of the end of the storm.

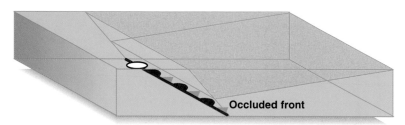

Occluded front

air from the north.

During the 19th century, scientists realized that storms were systems with circular winds. But they, like Franklin, believed the energy source was warm air rising straight up at the center — this is true of thunderstorms and hurricanes, but not extratropical storms.

By the 1920s, those Norwegian meteorologists who gave us the word "front" had also worked out an overall picture of storms. They recognized the role of temperature contrasts as the key energy source. Their picture of storms was so good we still use it every day. Each time you see a television weather map marked with "L"s and "H"s, and lines showing fronts radiating out from an "L," you're looking at the Norwegians' work: their World War I era picture of storms.

The changes fronts bring

The changes fronts bring depend on the season and location. A summer cold front in Oklahoma could push out 90°F air with 70°F air. A winter cold front in Colorado could replace 40°F air with 20°F air. Temperatures after a summer warm front arrives might be in the 90s in New York. A winter warm front might raise temperatures only into the 30s in Wisconsin.

While fronts are named for temperatures, they don't always bring big temperature changes. For example, in the summer a cold air mass that forms over northwest Canada won't be nearly as cold as a winter air mass. But it will be dry. As the air mass moves toward the Midwest or Northeast, it is warmed as it passes over ground heated by the summer sun. But it picks up very little water. Imagine that Detroit has been sweltering in 90°F afternoon temperatures, 60 percent relative humidity at noon and overnight temperatures in the 70s. After the summer cold front arrives, the afternoon temperature might still reach 87°F, but the afternoon relative humidity is only around 40 percent and the temperature that night falls into the 50s. It has brought comfort, not cold.

While the temperature might not change much, meteorologists have other ways of determining when a front has passed through. First, the wind changes direction. A cold front will usually cause a wind shift from southwest to northwest; a warm front brings a shift from southeast to southwest. A barometer shows falling pressure — it might not be much — before the front passes and then a rise after the front goes through.

The term "storm front" is often used on the West

Fierce storms can be 'meteorological bombs'

February 15, 1982, was the kind of day North Atlantic sailors dread. A storm's 100 mph winds whipped cold water into 50-foot waves. The giant oil-drilling rig *Ocean Ranger* sank about 200 miles southeast of St. John's, Newfoundland, killing all 84 aboard. The Soviet freighter *Mekhani Tarasov* also went down about 65 miles to the east of the *Ocean Ranger*, killing 18 of the 37 aboard.

The storm was what many meteorologists call a meteorological "bomb." The term was brought into common usage by now-retired MIT professor Fred Sanders in a 1980 article in *Monthly Weather Review*. The term was first used informally in the late 1940s for North Atlantic storms that develop "with a ferocity we rarely, if ever, see over land," Sanders says. Today, the formal definition of a "bomb" is a storm in which the center pressure drops an average of one millibar an hour for 24 hours.

"These are pretty good size storms with a horizontal span of 1,000 miles or more" with winds above 35 mph, Sanders says.

Fred Sanders studies 'bombs'

"They can grow to hurricane strength in 12 to 24 hours from something that had been very ordinary."

The storms' quick growth makes them especially dangerous to ships, including the fishing vessels that comb the North Atlantic off the United States coast. If a "bomb" explodes close enough to land, it can bury the East Coast in snow. Meteorological bombs also form in the western Pacific off the Asian Coast and occasionally in the eastern Atlantic and Pacific.

The ocean water is much warmer than the winter air blowing from the land. The resulting strong temperature contrasts along with the reduced friction of wind over the ocean — which allows the wind to blow harder — partly account for the storms' quick explosions into monsters.

Predicting "bombs" has been evolving since the '40s. Without weather satellites, forecasters had to rely on hit-or-miss radio reports from ships. By the 1970s computer models were developed to improve forecasting. Today scientists try to predict bombs by feeding information from satellites, offshore weather buoys and other sources into vastly improved computer models. "It's a question of how far ahead computer models can go," Sanders says. "Now the forecasts are good for a day or two ahead. But if you're figuring out the best route for a slow-moving ship, it would be nice to know five days ahead that a meteorological bomb is likely to explode off the coast of Newfoundland. "

Coast, but you won't find the term in meteorological glossaries. "Storm front" is usually used to describe what is, strictly speaking, an occluded front or a cold front moving in from the Pacific. The air that is moving in is usually chilly, humid Pacific Ocean air, which causes clouds and precipitation.

"Back-door cold front" is a weather term in the Northeast. Most cold fronts across the United States move northwest to southeast. But sometimes a cold air mass east of the Appalachians will move down the East Coast from the Northeast. It's coming from the "wrong" direction, through the back door.

The paths storms follow

Cold-weather storms are normally the strongest because the temperature contrasts are the greatest. While strong winds with below-zero temperatures are blowing snow across the Dakotas, Illinois and Indiana can be in the 50s and 60s. A strong winter storm literal-ly could affect the entire United States — including Alaska and Hawaii — in a single week of tracking across the country.

Since storms draw much of their energy from temperature differences, the boundary between warm and cold air is a favored path for storms. From the fall into the spring, storm centers move across the United States, often across the South as well as the North. In the summer, storm centers are more likely to track across Canada with the fronts trailing down across the northern United States. Thunderstorms, not extratropical cyclones, bring the United States much of its summer rain.

Tracking a storm

The picture of a typical storm and its movement (next four pages) helps explain the day-to-day weather.

A few hundred miles east of an advancing warm front, high, thin cirrostratus clouds will create a halo — a circle of light — around the sun or moon. As the front

LOOKING INSIDE A STORM

This look at a big, but typical spring storm moving across the United States over two days shows its effects on a wide area. Start with the storm already well formed, as shown on the weather map to the right. Now let's look at a vertical cross-section of the storm along a Des Moines-to-Albany line as the storm moves eastward.

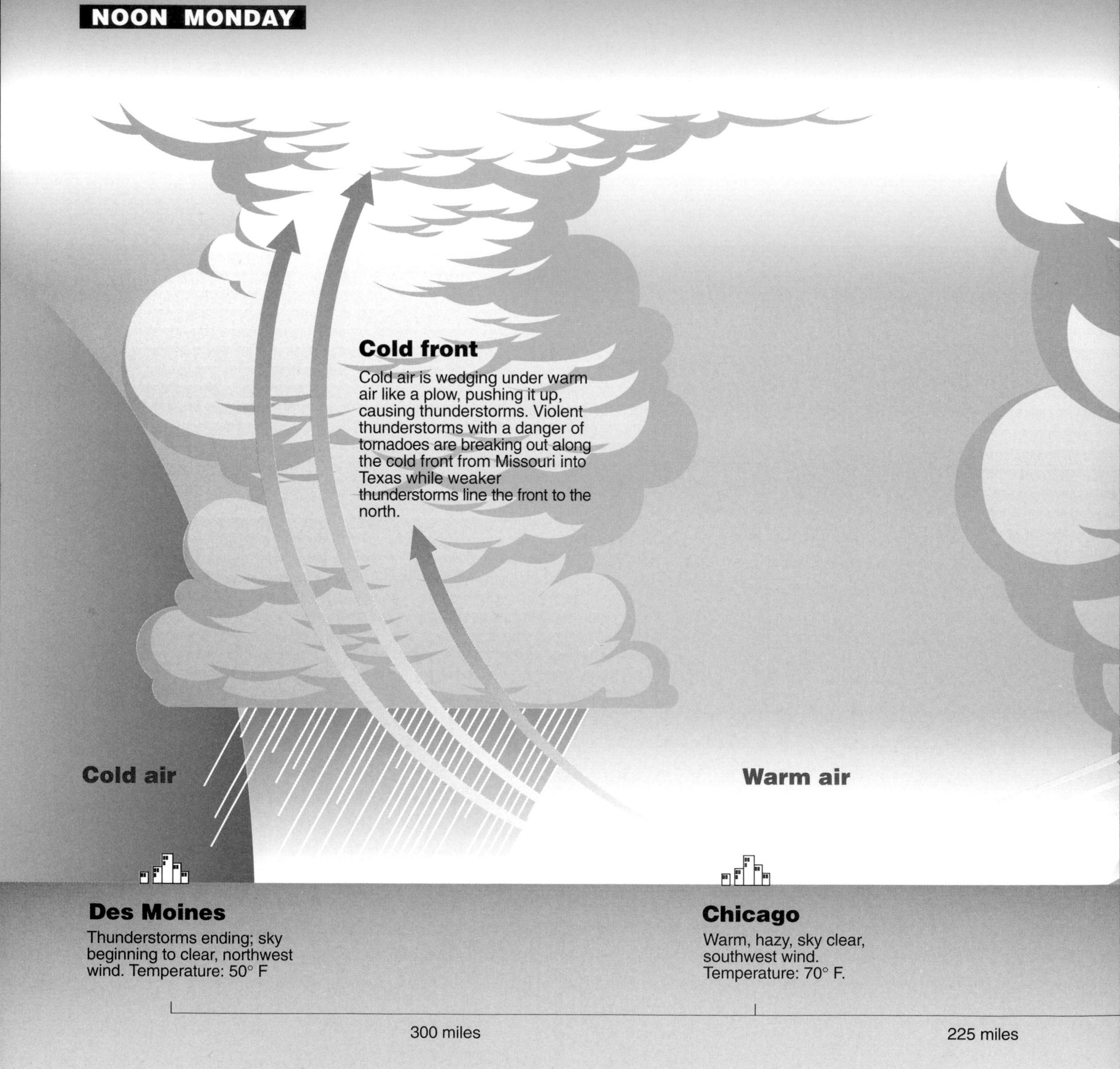

NOON MONDAY

Cold front

Cold air is wedging under warm air like a plow, pushing it up, causing thunderstorms. Violent thunderstorms with a danger of tornadoes are breaking out along the cold front from Missouri into Texas while weaker thunderstorms line the front to the north.

Cold air

Warm air

Des Moines

Thunderstorms ending; sky beginning to clear, northwest wind. Temperature: 50° F

Chicago

Warm, hazy, sky clear, southwest wind. Temperature: 70° F.

300 miles   225 miles

Low-pressure center

The storm's counterclockwise winds are pushing the fronts. Cold, moist air around the center and to its north is creating snow across northern Michigan, Wisconsin and Minnesota.

Weather map view

Cold front

Warm front

L

High, thin clouds stream out 500 miles ahead of the front where the warm-cool boundary is more than 20,000 feet up.

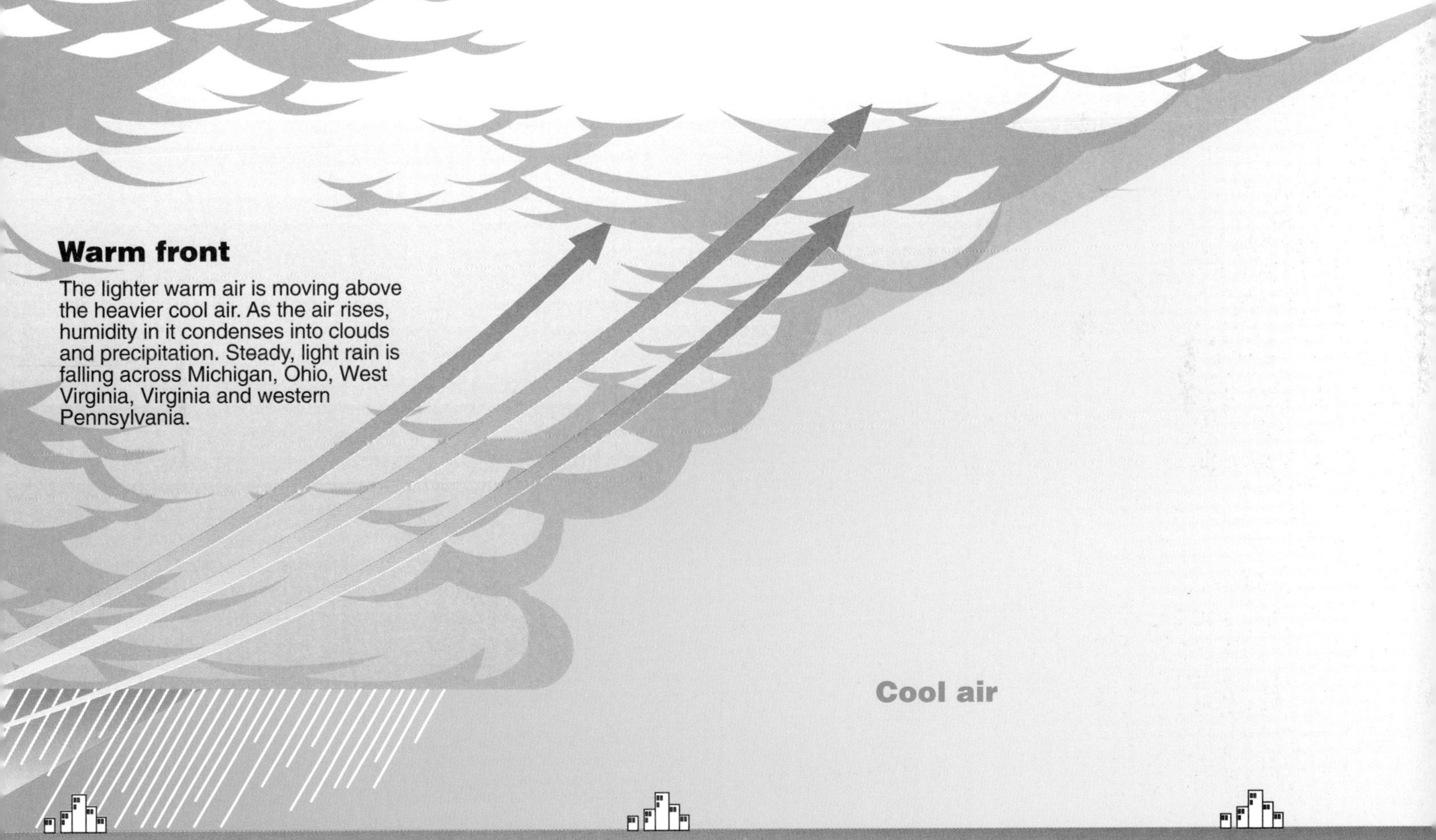

Warm front

The lighter warm air is moving above the heavier cool air. As the air rises, humidity in it condenses into clouds and precipitation. Steady, light rain is falling across Michigan, Ohio, West Virginia, Virginia and western Pennsylvania.

Cool air

Detroit
Rain, fog, low clouds, southeast wind.
Temperature: 54° F.

Buffalo
Cloudy, southeast wind.
Temperature: 48° F.

Albany
High, thin clouds with a halo around the sun, east wind.
Temperature: 40° F.

215 miles

250 miles

Turn page to see how this storm progresses in the next 24 hours.

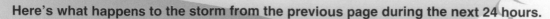
Here's what happens to the storm from the previous page during the next 24 hours.

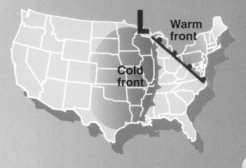

MIDNIGHT TUESDAY

The cold front is moving through Chicago. The warm front passed Detroit in the last two hours. Steady rain is moving eastward into New England. Thunderstorms along the cold front have weakened, but lightning is still flashing in the sky southward into Louisiana.

Des Moines

Clear, chilly night, northwest wind. Temperature: 25° F

Chicago

Thunderstorm, wind changing from southwest to northwest. Temperature: 30° F.

NOON TUESDAY

An occluded front has formed near the storm's center. Heavy snow is falling across northern Michigan into Canada. Lighter snow is falling west of the Great Lakes. Steady rain has spread eastward across the Northeast and New England. Thunderstorms are dumping heavy rain on the mountains of western Virginia and North Carolina. A severe thunderstorm watch is posted for South Carolina, Georgia and northern Florida.

Des Moines

Clear, chilly, west wind. Temperature: 40° F.

Chicago

Mostly clear, northwest wind. Temperature: 36° F.

MIDNIGHT WEDNESDAY

The occluded front now stretches as far south as the New Jersey coast. Rain is intense with thunderstorms and downpours mixed in with lighter but steady rain. Snow is falling in Canada and northern New England. The cold front and its thunderstorms have moved off the East Coast over the Atlantic. High, thin clouds about 600 miles ahead of a new storm to the west are moving over Des Moines.

Des Moines

High thin clouds, halo around the moon. Calm wind. Temperature: 33° F.

Chicago

Clear, light west wind. Temperature: 28° F.

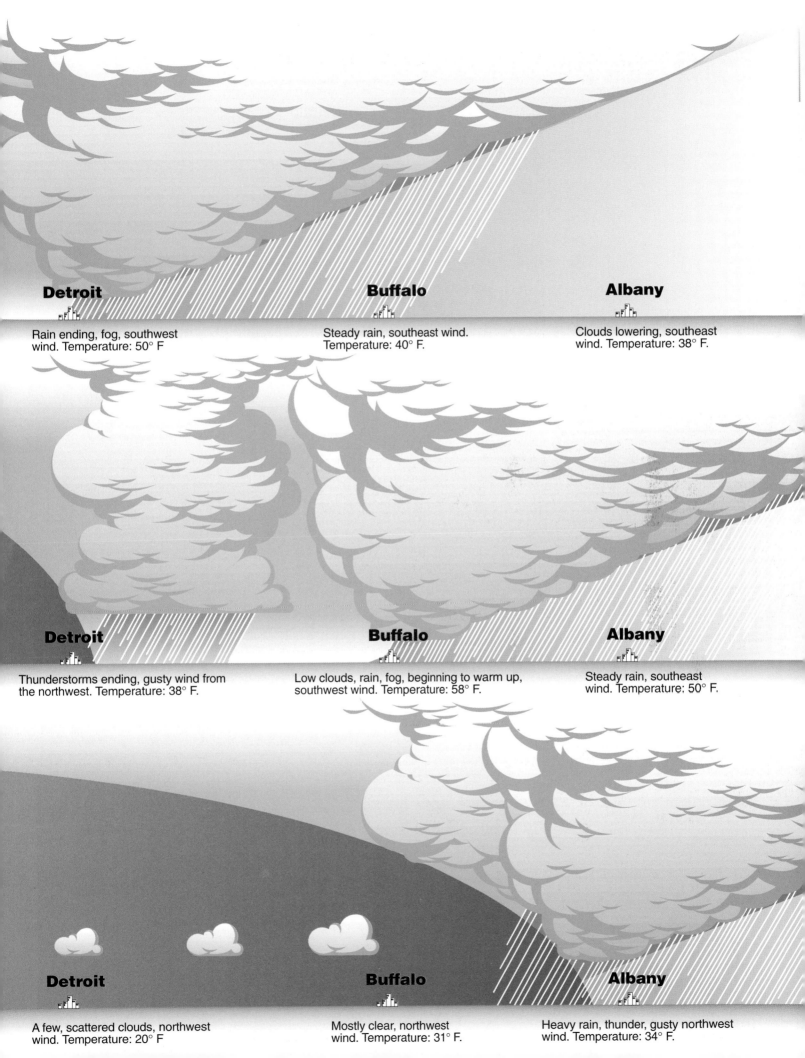

Detroit

Rain ending, fog, southwest wind. Temperature: 50° F

Buffalo

Steady rain, southeast wind. Temperature: 40° F.

Albany

Clouds lowering, southeast wind. Temperature: 38° F.

Detroit

Thunderstorms ending, gusty wind from the northwest. Temperature: 38° F.

Buffalo

Low clouds, rain, fog, beginning to warm up, southwest wind. Temperature: 58° F.

Albany

Steady rain, southeast wind. Temperature: 50° F.

Detroit

A few, scattered clouds, northwest wind. Temperature: 20° F

Buffalo

Mostly clear, northwest wind. Temperature: 31° F.

Albany

Heavy rain, thunder, gusty northwest wind. Temperature: 34° F.

When storms meet mountains

A simplified view of a storm as a spinning column of air shows why storms weaken as they move into high mountains, such as the Rockies, and then re-form or strengthen on the east side of the mountains.

1 As the storm approaches, it's a column of air from the ground to the bottom of the stratosphere, somewhat like an ice skater spinning with her arms over her head.

2 The bottom of the storm is pushed upward by the rising terrain while the bottom of the stratosphere acts like a lid on its top. The storm becomes a short, fat column of air — like a spinning ice skater with her arms extended, the storm's winds slow down.

3 As the storm moves across the mountains, its bottom follows the descending land. The storm turns back into a tall, thin, spinning column of air — like the ice skater pulling her arms back in — and wind speeds pick up.

comes closer, the clouds grow lower and thicker and turn the sky gray. A few hours later even lower clouds arrive with rain or snow.

What happens then depends on whether you're north or south of the storm's center as it passes by. When a storm's center passes by to the north, as shown for Des Moines, Chicago, Detroit, Buffalo and Albany in

Major storm tracks

Some paths followed by storm centers:

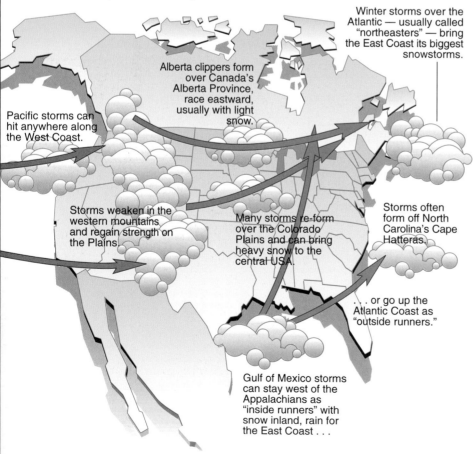

Winter storms over the Atlantic — usually called "northeasters" — bring the East Coast its biggest snowstorms.

Alberta clippers form over Canada's Alberta Province, race eastward, usually with light snow.

Pacific storms can hit anywhere along the West Coast.

Storms weaken in the western mountains and regain strength on the Plains.

Many storms re-form over the Colorado Plains and can bring heavy snow to the central USA.

Storms often form off North Carolina's Cape Hatteras.

. . . or go up the Atlantic Coast as "outside runners."

Gulf of Mexico storms can stay west of the Appalachians as "inside runners" with snow inland, rain for the East Coast . . .

the illustration on the previous pages, clouds and precipitation arrive ahead of the warm front. The weather turns warmer and may clear up. More clouds, perhaps thunderstorms, arrive. Finally the weather turns colder and begins to clear.

If the storm's center is south of you, clouds will become lower and thicker with rain or snow lasting around 12 to 24 hours. The precipitation will taper off to showers or snow flurries, the air will remain cold and the sky will clear. You'll never be in the storm's warm air.

During some winters, storms seem to follow the same paths. They are being guided by jet stream patterns that become established for long periods. A winter with most storms east of the Mississippi, moving from Texas to the Great Lakes, will give warm, rainy days to the East Coast. A year with winter storms moving along the Atlantic Coast east of the Appalachians will be cold and snowy for the mid-Atlantic and Northeast states. A winter that sees most Pacific storms hitting Canada's west coast, instead of California, will be dry in the Sierra Nevada and could leave ski areas in the southern Rockies wishing for snow.

Upper atmosphere triggers

To understand storms and forecast when they will form and what they will do, you have to know what's going on above 10,000 feet.

As we saw in the drawings on Page 49, fronts — and the storms that they are part of — extend high into the atmosphere. It's not surprising that high-altitude winds help supply energy for such storms. They also help determine where and when storms form, how they grow and decay, and where they travel.

Shifting winds, changing weather

At 7 a.m. on Dec. 24, 1989, winds about 18,000 feet above the Earth were blowing up to 100 mph directly from northwestern Canada into Florida.

This was the height of one of the worst cold snaps on record. On Dec. 22, 23 and 24 nearly 300 cold temperature records fell. In addition to hauling in cold air, the upper air pattern also helped trigger a storm that spread snow across the Southeast from Mobile, Ala., to Wilmington, N.C., where 15 inches fell. Miami was 31° F.

A little over a week later, Jan. 3, 1990, the winds at 18,000 feet were almost as strong, but were blowing from over the tropical Pacific.

By that afternoon, temperatures reached 79° F in Miami and 52° F in Washington, D.C.

In the 1920s and 1930s, scientists began exploring the upper atmosphere with kites, balloons and airplanes. By World War II weather stations around the globe were making upper-air observations. Today no meteorologist would think of trying to make a weather forecast without extensive upper-air measurements.

Jet streams are the cores of the general west-to-east flow of winds in the upper atmosphere. As the drawing to the right shows, these winds not only flow west to east, but they also often make excursions to the north and south. In the drawing we see that these winds make wavy patterns. A closer look shows "long waves" with crests and troughs thousands of miles apart and "short waves," ripples on the long waves.

Let's look at how both affect the weather.

Long waves and short waves

The wavy course followed by upper-altitude winds, including jet streams, is important to weather. Sometimes a global pattern, like that shown to the right, will stay in place for days. The long-wave troughs of cold, low-pressure air are associated with generally cool, cloudy weather. The warmer, high-pressure ridges usually bring warm, clear weather. The pattern shown here would bring the West good weather while the East is cloudy and chilly.

Most of the time the long-wave pattern drifts eastward under 10 mph. In this case, the nice weather from the West would move eastward while the trough over the Pacific would move into the West with clouds and maybe a storm. On rare occasions the pattern drifts westward. This is called "retrograding."

The long waves generally guide storms and determine whether temperatures will be cold or warm. Short waves help stir up storms and precipitation.

Also remember, while the waves are standing still or moving slowly, the winds that create the wavy patterns can be roaring at jet-stream speeds.

When forecasters talk about an "upper-air disturbance" or a "package of upper-air energy," they are often referring to a short wave that's moving along as a ripple on the long-wave pattern.

But sometimes forecasters might be talking about a trough passing overhead. As the drawing below shows, an upper-air trough is where cold, low-pressure air dips toward the equator. Even though the surface weather map might not show any fronts, a trough can act like a front in many ways. The sky could turn cloudy; rain or snow might fall. The wind might shift as it does when a front passes. Upper-air troughs can bring bad weather even though the surface weather map doesn't show any storms or fronts.

Old weather sayings are sometimes based on good science. Take for instance:

"Circle around the moon, rain or snow soon" and "When the sunset is gray, tomorrow may be a rainy day."

Thin, cirrostratus clouds far ahead of a storm can create a halo — a circle of light — around the sun or moon. As the front advances, the clouds lower, making the sky gray before rain arrives.

How upper-air patterns affect the weather

Here's a typical pattern of winds 18,000 feet above the surface. The distance from the crest of the wave over the West — a ridge — to the bottom over the East — a trough — shows this is the long wave pattern. Ripples on the waves are short waves. The global pattern usually has four to seven long waves.

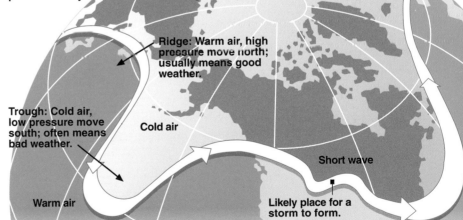

Ridge: Warm air, high pressure move north; usually means good weather.

Trough: Cold air, low pressure move south; often means bad weather.

Cold air

Warm air

Short wave

Likely place for a storm to form.

Sometimes the tip of a low-pressure trough becomes pinched off from the flow. Such "cut off lows" can stay in place for days or drift slowly eastward causing clouds and precipitation.

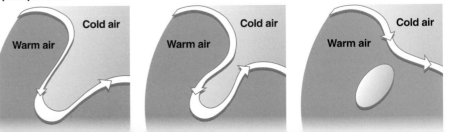

Tides, winds combine to create a super storm

For people who live along the ocean, storms become landmarks in time. At Folly Beach, S.C., local history is often divided into the time before or after Hurricane Hugo. At New York's Fire Island, residents may date events from a northeaster that left water two feet deep around their cottages. Usually storms hit just a few states, maybe only one or two, while the rest of the coast escapes harm.

As long ago as the 4th century B.C., the Greeks realized tides were linked to the moon and sun. But the link wasn't understood until Isaac Newton worked out his theory of gravity. In *Principia Mathematica,* published in 1687, Newton calculated the moon's and sun's effects on the tides.

But one 20th-century storm — not a hurricane, but an extratropical cyclone — ripped the East Coast from Charleston to Boston. For 65 hours beginning March 6, 1962, huge waves wiped out entire beach communities from the Carolinas to New England, killing more than 40 people and destroying an estimated $500 million worth of homes, stores, piers and roads.

Government reports called it the Great Atlantic Coast Storm. Many people refer to it as the Ash Wednesday Storm. "This one left the public aghast. It came out of the dark of the night and the dark of the moon," says Fergus J. Wood, a retired government scientist who detailed how tides added to the storm's destruction.

Unlike hurricanes that come and go in less than a day, the March 1962 storm sent its worst wind and waves crashing against beaches through the cycles of five high tides over more than two days. As the water rose for each successive high tide, storm waves washed away sand dunes, buildings and roads left defenseless by the previous high tide.

A strong high-pressure area to the north blocked the storm, keeping it in place to batter the East Coast hour after hour. But more was involved.

In the early 1970s, Wood, then with the National Ocean Survey, began studying this powerful, flukey storm. With the help of the U.S. Naval Observatory, he obtained computerized data on the moon's and sun's orbits at the time of the 1962 storm and the times of other destructive coastal storms going back to early colonial days.

Nags Head, N.C., in the aftermath of the 1962 Ash Wednesday Storm. The storm washed away beaches along hundreds of miles of the East Coast.

To see what Wood saw, you need to understand tides. The moon's and sun's gravity pull the oceans' water, causing the level to rise and fall regularly. In some places the difference between high and low tide is only a couple of feet. In rare cases, such as the Bay of Fundy in Canada, the water rises more than 43 feet from low to high tide.

While the sun and moon — especially the moon — set the tides in motion, scores of factors such as water's inertia and the shapes of the ocean bottom and the shoreline determine what the tides actually do.

Twice each 29.5 days, at new moon and full moon, the sun, Earth and moon are in a nearly straight line — this is called syzygy (pronounced SIZZ ah gee). The moon's gravity and the sun's gravity combine to create a stronger tidal pull than at other times. This stronger pull creates "spring tides" — named not because of any seasonal link, but because the tides "spring up." The high tides are higher and the low tides lower, making the range between high and low greater than at other times.

Fergus Wood investigates the links between tides and storms.

The March 1962 storm hit when the moon was new — "the dark of the moon" — which means spring tides helped make it worse.

But that wasn't the whole story. The moon doesn't trace out a perfect circle as it races around the Earth. The orbit is slightly egg shaped with the Earth closer to one end of the oval than the other. Once in each orbit, the moon reaches perigee, its closest approach to the Earth. This can be as close as 221,428 miles compared to the average distance of 238,855 miles. All perigees aren't equal; some are closer than others. Since the pull of its gravity on the oceans becomes stronger as the moon draws nearer the Earth, tides will tend to be higher at perigee and will be highest at the closest perigees.

Most of the time, perigee and syzygy happen at different times. But at least twice a year, and no more than five times a year, they occur within 36 hours of each other. At such times the tides rise prominently. In 1609, Johannes Kepler defined this alignment as "in syzygiis perigaeam." With new knowledge of the times when the moon and Earth approach very closely at syzygy, Wood coined the terms "proxigee-syzygy" and "proxigean spring tides" for the highest tides.

Sure enough, the March 1962 storm hit at one of these times of proxigean spring tides. Proxigee and syzygy were only 31 minutes apart. That, Wood points out, would not be enough to cause widespread trouble without a strong storm at the same time.

That is the secret of the Ash Wednesday Storm. An extremely strong storm hit just when the astronomical tides were already at about their highest levels.

When he looks back on it, Wood says his entire scientific career seemed to lead to his study of tides and storms. With a background in astronomy, meteorology

and oceanographic research, "I joined knowledge of gravitational forces of the moon and sun with aspects of meteorology — how winds form over and react with the sea surface — and oceanography, including tides and ocean currents."

Wood was able to plot times of other perigean and proxigean spring tides and match them to other storms. The country's first recorded perigean spring tide storm hit the Massachusetts Colony in 1635, only five years after the Pilgrims arrived. A contemporary account by Nathaniel Morton says the storm and tides "caused the sea to swell in some places to the south-ward of Plymoth, as that it arose to 20 feet ... , and made many of the Indians to climb into trees for their safety." He also noted "the moon suffered a great eclipse 2 nights after it." All eclipses of the sun or moon happen at syzygy, but syzygy doesn't always bring an eclipse.

Astronomical tides can be predicted with extreme accuracy decades ahead because even though the orbits of the sun, Earth and moon are complicated, scientists can calculate where they will be at any time in the future. But the tide tables can't say whether a storm will make the water rise higher than the tide alone will, or whether high pressure will keep the water level lower. Only weather predictions made three days or so ahead of time, combined with the tide table listings, can show how high the water actually will be.

Proxigean spring tides can be predicted far ahead by programming a computer with the necessary lunar and solar data from the U.S. Naval Observatory. Wood says proxigean tides will occur Sept. 16, 1997; Nov. 4, 1998; Dec. 22, 1999; Feb. 8, 2001; and Oct. 6, 2002.

As any of these times approach, you might start hearing "predictions" of earthquakes and volcanic eruptions, but geologists have never been able to relate such events to tides, even though the tides pull on the earth. If any disasters happen at these times, they're likely to be a repeat of the Ash Wednesday storm, but not necessarily on the East Coast.

Of course these aren't the only dates to worry about ocean storms. Beach houses have been washed away by storms when the tides weren't especially high. But if the moon is new or full — the times of spring tides — and a storm is predicted, watch out. And if a storm is expected near the time of a proxigean spring tide, be especially careful if you live on the coast.

"Neap tides" occur twice a month, at the moon's first and third quarters. The sun's and moon's gravities are pulling on the Earth at right angles to each other. Neap tides always have a small range between high and low tides; often the month's smallest. But other factors can sometimes bring lower tides. "Neap" comes from the Anglo-Saxon word for "scant" or "lacking."

How upper-air patterns affect storms

Air is flowing into an upper-air trough. The blue lines represent streams of air.

1 The streams of air are coming together; they're converging.

4 Here, the streams of air are pulling apart; they're diverging.

2 At Point 1, where the air is converging, air is piling up; the pressure is increasing. The bottom of the stratosphere acts like a lid, keeping air from going up. But, there's nothing to keep the air from going down. In fact, that's what happens — air descends from an area where air is converging aloft.

5 At Point 4, where the air is diverging, the pressure is decreasing. The lower pressure can't pull air down from above, but air does come up from below toward the low pressure. Air rises into an area where air aloft is diverging.

3 Since air is descending from above, the pressure on the ground is increasing. An area of high pressure is growing and air is flowing out from it in a clockwise direction. This turns out to be the area of high pressure found to the north and west of a storm center.

6 Air is going up into the area of divergence at Point 4. This means low pressure is being created, or made stronger at the ground. Winds are flowing counterclockwise into the surface low-pressure area and going up. This is the center of a storm and the winds are pushing the cold and warm fronts.

Convergence and divergence

Air coming together at any place is said to be "converging." Air that's flowing apart is "diverging." Areas of convergence and divergence in the upper atmosphere are keys in determining where high and low pressure will form and grow on the surface.

Changes in the speed or the direction of winds in the upper atmosphere can cause convergence or divergence. One place where convergence occurs is that point where the jet stream changes from a clockwise to a counterclockwise flow, such as on the east side of a ridge. Air converging in the upper atmosphere sinks into high pressure at the surface below.

On the east side of a trough, on the other hand, the jet stream changes from a counterclockwise to a clockwise flow. This causes air to diverge. Air diverging in the upper atmosphere causes air to flow upward, strengthening low pressure at the surface. This is why storm centers tend to form on the east side of upper-air troughs.

Other factors can cause upper-air convergence or divergence, and all of them can affect high- and low-pressure areas below.

The path and speed of the winds at around 18,000 feet help forecasters figure out where storms are heading. In general, surface storms will move in the same direction at about half the speed of these winds. This isn't as simple as it sounds, however. As a storm strengthens, the upper-air winds can begin blowing faster, which helps change the pattern of troughs and ridges. That is one reason why storm forecasts sometimes go wrong.

Weather maps and models

Maps of storms, such as those developed by the Norwegians in the early 1920s, are idealized concepts. They help people organize and make sense of what otherwise would be a confusing jumble of phenomena seen in the sky and gleaned from weather reports at distant places.

Scientists call such drawings "models." They not only help us make sense of the world, but they also help us see things we might otherwise miss. In the past, you might hardly have noticed thin, high clouds slowly covering the sky from the west. The illustrations on Pages 52-55 will help you recognize high, thin clouds as part of a storm system.

Such simple models are the first step toward understanding and forecasting. But they are only a first step. Since the 1920s scientists have been building more complicated models using mathematics and physics rather than drawings and maps.

Mathematical equations must describe the big picture of an atmosphere bounded on the bottom by oceans, plains and mountains and on the top by the vacuum of space. Equations must account for the Earth's rotation and its tremendous temperature contrast, swirling air motions, heat added and lost, effects of water changing among its liquid, solid and vapor forms and the differences in winds and air pressures from the surface upward. These equations must also include interactions between the air and the ocean, and the air and land that's wet in some places and dry in others; smooth in some places, rugged in others. All these factors — and others not mentioned here — are pushing and tugging on each other, making the chain of cause and effect extremely tangled.

It is no surprise that knowledge often outstrips models in meteorology, forcing scientists to refine the old models or create new ones.

A more detailed picture

Newspaper and television weather maps show what's known as the "synoptic scale." A synoptic weather map covers a large area, such as the entire United States, and is a snapshot of either the weather at a particular time or a forecast showing what's expected. Synoptic maps show large features, such as storms hundreds of miles across, but can only hint at smaller features, such as thunderstorms.

Observations taken at widely scattered weather stations are a good net for catching synoptic scale events such as fronts. However, they miss the details of small disturbances.

After World War II, meteorologists began using radar to see details of storms that previously had been hidden. They also began using airplanes to take scientific measurements in storms, collecting detailed data that the World War I Norwegian meteorologists could have only dreamed of. In the 1970s, satellites began showing not only the big picture of storms but also details that had never before been seen.

As scientists collected more data on small-scale weather events, they started using the term "mesoscale" to describe weather systems between two miles and 250 miles across. Mesoscale weather studies offer the promise of improved forecasts for thunderstorms, tornadoes, and snowstorms. They also are helping scientists refine their models of extratropical cyclones.

The Norwegians developed their models by studying storms that were coming into northwestern Europe

from the Atlantic Ocean. Obviously, the model is a good one. We're still using it to describe weather systems in the middle latitudes of both the northern and southern hemispheres. The big breakthrough the Norwegians made was realizing that boundaries between different air masses — fronts — are major weather makers.

In the 1920s and 1930s, as American meteorologists started using the Norwegian model of storms, they began seeing weather that didn't quite fit the picture.

For example, some cold fronts on the Plains do not touch the ground. The air between the bottom of the front and the ground does not have the temperature contrast needed to fit the Norwegian description of a front.

We call these "upper air fronts," and they show why the Norwegian model is too simple to explain all of the weather in all parts of the world.

As a cold front from the west crosses the Rockies, air at the lower levels is warmed as it flows downhill on the east side of the mountains. (In Chapter 5, we'll see why air warms as it flows downhill.) This warming erases the cold-warm air boundary near the ground but leaves it in place a few thousand feet up. The resulting upper air cold front can act much like a surface cold front pushing warmer air up to create clouds and precipitation in locations that don't fit the Norwegian model.

Air that warms as it flows down the east side of the Rockies into a storm, or a developing storm, adds another complication that the Norwegian model can't accommodate.

This warmed air from the west is dry. At the same time, warm, humid air is flowing into the storm from the Gulf of Mexico. Now, instead of the Norwegian model of a storm with warm air flowing in from the south, the storm has warm, humid air from the south and warm, dry air from the west.

This is just one example of how the Norwegian model can't accommodate all storms.

Another recent change in understanding how storms move, the shapes they show on satellite photos, and whether they bring rain or snowfall is the concept of "conveyer belts" of air inside storms. The graphic on Page 62 is based on a study of U.S. East Coast snowstorms by Paul J. Kocin and Louis W. Uccellini, both of the National Weather Service. It's typical of storms that bring heavy snow to the eastern USA.

The conveyor belts are not like jet streams, which span continents and stay at the same altitude. Instead,

Early mathematical models

Lewis Fry Richardson (1881-1951), a British weather observer, studied the theories of Norwegian meteorologists. During World War I, Richardson worked out the basics of computer modeling. His *Weather Prediction by Numerical Process* showed how to calculate a sample forecast. But his method wasn't practical in those pre-computer days: He estimated that 64,000 people working together would be needed to make forecasts.

John von Neumann (1903-1957) and his research team at the Institute for Advanced Study at Princeton, N.J., made the first computer weather forecast in April 1950. It was the beginning of today's computer forecasting models.

'Conveyor belts' give a more complete picture of storms

The storm graphics on Pages 52-55 are generalized and focus on what's happening at the ground. In recent years, satellites, radars, and computers have helped scientists develop more detailed pictures of storm airflows at all altitudes, including "conveyor belts" of air. The cold front and warm front in the graphic here are the same as in the Pages 52-55 graphics. The conveyor belts are an added feature. The airflows shown are typical of northeastern USA snowstorms.

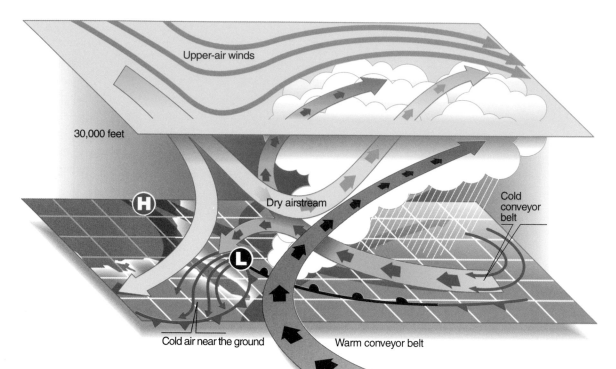

the conveyor belts are much smaller streams of air that rise or fall within a storm. The cold conveyor belt shown in the graphic, for example, brings in cold air that picks up moisture as it skims southward across the ocean to wrap around the north side of the storm's low-pressure center and rise. The warm conveyor belt brings in humid air from the south. The dry airstream actually brings jet stream air down, drying out clouds in part of the storm on its way down. Forecasters use this more complete model to help predict where a storm's snow will fall.

Knowing about upper air fronts and conveyer belts, which are part of more refined models, helps meteorologists prepare more accurate forecasts. A forecaster with only the classic Norwegian model can say a big storm is coming but has a much harder time predicting where the storm's strong thunderstorms, rain, snow, sleet, or freezing rain are likely to hit.

Chicken and egg

No simple picture tells the whole story of storms. The Norwegian model is a good starting point for understanding the large sweep of weather across the middle latitudes, the parts of the Earth outside the polar regions and tropics. The highs, lows, and fronts shown on television, newspaper, and Internet weather maps will give a good general idea of what kind of weather to expect. But the details aren't always likely to fit neatly into the simplified model these standard weather maps represent.

Meteorologists must use verbal and visual shorthand to communicate with people who aren't atmospheric scientists. Sometimes, for instance, a forecaster may say, "It's going to turn cold because the jet stream is going to dip south." But is the jet stream going to bring cold air south? Or will the jet stream dip south because that's where the cold-warm boundary is going to be?

For while jet streams help guide hot and cold air masses, temperature differences create the air-pressure differences that make jet streams in the first place.

The relationship among upper-air winds, locations of cold and warm air, and what storms are doing is like the chicken and the egg: Which came first? Still, you don't have to understand complex models to appreciate how jet streams and storms build the framework for other weather events that play themselves out on the atmosphere's stage.

If our air were perfectly dry, winds created by the sun's unequal heating of the Earth would blow dust storms across an unlivable planet.

The Earth supports life because, even in the driest places, the air always carries at least a little water. Not only does water in the air bring needed rain, but the changes among water's three phases — ice, liquid, and vapor — also help create much of our weather.

Water vapor, the invisible gaseous form of water that's always in the air, supplies the moisture needed for dew, drizzle, fog, frost, rain, sleet, snow, thunderstorms, tornadoes, hurricanes, and rainbows. It also makes clouds.

When water changes among its gas, liquid, or solid phases, heat is added or taken from the surrounding air. Air warmed or cooled by phase changes is a source of weather energy.

What happens in phase changes

To understand how water creates clouds and precipitation and how water adds or subtracts heat from the air as it changes among its three phases, start by looking closely at water molecules to see how they interact as ice, as liquid, and as vapor.

Molecules are always moving. A substance's temperature depends on how fast its molecules are moving. The faster the molecules, the hotter the substance. Temperature is one of the factors that determine whether a substance is a solid, liquid, or gas.

When we talk about speeds of molecules changing with temperature, we're talking about averages. In a glass of cold water the average speed of molecules is slower than in a glass of hot water. But if you could beam a state trooper's radar gun at each molecule in both glasses, you'd find a variety of speeds with some molecules mov-

USA's highest normal annual precipitation (in inches)

Yakutat, Alaska	134.96
Hilo, Hawaii	128.15
Annette, Alaska	115.47
Quillayute, Wash.	104.50
Kodiak, Alaska	74.24
Astoria, Ore.	69.60
Blue Canyon, Calif.	67.87
Mobile, Ala.	64.64
Tallahassee, Fla.	64.59
Pensacola, Fla.	61.16

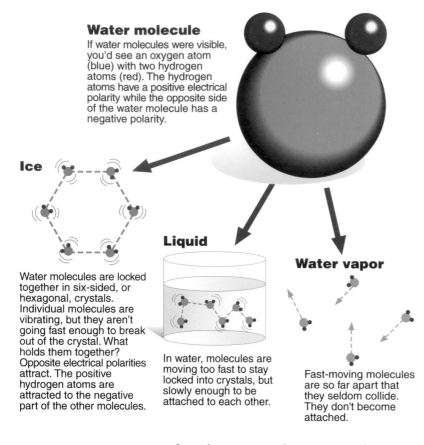

Water molecule
If water molecules were visible, you'd see an oxygen atom (blue) with two hydrogen atoms (red). The hydrogen atoms have a positive electrical polarity while the opposite side of the water molecule has a negative polarity.

Ice
Water molecules are locked together in six-sided, or hexagonal, crystals. Individual molecules are vibrating, but they aren't going fast enough to break out of the crystal. What holds them together? Opposite electrical polarities attract. The positive hydrogen atoms are attracted to the negative part of the other molecules.

Liquid
In water, molecules are moving too fast to stay locked into crystals, but slowly enough to be attached to each other.

Water vapor
Fast-moving molecules are so far apart that they seldom collide. They don't become attached.

ing faster than average and some moving slower.

Imagine you are measuring the speed of a single water molecule. If it moves faster than the average speed of all the molecules in the glass, it might break free of the liquid water and fly into the air. It would have enough energy to break the attractive force that had been holding it in the liquid.

Warm up the glass of water and more faster-moving molecules will gain enough energy to break free and move into the air to become water vapor. This is evaporation. The molecules in the air, water vapor, also are

moving at various speeds. Some move slowly enough to join the water in the glass. This is condensation, the opposite of evaporation.

Understanding what happens to the molecules in our two glasses of water is the basis of learning what happens in clouds and storms.

If evaporation continues long enough, the number of vapor molecules rejoining the water becomes equal to the number leaving the water. The air is "saturated."

As the drawings below show, when the air and water are warm, more vapor molecules are needed to reach this saturation point than when the air and water are cold. This is why warm air is said to "hold" more water vapor than cold air.

Some scientists don't like to talk about warm air "holding" water vapor because that can give the wrong idea: that air "holds" water like a basket "holds" apples. But in the atmosphere's "basket," the "apples" are all moving. Warm air can "hold" more water vapor than cool air because in warm air more molecules are going fast enough to remain as vapor instead of condensing into liquid water.

Keep the term "saturated" in mind. It's needed to explain a lot of things about humidity. Air is saturated when any additional vapor will condense if it gets the chance. Air that's saturated won't "hold" any more water vapor.

If we took the top off the container in the drawing below and blew dry air over it, more water would evaporate because the wind would blow away air above the container before it became saturated. This is why water evaporates faster on dry, windy days than on humid, still days.

Looking at water and vapor on the molecular level

How air becomes 'saturated'
A sealed container is about half full of water and half full of perfectly dry air. The water and air are at the same temperature.

Even though the air starts out perfectly dry, some of the liquid water molecules are going fast enough to break free into the air: They've evaporated into the air, becoming water vapor.

After a while water molecules are going from the air to the water as well as from the water to the air. Eventually, as many molecules will be going back into the water as leaving it. When this happens we say the air is saturated with water vapor. At this temperature one water molecule will leave the air for every one entering the air — the air can't "hold" any more water vapor.

If the container is put into the sun, it heats up. Water molecules in both the air and water speed up, which means more molecules are going fast enough to leave the water at this warmer temperature. Eventually the air becomes saturated with more vapor molecules.

If the container is cooled, the molecules' average speed slows, and more vapor molecules return to the liquid water than leave it. But no matter how cool the container becomes, a few molecules will still be going fast enough to break free and become vapor. At cooler temperatures fewer water molecules are in saturated air.

Taking photographs of raindrops to figure out how clouds work

Sometimes serious science doesn't look very serious.

Anyone visiting Duncan Blanchard's lab in 1949 might well have wondered about the hodgepodge of wood, construction paper, plastic and metal screening hanging from the ceiling. The wobbly affair was part of a serious scientific study, an attempt to take photos of raindrops.

Blanchard would release one drop of water at a time from the top. The air blowing from below would hold the drop in place in front of his camera. The study slowly unfolded secrets of raindrops. Among other things, Blanchard's wobbly wind tunnel photos showed that small raindrops are round and large ones are shaped like hamburger buns, as shown in the graphic on Page 66.

Blanchard's work took place at the General Electric Research Laboratory in Schenectady, N.Y. It was part of Project Cirrus, a government project to develop cloud seeding — to make rain or snow fall when and where it was needed. The project was directed by Irving Langmuir, the 1932 Nobel Prize winner in chemistry.

There was work outside the laboratory as well. Military planes flew to many parts of the USA to seed clouds by dropping dry ice while researchers recorded the results. But to make it rain, you have to know details of how rain forms. The raindrop photos showed how falling drops act, including how large they grow before breaking up.

Blanchard says that unlike other scientists he knows, he didn't build radios or blow things up with chemistry sets as a youngster growing up in New Lenox, Mass.

In 1942 after graduating from high school "with mediocre grades and no money," he became an apprentice machinist at General Electric in Pittsfield, Mass. The next year the Navy sent him to Harvard University and later

Blanchard is working on a new book about snowflake photography.

to Tufts University under a World War II program that sent promising youngsters to college to become officers. After commissioning him as an ensign, the Navy sent him to Guam just after World War II ended in 1945. "I was just 21 and it all seemed just marvelous," he says. "I had decided on a Navy career."

"All this changed one evening after I had been there three or four months. I was walking along a path through the jungle to the officers' club when I noticed something off the path. Here was a book someone had tossed away." The book was a soaking wet copy of *The Human Side of Science* by Grove Wilson. Blanchard picked it up, took it back to his quarters, dried it out and began reading. "I saw that science was a human activity, trial, errors, hits and misses. This intrigued me. I decided 'I'm leaving the Navy, going back to school and somehow become a scientist.' It was the nearest thing to a religious conversion

I could imagine."

When Blanchard joined Project Cirrus he had just a bachelor's degree and no experience as a scientist. Before meeting Langmuir for his first assignment, Blanchard, as a rank scientific beginner, expected to be told exactly what to do and how to do it. He figured Langmuir would be like the factory foremen he worked for as an apprentice.

Instead, Langmuir "suggested I take on the job of designing and building a vertical wind tunnel." Blanchard discovered that Langmuir was a colleague, not a foreman.

After leaving Project Cirrus, Blanchard earned a master's degree in physics from Penn State University and a Ph.D. in meteorology from Massachusetts Institute of Technology. "After getting my union card, a Ph.D., I was able to root contentedly in the citadel of science, sniffing out a variety of interesting research problems," he says.

In 1967, while working in Massachusetts with the well-known oceanographer Alfred Woodcock at the Woods Hole Oceanographic Institution, Blanchard wrote *From Rain Drops to Volcanoes*. The raindrops are those he photographed; the volcanoes refer to studies he made of lightning created by eruptions. His research also showed how bubbles that are formed by breaking ocean waves send salt particles into the air to become condensation nuclei for raindrops. In a 1996 article in the *Bulletin* of the American Meteorological Society, Blanchard wrote about the importance of "serendipity" in science, which Langmuir defined as "the art of profiting from unexpected occurrences."

Blanchard likes the way Albert Szent-Gyorgyi, the 1937 Nobel Prize winner in physiology or medicine, summed it up: A scientist has "to see what everybody else has seen, and think what nobody else has thought."

What falling raindrops look like

High-speed photos show raindrops don't look like "teardrops." Water's surface tension pulls drops into a sphere.

A drop smaller than about .08 of an inch in diameter remains spherical as it falls.

As a larger drop falls, air pressure flattens its bottom. The sides bulge out because air pressure there is lower.

When a drop grows larger than about a quarter inch across it begins breaking up into smaller drops.

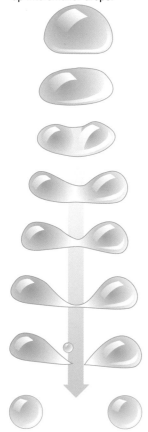

also helps explain why evaporation cools things. To break away and become vapor, a molecule has to acquire energy. Surrounding heat supplies that energy. In effect, a molecule that evaporates is carrying some of the heat from the liquid it has just left. When you perspire, the water your sweat glands deposit on your skin has been warmed by your body. When this water evaporates, it carries away heat.

Exercising when the air is nearly saturated is uncomfortable and sometimes dangerous. Little perspiration evaporates into the nearly saturated air, which means little heat is carried away. You aren't cooled.

Just the opposite happens when the air is dry, especially if it's windy. After a swim on a hot, dry, windy day, water quickly evaporates from your body. It can carry away so much heat that you get goose bumps even if the temperature is 90°F.

Measuring humidity

The measure of humidity used most often is "relative humidity." To explain it, we'll start with some other ways of measuring the air's humidity and then show what relative humidity is all about.

Meteorologists use various ways to describe how much water vapor is in the air. We will use the "saturation mixing ratio" in the chart above. It describes how many grams of water vapor are in the air for each kilogram of dry air when air at different temperatures is saturated.

How much air is a kilogram? If the air is at a normal sea-level pressure and a temperature around 60°F, a kilogram of air would fill about a cubic yard of space; that is, a 3-by-3-by-3-foot box. Using this information, we can see just how little water is in even the most humid air. If you condensed all of the water vapor in a kilogram of saturated 95°F air, you'd have only 37.25 grams of water — not even a quarter cup.

We'll use the chart to explain one of the most important measures of humidity, the dew point. Let's say that at 3 p.m. on a particular day, a weather observer measures the temperature at 86°F. The observer also finds that the air happens to have 10.83 grams of water per kilogram of dry air.

As the sun sets, the temperature goes down. What happens when the air cools to 59°F? The chart shows that at 59°F air is saturated if it has 10.83 grams of water vapor per kilogram. If it gets any cooler, the water vapor will begin condensing into liquid water. Dew will form

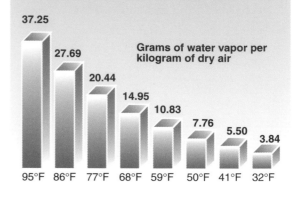

Saturation amounts

The amount of water needed to saturate air increases as the air's temperature increases. This chart shows how many grams of water vapor are in the air for each kilogram of dry air when it's saturated.

Grams of water vapor per kilogram of dry air

95°F	86°F	77°F	68°F	59°F	50°F	41°F	32°F
37.25	27.69	20.44	14.95	10.83	7.76	5.50	3.84

on the grass.

Back at 3 p.m. when we saw how much water was in the air, we could have said: "If the air is cooled below 59°F, it will be saturated and dew will form." In other words, 59°F is the air's dew point.

Hourly weather observations from around the world don't give the relative humidity; they give the dew point. The dew point temperature is based on how much water vapor is in the air. So while dew point is given in terms of temperature, it's really a measure of humidity.

What relative humidity tells you

The chart above also helps explain what relative humidity means. Go back to the 3 p.m. temperature of 86°F with 10.83 grams of water vapor per kilogram of air.

The chart shows that 86°F air needs 27.69 grams of water vapor per kilogram of dry air to be saturated. Divide the actual vapor in the air (10.83 grams) by the amount the air could hold (27.69 grams) and multiply by 100. That percentage shows what share of the vapor the air could hold is actually in the air. This is relative humidity. In this case, the air has 39 percent of the water vapor it could hold at its current temperature.

Relative humidity depends not only on how much water vapor is in the air but also on the air's temperature. The relative humidity table on Page 67 shows how this works. The amount of vapor in the air was measured at 3 p.m. when the temperature was 86°F. As the air cools, no water vapor is being added or taken away. The table shows how the relative humidity increases as the air cools, even though the amount of water vapor in the air stays the same.

Why your house dries out in winter

Imagine it's a cold, damp day with the outside air saturated at 32°F. According to the chart, saturated 32°F air has 3.84 grams of water vapor per kilogram of dry air. Now bring some of this 32-degree outdoor air inside and warm it up to 68°F. What's its relative humidity? Take the 3.84 grams of vapor actually in the air and divide it by the amount of vapor 68°F air can hold, 14.95. Multiply by 100 and you get a relative humidity of about 26 percent. A comfortable range is usually 40 to 70 percent.

The colder the outside air, the worse it gets because colder air contains even less water vapor. Air at 5°F can hold only 1.19 grams of water vapor per kilogram of dry air. Bringing that air inside and warming it to 68° F — without adding any humidity — gives a relative humidity of a very uncomfortable 8 percent.

Heat and water's phase changes

A liquid molecule needs energy to break its bonds and become vapor. Imagine a water molecule evaporating from your skin on a hot day. It gains the needed energy by taking a tiny amount of energy or heat from your skin. That helps cool you off. This heat doesn't disappear; it's converted into the vapor molecule's added energy. As the vapor molecule moves away from your body, it takes this energy — called latent heat — with it. When the vapor molecule condenses back into liquid, it releases the latent heat, warming its surroundings. Imagine the molecule that evaporated from your body rising to join billions of water vapor molecules condensing into the tiny drops of a cloud. When the molecule condenses, it releases the latent heat carried from your body, helping to warm the air slightly in the growing cloud. Latent heat might not seem like much, but on a much larger scale it's the major source of energy for hurricanes.

Latent heat also is involved when water freezes or ice melts. To turn water into ice, a refrigerator must pump away heat to cool the water to the freezing point and then continue pumping away the latent heat released as the water freezes.

To melt, ice must take in enough heat for its molecules to gain sufficient energy to break their crystal bonds. In a glass of iced tea, for instance, this energy comes from the warm tea. Energy going from the tea into the ice accounts for most of the tea's cooling.

Water makes two other kinds of phase changes: directly from vapor into ice and directly from ice to vapor. In the United States, meteorologists usually call both changes "sublimation."

Sometimes after snow falls, the sun comes out but the air stays very cold and dry. Piles of snow just seem to disappear without any melting; the snow is sublimating directly to vapor in the dry air. Frost is a good example of vapor sublimating directly into ice.

Sublimation, like the other phase changes, also involves latent heat. When vapor sublimates into ice, heat is added to the surroundings. When ice sublimates into vapor, heat is taken from the surroundings.

Effects of water vapor in the air

Understanding how water acts as it evaporates into the air or condenses out of the air helps explain why a bathroom mirror fogs up when you take a hot shower. It also explains how dew, frost, fog, and clouds form.

The bathroom mirror fogs because some of the hot water spraying from the shower evaporates into the air, increasing its humidity, and therefore the air's dew-point temperature. The mirror's surface is cooler than the dew point of the now humid air in the bathroom. Some of the vapor in the air that touches the mirror condenses onto it, making tiny water drops. Dew has formed on the mirror. If the shower is hot enough and the air already humid enough, a light fog might form in the room. The vapor is condensing into tiny drops that float in the air.

As we saw in Chapter 2, fog or dew or frost is more likely to form on clear nights than on cloudy ones. The Earth is always radiating away infrared energy into the atmosphere. After the sun goes down, solar energy is no longer warming the ground, but the infrared energy keeps sending heat upward. The ground cools. On cloudy nights, the clouds absorb infrared energy from the Earth and radiate it back down. On clear nights, infrared energy is lost to space; the ground becomes

Relative humidity

Table shows the relative humidity of air with 10.83 grams of water vapor per kilogram of dry air at various temperatures.

Temperature of the air	Vapor air can hold	Vapor actually in the air	Relative humidity
86°F	27.6	10.83	39%
77°F	20.4	10.83	53%
68°F	14.9	10.83	72%
59°F	10.8	10.83	100%

As air cools, its molecules — including water molecules — lose energy. More and more water molecules condense, forming clouds or fog.

Cloud drop

Water vapor

What happens as water changes phases

We'll use a pile of snow in a puddle, one side in the sun and the other in the shade, to sum up how water changes from one phase to another and whether the change cools or warms the surroundings.

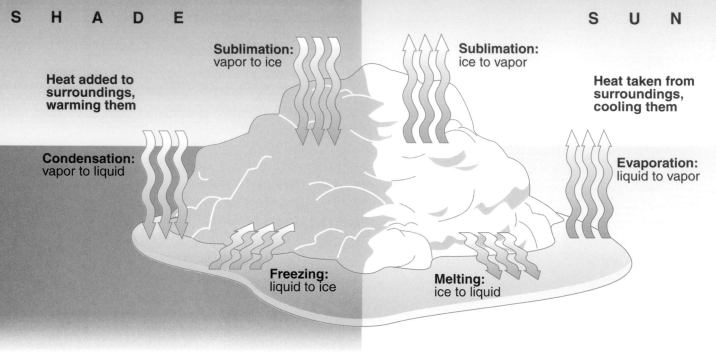

SHADE

SUN

Sublimation: vapor to ice

Sublimation: ice to vapor

Heat added to surroundings, warming them

Heat taken from surroundings, cooling them

Condensation: vapor to liquid

Evaporation: liquid to vapor

Freezing: liquid to ice

Melting: ice to liquid

cooler. The cold ground cools the air next to it. If the air cools to its dew point, dew, frost or fog forms.

What the dew point tells you

When water vapor changes into liquid or directly into ice, it releases latent heat and warms the air a little. This means that overnight as the air cools to its dew point, condensation will begin slowing the temperature fall. As a result, the air is not likely to get colder than its initial dew point anytime during the night.

Of course, the air doesn't always cool to the dew point. And a mass of new cold air could move in during the night making the temperature plunge after the front passes and colder, drier air arrives. But if no fronts are expected, the afternoon's dew point can give you an idea of what minimum temperature to expect that night. For instance, if you're vacationing in the Florida Keys and the afternoon temperature is 89°F and the dew point is 72°F, you better make sure your motel has good air conditioning. That night's low temperature is going to be in the 70s and the relative humidity is going to be around 100 percent.

If the temperature and dew point are close together in the late afternoon as the air turns cooler, fog is likely during the night. This is why weather observations from airports include both the temperature and dew point. Pilots know that if the two are only 4°F or 5°F apart, fog

is likely. Pilots might not have enough visibility to land.

Condensation in midair

Getting condensation going in the air isn't as automatic as it might seem. We introduced evaporation and condensation on Page 64 with the special case of water in a sealed container. After saturation was reached, vapor molecules began condensing back into the water. But in the atmosphere things do not happen so simply.

In clean air, condensation doesn't take place easily. Colliding water molecules have to stick together. That is unlikely. Water vapor molecules need something to stick to. These "somethings" are called "condensation nuclei." These are tiny particles in the air. They are so tiny that they are measured in microns. A micron or micrometer is one-millionth of a meter. A human hair is about 100 microns thick, about .004 inch. Most condensation nuclei are about 0.1 micron in diameter; the largest are 10 microns.

Condensation nuclei can come from sand or dust in the air, from particles in smoke from fires, from material ejected by volcanoes, even from tiny particles left from shooting stars.

Many condensation nuclei come from sea salt. Salt shoots into the air when bubbles burst in the ocean. Wind picks up salt when it blows foam off breaking waves. Winds can carry sea salt 100 miles inland. Relatively large salt particles explain the haze often seen at the beach. The haze is caused by water collecting on the many salt nuclei in the air. Anyone who has lived near the ocean knows how the salty mist and dew quickly rust metal.

Pollution also adds condensation nuclei to the air. A study of rainfall in the St. Louis area from 1971-1975 showed that most rain fell downwind from the city. Researchers concluded that the city's pollution added extra condensation nuclei to air flowing over the city. Similar effects have been found in other areas.

Even when they don't create more clouds and rain, added condensation nuclei can make polluted air hazy, but all hazy air isn't polluted. In many places tiny particles of natural substances from trees and other vegetation combine with ozone to form particles that scatter blue light in all directions. This haze gives the Blue Ridge Mountains of Virginia their name. Larger particles, from sources such as sea salt as well as pollutants, create a white haze.

When does water freeze?

You often hear that water freezes at 32°F, but that's not always true — another complication of how water acts in the atmosphere. As water cools, its molecules begin going slowly enough to line up as six-sided ice crystals. But they need a pattern on which to form.

Molecules can come together by chance into the six-sided form, supplying the needed pattern. Or a "freezing nucleus," a bit of foreign matter such as clay or vegetation debris in the water, may provide the ice pattern.

The amount of water in a freezer's ice tray doesn't seem like much. But compared with even the largest raindrop, an ice tray holds a huge amount of water. In an ice tray, odds are extremely high that some water molecules will come together to form a six-sided ice crystal. That is enough to start the ice tray's water freezing. But in the atmosphere, the odds are astronomical against molecules in tiny drops coming together by chance to form a six-sided crystal. In fact, scientists have found that tiny water drops aren't sure to form ice until the temperature drops to -40°F.

In the atmosphere, though, water doesn't have to be cooled to -40°F to start making ice crystals. Freezing nuclei make the difference. Some, but not all of these particles, have a structure much like ice. Vapor or liquid water molecules hook on and grow into ice crystals.

Life becomes even more complicated when both ice crystals and water drops form in air at the same time. Once ice starts forming, such as in a cloud, it begins stealing water from the water droplets. Water molecules are coming and going between the liquid and vapor forms. The same is true for ice crystals, but there's an important difference. Once a water molecule hooks on as part of an ice crystal, it has a harder time escaping than another molecule that's become part of a drop of liquid water. As a result, more molecules leave the water

At about the time he was working out the theory of continental drift, German geophysicist and meteorologist Alfred Wegener proposed that the question of how droplets could grow to raindrop size in clouds could be answered by the growth of ice crystals.

Wegener's rain theory was developed further by Tor Bergeron, a Swede, and Walter Findeisen, a German. It was named after all three scientists.

Official definitions of liquid precipitation

Drizzle
Drops with diameter less than .02 inch, falling close together. They appear to float in air currents, but unlike fog, do fall to the ground.

Light drizzle
Visibility more than 5/8 mile.

Moderate drizzle
Visibility from 5/16 to 5/8 mile.

Heavy drizzle
Visibility less than 5/16 mile.

Rain
Drops larger than .02 inch or smaller drops that are widely separated.

Light rain
0.1 inch or less in an hour. Individual drops easily seen.

Moderate rain
0.11 to 0.30 inches per hour. Drops not clearly seen.

Heavy rain
More than 0.30 inches per hour. Seems to fall in sheets, reducing visibility.

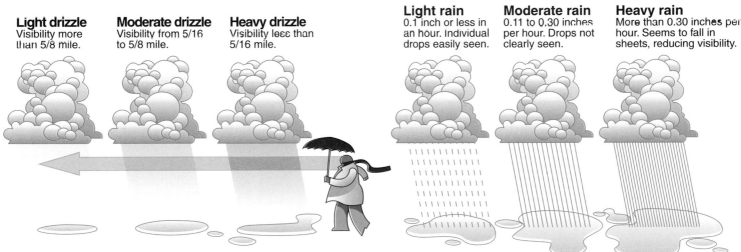

How fog forms

Fog is a cloud on the ground. The most common kinds of fog form when humid air is cooled to its dew point, causing water vapor to begin condensing into tiny drops. Sometimes fog forms when extra water evaporates into the air, increasing the dew point enough to match the temperature.

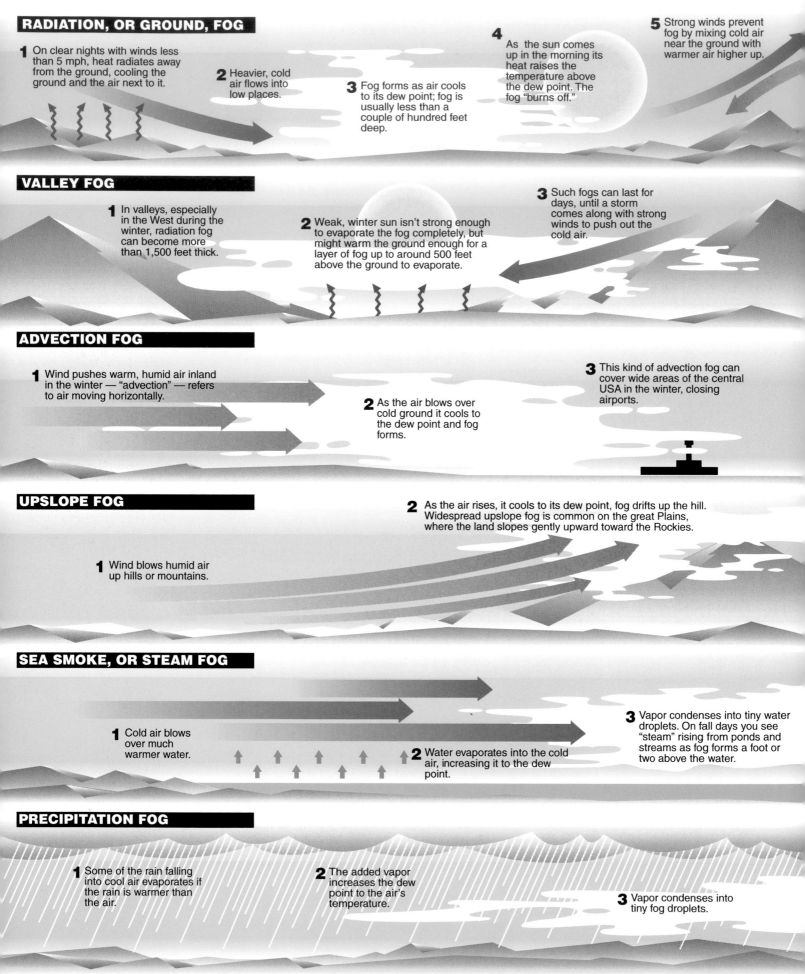

RADIATION, OR GROUND, FOG

1 On clear nights with winds less than 5 mph, heat radiates away from the ground, cooling the ground and the air next to it.

2 Heavier, cold air flows into low places.

3 Fog forms as air cools to its dew point; fog is usually less than a couple of hundred feet deep.

4 As the sun comes up in the morning its heat raises the temperature above the dew point. The fog "burns off."

5 Strong winds prevent fog by mixing cold air near the ground with warmer air higher up.

VALLEY FOG

1 In valleys, especially in the West during the winter, radiation fog can become more than 1,500 feet thick.

2 Weak, winter sun isn't strong enough to evaporate the fog completely, but might warm the ground enough for a layer of fog up to around 500 feet above the ground to evaporate.

3 Such fogs can last for days, until a storm comes along with strong winds to push out the cold air.

ADVECTION FOG

1 Wind pushes warm, humid air inland in the winter — "advection" — refers to air moving horizontally.

2 As the air blows over cold ground it cools to the dew point and fog forms.

3 This kind of advection fog can cover wide areas of the central USA in the winter, closing airports.

UPSLOPE FOG

1 Wind blows humid air up hills or mountains.

2 As the air rises, it cools to its dew point, fog drifts up the hill. Widespread upslope fog is common on the great Plains, where the land slopes gently upward toward the Rockies.

SEA SMOKE, OR STEAM FOG

1 Cold air blows over much warmer water.

2 Water evaporates into the cold air, increasing it to the dew point.

3 Vapor condenses into tiny water droplets. On fall days you see "steam" rising from ponds and streams as fog forms a foot or two above the water.

PRECIPITATION FOG

1 Some of the rain falling into cool air evaporates if the rain is warmer than the air.

2 The added vapor increases the dew point to the air's temperature.

3 Vapor condenses into tiny fog droplets.

Why rising air cools

To see why air cools and warms at a regular rate as it rises or sinks, we'll imagine blowing up an ordinary balloon; putting a thermometer in it, and taking it up 1,000 feet from the ground and down again. The balloon is being carried, it is not rising and sinking on its own.

4 At 1,000 feet the temperature inside the balloon has fallen 5.4° F. The decrease doesn't depend on the original temperature inside the balloon or air temperatures outside it; only on how much the air pressure has decreased.

3 Thermometer shows the air temperature in the balloon is decreasing. Why? Expanding the balloon means the air inside is doing work, which requires energy. The energy comes from the air molecules, slowing them down and therefore decreasing the temperature.

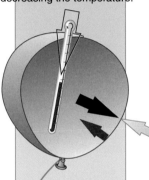

2 The balloon rises into decreasing outside air pressure, but no air escapes to equalize inside pressure. Air inside expands, equalizing pressures.

5 On the way down, increasing outside air pressure squeezes the balloon, compressing the air inside. Instead of doing work, the inside air is having work done on it, increasing its temperature.

1 On the ground the balloon's inside air pressure equals outside air pressure plus balloon's elastic pull, which tries to return it to its uninflated shape and size.

6 Back on the ground, the balloon has regained the 5.4° F in temperature it lost on the way up.

---

drops than the ice crystals. If this goes on long enough, a cloud of mixed ice and water will turn into a cloud of ice crystals.

How frost forms

Near the ground, at temperatures from 0°F to 32°F, water vapor condenses as supercooled dew and then turns to ice and frost. After that, vapor molecules begin sublimating directly from the air onto the frost.

This process of condensing as supercooled dew, turning to ice, and then growing by sublimation accounts for many of the frost patterns we see on windows. Sometimes dew condenses above 32°F, but then the air turns colder, freezing the dew. This frozen dew creates solid ice drops or a glaze of clear ice. Frost forms inside windows when the glass cools to the frost point of air inside the house or between the panes of a double window.

What shapes the clouds?

Air below the temperature at which it's saturated is "supersaturated." Water vapor begins condensing into liquid or sometimes sublimating into ice. Dew, frost, or fog forms in supersaturated air at the ground.

Air also cools as it rises, as shown in the graphic on this page. Most clouds are created by air rising until it cools to the dew point. How the air happens to rise helps determine what the clouds look like. We'll spend the rest of this chapter focusing on clouds. We'll see why some clouds are puffy, often billowing high into the sky, while others are flat and form in layers. We'll look at why rain or snow falls in showers some days, steadily on others.

Atmospheric stability is the crucial factor that determines cloud shapes and whether precipitation — rain or snow — is showery or steady.

Anything that's stable tends to return to the state it was in before it was disturbed. When we talk about the atmosphere, or the air, being stable, we're saying air that's forced upward tries to return to its original altitude

How dew and frost form

If the air cools to its dew point — the point at which the air is saturated with water vapor — either dew or frost will form. Which one depends on the air's temperature.

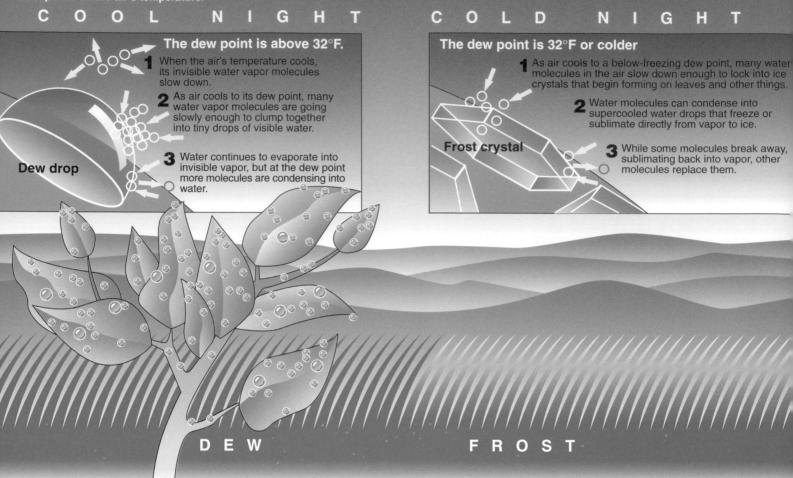

COOL NIGHT

The dew point is above 32°F.

1 When the air's temperature cools, its invisible water vapor molecules slow down.

2 As air cools to its dew point, many water vapor molecules are going slowly enough to clump together into tiny drops of visible water.

3 Water continues to evaporate into invisible vapor, but at the dew point more molecules are condensing into water.

Dew drop

COLD NIGHT

The dew point is 32°F or colder

1 As air cools to a below-freezing dew point, many water molecules in the air slow down enough to lock into ice crystals that begin forming on leaves and other things.

2 Water molecules can condense into supercooled water drops that freeze or sublimate directly from vapor to ice.

3 While some molecules break away, sublimating back into vapor, other molecules replace them.

Frost crystal

DEW

FROST

---

after the disturbance ends. In an unstable atmosphere, air continues in the path along which it was pushed. It continues rising after the original force stops.

To begin understanding why the atmosphere is stable sometimes and unstable at other times, take a closer look at why and how air cools as it rises. The drawings on Page 71 explain why air cools as it rises and warms as it descends. In these drawings the balloon isn't just floating up; it is being forced to go up or down. Imagine someone is carrying it up and down a hill.

A bubble of air rising without a balloon around it acts as if it were a balloon. Instead of mixing with the air around it, the rising air bubble expands, pushing on the surrounding air. That pushing uses up energy and the air bubble cools.

A key point: The temperature changes in rising or sinking air depend only upon how much the pressure increases or decreases. The temperature of the surrounding air doesn't affect the temperature of the air rising or sinking through it.

You can use a bicycle tire and tire pump to feel air cooling as it moves from higher pressure to lower pressure and warming as its pressure increases. Let air out of the tire; it feels cool as it rushes out because its pressure is decreasing. Now pump up the tire and notice how the pump warms. Little of that heat comes from friction in the pump; most comes from air being warmed as the pump compresses it.

Remember, the amount of cooling or heating depends only on the pressure change the air undergoes. For rising air, pressure change depends on how high it rises. Rising air cools 5.4°F per 1,000 feet; sinking air warms at the same rate.

This 5.4°F figure is the "dry adiabatic lapse rate." "Dry" means water isn't changing from one phase to another. "Adiabatic" means heat isn't being added or taken away; temperature changes result only from what the air itself is doing. "Lapse rate" means a rate of temperature change with altitude.

The fact that rising air cools explains why low-pressure areas normally bring clouds and precipitation. As we saw in Chapter 4, surface air rises in low-pressure areas. When this rising air cools to its dew point, clouds begin forming.

High-pressure areas usually bring good weather. In Chapter 4, we saw how winds converging at middle and high levels of the atmosphere force air down, building high pressure at the surface. As the air descends, it warms. That keeps clouds from forming and evaporates any existing clouds. This is the main reason why places under large, weekslong high-pressure areas — such as the West Coast in the summer — are mostly sunny and dry.

Winds grow warmer and drier as they blow down mountains. Examples are Santa Ana winds in California and Chinook winds in the Rockies. As the wind comes down the mountains, its temperature increases, which means it can "hold" more water vapor. But there's hardly any water to evaporate into the air. The relative humidity drops.

California's Santa Ana winds often contribute to terrible grass fires. The winds, which come from the deserts and blow down into places such as the Los Angeles Basin, start dry and end with relative humidities in the single digits after warming as they blow downhill. This dries out vegetation. As the winds squeeze through canyons, their speeds increase. These winds can whip even tiny fires into roaring conflagrations.

Keys to air stability

With the knowledge that rising air cools at the rate of 5.4°F per 1,000 feet — as long as water vapor isn't condensing — we're almost ready to see how meteorologists determine where and when the atmosphere will be stable or unstable.

The second important piece of information is this: Warm air is less dense than cold air. The same volume of warm air weighs less than cold air at the same pressure.

This is why a hot-air balloon rises. As long as the pilot keeps the air in the balloon's envelope warmer than the surrounding air, the balloon will float. The pilot can make the balloon rise faster by turning on a burner under the open balloon. This warms the air in the envelope. The warmed air, now lighter than the surrounding air outside the balloon, floats upward carrying the balloon with its pilot and passengers.

But what makes a bubble of air float upwards? It

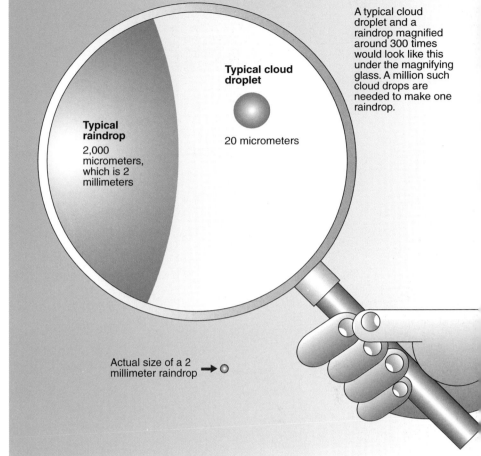

Relative sizes of cloud, raindrops

Typical cloud droplet

20 micrometers

Typical raindrop
2,000 micrometers, which is 2 millimeters

A typical cloud droplet and a raindrop magnified around 300 times would look like this under the magnifying glass. A million such cloud drops are needed to make one raindrop.

Actual size of a 2 millimeter raindrop ➙ ○

has no pilot to turn on a burner.

When we say the air inside the balloon or the bubble is warmer than the outside air, that's the same thing as saying the outside air is colder than the air in the balloon or the bubble. And that's how a bubble of air, with no pilot to turn on a burner, rises. The outside air has to be cold enough. How cold? Only a fraction of a degree colder would be enough.

There's another requirement. Remember that the rising air always cools at the rate of 5.4°F for each 1,000 feet it rises. For the bubble to rise, the surrounding air has to keep getting colder as the bubble goes higher.

So, to determine whether the air is stable or unstable, you need to know the temperature of the air at regular intervals from the ground up. Meteorologists put together these vertical profiles of air temperature by sending up weather balloons with instruments that radio back temperatures at particular altitudes.

Imagine a bubble of rising air cooling at the rate of 5.4°F per 1,000 feet and compare its temperatures with

Human hair grows longer as the humidity rises and shorter as it falls. The hair absorbs more water when there's more vapor in the air, which pushes the hair molecules apart.

That's why naturally curly hair becomes frizzy and straight hair becomes limp as humidity increases. It is also the principle behind the hair hygrometer, an instrument for measuring humidity.

Examples of stable, unstable air

Air is unstable when rising air — always cooling 5.4°F for each 1,000 feet it ascends — stays warmer than the surrounding air. This depends on the temperature of the surrounding air. The drawing below shows how this works.

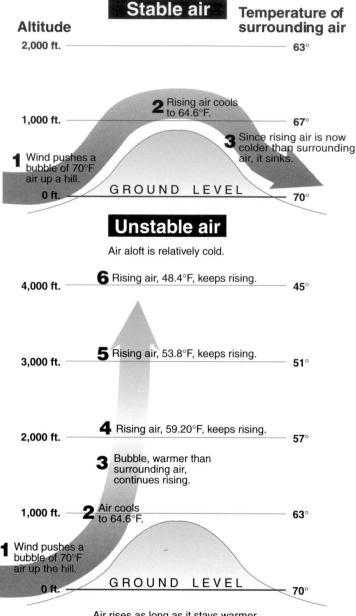

Stable air

Altitude		Temperature of surrounding air
2,000 ft.		63°
1,000 ft.	**2** Rising air cools to 64.6°F.	67°
	3 Since rising air is now colder than surrounding air, it sinks.	
	1 Wind pushes a bubble of 70°F air up a hill.	
0 ft.	G R O U N D L E V E L	70°

Unstable air

Air aloft is relatively cold.

4,000 ft.	**6** Rising air, 48.4°F, keeps rising.	45°
3,000 ft.	**5** Rising air, 53.8°F, keeps rising.	51°
2,000 ft.	**4** Rising air, 59.20°F, keeps rising.	57°
	3 Bubble, warmer than surrounding air, continues rising.	
1,000 ft.	**2** Air cools to 64.6°F.	63°
	1 Wind pushes a bubble of 70°F air up the hill.	
0 ft.	G R O U N D L E V E L	70°

Air rises as long as it stays warmer than surrounding air. At times it rises more than 50,000 feet.

those of the surrounding air. As long as the bubble stays warmer, it will keep on rising, and we say the atmosphere is unstable. If it grows cooler than the surrounding air, the atmosphere is stable.

To make a forecast of what will happen as the day goes on, forecasters have to predict the stability of the air. Here are two facts and two rules of thumb:

Fact 1: If cold air moves in at higher altitudes, the air will be come more unstable.

Fact 2: As we saw in Chapter 4, a short wave or upper-air disturbance can move over an area causing surface air to rise. This air cools as it rises and eventually generates a pool of cooler air.

Rule 1: Relatively cold air on top of relatively warm air creates an unstable atmosphere.

Rule 2: Relatively warm air atop relatively cold air creates a stable atmosphere.

On clear nights the ground cools the air near it, but it has little effect on the air higher up. By dawn, it's not unusual to have the air near the ground colder than the air a few hundred feet above.

But sometimes, a layer of warm air can move in above cold air. This is an "inversion" and is the opposite of the usual case of the air being cooler at higher altitudes. Inversions are extremely stable. Rising air will always be cooler — and therefore heavier — than the warmer air above. Since inversions block air from rising, they can trap pollution close to the ground.

How humidity affects air density

Hard as it is to believe, humid air is lighter than dry air. The reason is an important law of nature discovered in the early 1800s by the Italian physicist Amadeo Avogadro: A fixed volume of gas — say one cubic foot — at the same temperature and pressure will always have the same number of molecules no matter what gas is in the container. This is one of the keystones of chemistry.

Imagine a small container of perfectly dry air. It contains about 78 percent nitrogen; each molecule has an atomic weight of 28. Another 21 percent is oxygen; each molecule has an atomic weight of 32. The final one percent is various other gases, mostly argon.

The chart on the next page shows what happens if we add 10 water vapor molecules to the container. Eight molecules of nitrogen and two of oxygen leave, keeping the total in the box the same. In this case, the total atomic weight would decrease by 108 as water replaces heavier air molecules.

Humid air is lighter than dry air at the same temperature and pressure because water molecules are lighter than the nitrogen and oxygen they displace.

How humidity affects stability

The amount of water vapor in the air does have an important effect on stability, however. Unless rising air is extremely dry, such as over deserts, it is likely to cool to its dew point while rising. When air reaches its dew point, water vapor begins condensing onto condensation nuclei in the air, forming tiny drops. A cloud begins forming.

As we've seen, when water vapor condenses it releases latent heat into its surroundings. Now there's a tug of war going on. The air with the condensing water vapor is still rising, which means it's cooling. But the latent heat being released is trying to warm the air.

While the heat being added as water vapor condenses is never enough to prevent the air from cooling as it rises, it slows the rate of cooling. Scientists call the rate at which rising air cools during condensation the "moist adiabatic lapse rate." This rate varies, but on average it is about 3.3°F per 1,000 feet.

The slower cooling rate means the rising air is even more likely to remain warmer than the surrounding air and will continue rising. In other words, latent heat helps the air rise. It's like a fuel for thunderstorms. So the humidity that makes spring and summer days uncomfortable also supplies energy to thunderstorms.

Stability and the shape of clouds

On a day when the atmosphere is unstable, individual "bubbles" of air accelerate after they begin moving upward. Perhaps they've been shoved by an advancing cold front. Unless the air is very dry, water vapor will condense into clouds. Since individual bubbles of air are rising, the clouds will be bubbly or puffy. Their precipitation will be showery. If the layer of unstable air is only three or four thousand feet deep, it will produce small, puffy clouds. But if the air is unstable up to 20,000 or more feet above the ground, towering cumulus clouds or thunderstorms form.

When the air is stable, bubbles won't rise. Instead, entire layers of air may be pushed upward. A warm front is a good example; warm air is being pushed up over cold air. Since the whole layer of air is being pushed upward, clouds that form are flat with a layered look.

Drawings on Pages 76 and 77 show the key differences air stability makes on the weather.

The distinction between stable and unstable air isn't always sharp. Sometimes one layer of air is stable while another above or below it is unstable. Or the air can be on the borderline between stable and unstable This is why sometimes flat clouds typical of stable air and billowing clouds that mark unstable air can appear in the sky at the same time.

What goes on in clouds

To someone who pays little attention to the sky, clouds might look simple, merely white puffs or gray smudges. When you begin looking closely at the sky, however, clouds present a bewildering variety of shapes and colors. As scientists came to learn more about clouds, they found them to be far more complicated than they first appear. Not many years ago key questions about clouds seemed unanswerable.

Today, however, many of those questions are being answered by probing clouds with instrument-laden aircraft and Doppler radars that measure hidden, internal motions. Scientists need to learn more about clouds so they can make computer models, which are central to weather forecasting and understanding Earth's climate.

Clays are the most commonly identified ice nuclei. Bits of clay are found in snowflakes even at the South Pole, at least 2,400 miles from the nearest possible source of clay dust, the tip of South America.

"Showers" do not mean light precipitation. Showers can be heavy. Rain showers or snow showers do not fall steadily; they come and go. A day with showers can have clear skies between the clouds.

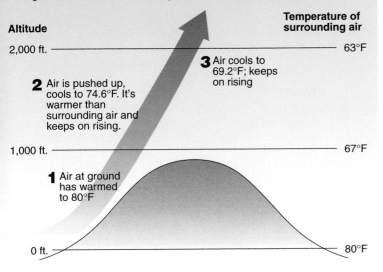

How stable air can become unstable

Heating of air near the ground can turn a stable day into an unstable one. The drawing below shows what happens if the stable air in the drawing on Page 74 is heated 10° F at the ground. It becomes unstable.

Altitude

Temperature of surrounding air

2,000 ft. — 63°F

3 Air cools to 69.2°F; keeps on rising

2 Air is pushed up, cools to 74.6°F. It's warmer than surrounding air and keeps on rising.

1,000 ft. — 67°F

1 Air at ground has warmed to 80°F

0 ft. — 80°F

What happens when humidity increases

The table below shows what happens when you add 10 water molecules to a fixed volume of air without changing the pressure or temperature. Eight nitrogen and two oxygen molecules leave to allow 10 water molecules in.

Gas	Atomic weight	Number of molecules added or lost	Atomic weight added or lost
Water moves in	18	10 added	180 added
Nitrogen leaves	28	8 lost	224 lost
Oxygen leaves	32	2 lost	64 lost
Atomic weight lost (nitrogen and oxygen)			288
Atomic weight added (water vapor)			180
Total change			**Atomic weight of 108 lost**

HOW AIR'S STABILITY
AFFECTS CLOUDS

What happens when air rises depends on whether the air is stable or unstable and how humid it is. A look at the kinds of clouds in the sky can often tell you whether the air is stable or unstable.

UNSTABLE AIR

To see what happens when the air is unstable we'll look at a day when thermals form. Unstable air can also be forced upward in other ways.

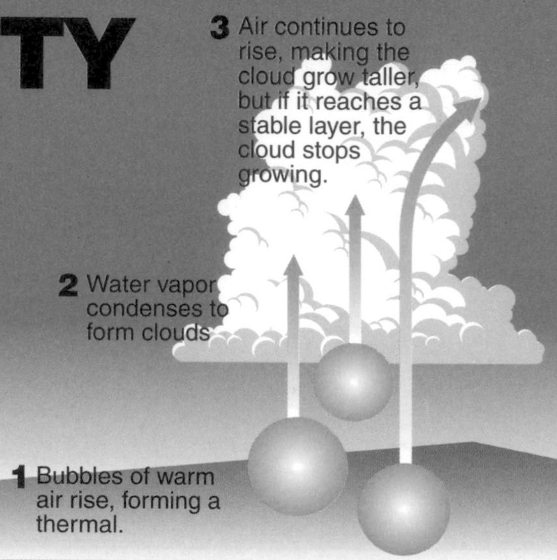

3 Air continues to rise, making the cloud grow taller, but if it reaches a stable layer, the cloud stops growing.

2 Water vapor condenses to form clouds.

1 Bubbles of warm air rise, forming a thermal.

UNSTABLE AIR WITH SHOWERS

Sometimes, of course, the cloud can grow big enough and last long enough for precipitation to form and fall. We'll look at a typical summer shower over the mainland United States.

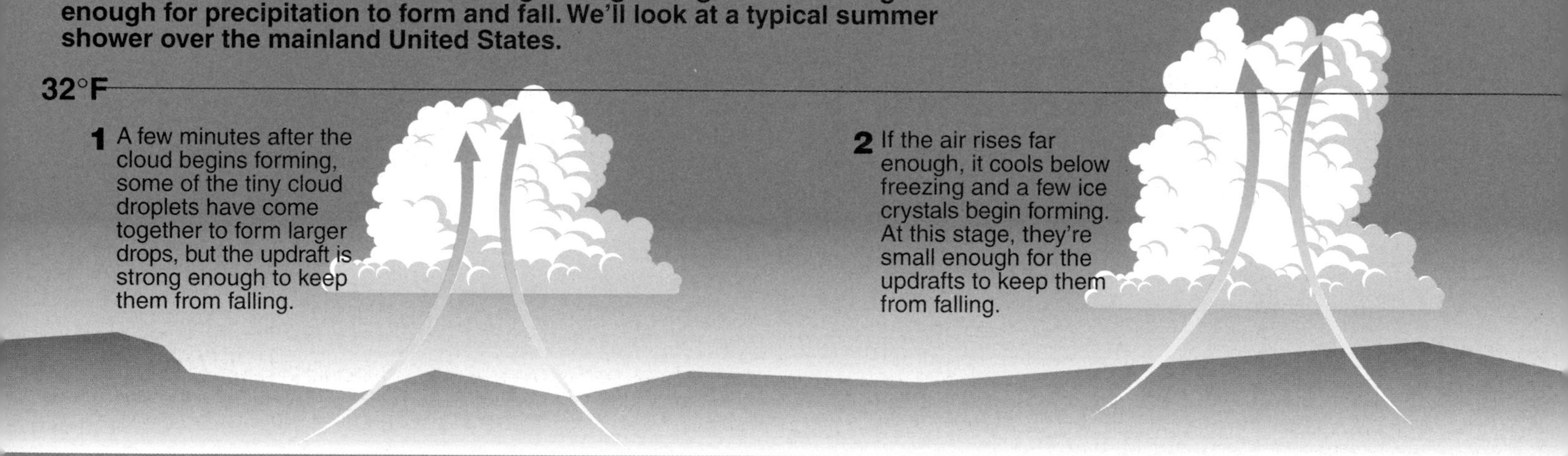

32°F

1 A few minutes after the cloud begins forming, some of the tiny cloud droplets have come together to form larger drops, but the updraft is strong enough to keep them from falling.

2 If the air rises far enough, it cools below freezing and a few ice crystals begin forming. At this stage, they're small enough for the updrafts to keep them from falling.

STABLE AIR

For clouds to form in stable air, something has to keep pushing the air upward until it cools to the dew point. This can happen in a low-pressure area where the air is slowly rising. It can cover an area the size of New England or a couple of Midwestern states. It's also common when warm air is pushed over cold air as a warm front advances.

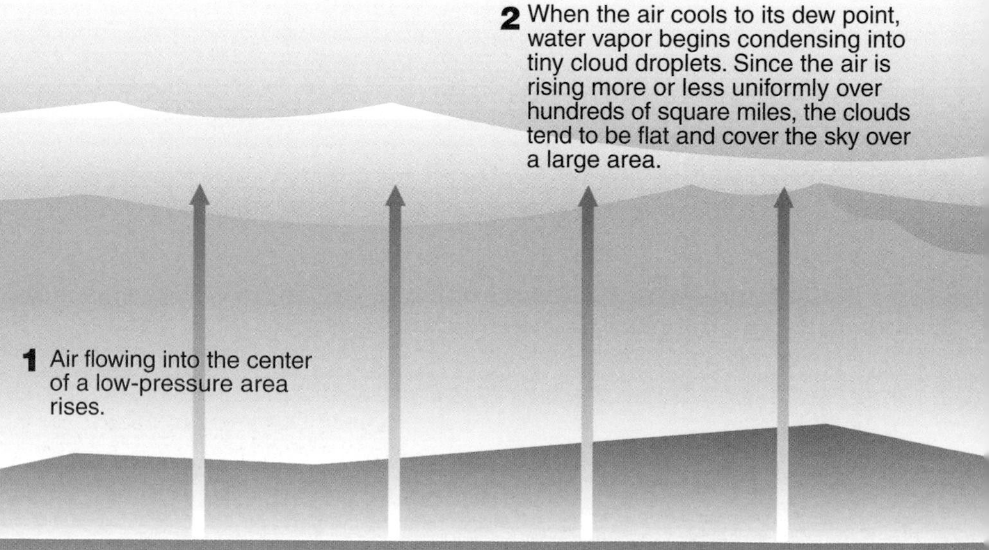

2 When the air cools to its dew point, water vapor begins condensing into tiny cloud droplets. Since the air is rising more or less uniformly over hundreds of square miles, the clouds tend to be flat and cover the sky over a large area.

1 Air flowing into the center of a low-pressure area rises.

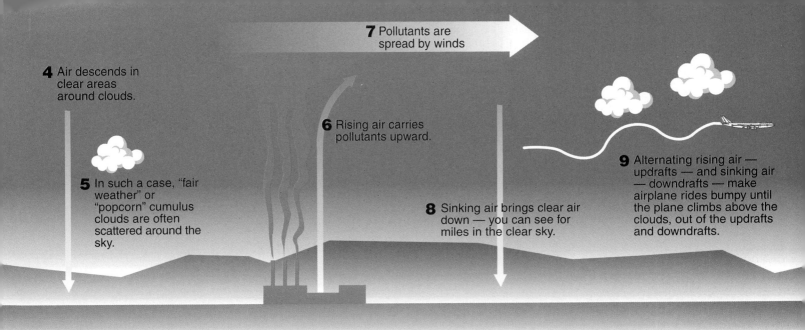

4 Air descends in clear areas around clouds.

7 Pollutants are spread by winds

5 In such a case, "fair weather" or "popcorn" cumulus clouds are often scattered around the sky.

6 Rising air carries pollutants upward.

8 Sinking air brings clear air down — you can see for miles in the clear sky.

9 Alternating rising air — updrafts — and sinking air — downdrafts — make airplane rides bumpy until the plane climbs above the clouds, out of the updrafts and downdrafts.

32°F

3 The amount of moisture in the air, how unstable the air is and other conditions determine how high and fast the cloud grows. If conditions are right, strong thunderstorms form. In an ordinary shower, ice crystals can grow big enough to start falling 20 to 30 minutes after the cloud began forming. In warm weather the ice crystals melt to fall as rain.

4 If the cloud hasn't developed into a larger thunderstorm, the rain begins easing up and the cloud is dissipating less than an hour after the cloud first began forming. But other clouds might be forming nearby.

3 Visibility under the clouds is likely to be poor since pollutants, haze or fog aren't being cleared out by the slowly rising air.

5 Without strong updrafts and downdrafts, airplanes are likely to have a smoother ride than on an unstable day.

4 When precipitation does fall it's likely to be light, but will fall over a wide area and last for hours.

How cloud seeding works

Cloud seeding is a way to encourage tiny cloud droplets, which aren't growing or are growing slowly, to grow big enough to fall as rain or snow. The most common use is to squeeze more snow out of winter clouds over Western mountains. Cloud seeding also is used to increase rain from summer showers and to turn water in thunderstorms into rain before it has time to form damaging hail.

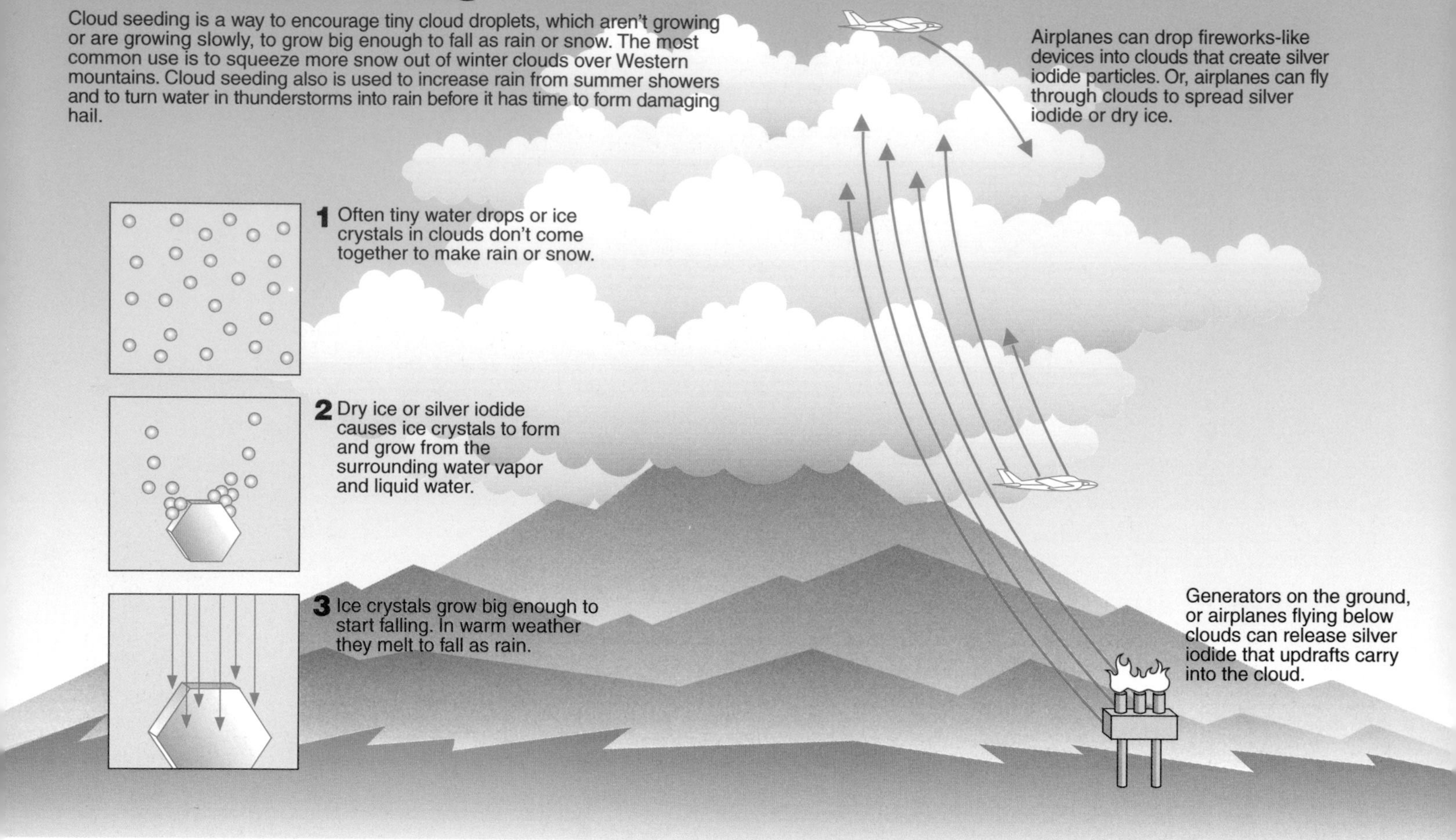

Airplanes can drop fireworks-like devices into clouds that create silver iodide particles. Or, airplanes can fly through clouds to spread silver iodide or dry ice.

1 Often tiny water drops or ice crystals in clouds don't come together to make rain or snow.

2 Dry ice or silver iodide causes ice crystals to form and grow from the surrounding water vapor and liquid water.

3 Ice crystals grow big enough to start falling. In warm weather they melt to fall as rain.

Generators on the ground, or airplanes flying below clouds can release silver iodide that updrafts carry into the cloud.

American meteorologists generally use the word "sublimate" as both a noun and a verb describing the change of water vapor directly to ice or ice back directly to vapor. British meteorologists and other U.S. scientists, such as chemists, generally call the process of going from vapor to solid "deposition." And, dictionaries list "sublime" as the verb for sublimation.

Clouds are made of tiny water droplets, tiny ice crystals, or both. Precipitation that falls from clouds is made up of larger water drops, ice crystals that have melted, or still-frozen crystals. Sometimes all three fall together.

Gravity pulls the cloud droplets toward the Earth like everything else. But, also like everything else, cloud droplets have a "terminal velocity," which is the speed they reach when the upward force of air resistance matches the downward pull of gravity. Small water drops have a slower terminal velocity than large drops. For an average-sized cloud droplet, the terminal velocity is less than two feet per minute, about 0.02 miles per hour.

The question of how a cloud made of water could float in the air was a big puzzle through the early 19th century. Many scientists argued that clouds must be composed of bubbles because bubbles would be able to float in the air. They didn't have a good understanding of terminal velocity, which explains how air can hold up water drops. Even if the updrafts supporting the tiny drops stop and the droplets start falling, they're so tiny they quickly evaporate into the surrounding air.

For rain to begin falling, the drops must grow much larger. The drawing on Page 73 shows just how much bigger the average raindrop is compared to a cloud droplet. While the average cloud droplet has a terminal velocity of only 0.02 miles per hour, an average raindrop falls at 14 miles per hour. As drops grow larger, their terminal velocity increases.

It might appear that condensation continues until the drops become big enough to start falling. But early in the 20th century scientists realized that condensation alone is too slow and could never produce rain-sized drops in a half-hour. Yet showers often began falling 30 minutes or less after the first wisps of a new cloud show up.

Between 1911 and the 1930s, scientists developed the theory that ice is the answer. As we see above, if ice crystals and water drops are mixed together in a cloud, the ice crystals will grow at the expense of the water droplets. In such cases, researchers found, the ice crystals quickly could grow large enough to begin falling, melt on the way down, and hit the ground as rain.

Until World War II, meteorologists accepted the ice-crystal theory as central to most precipitation. During the war meteorologists were sent to the tropics to

Rainmaking: How we try to fool Mother Nature

When Thomas Henderson was just starting in the cloud-seeding business in the early 1960s, many potential customers classified him with the quacks of earlier days.

"Farmers and people from irrigation districts would say, I know, you're one of those rainmakers like Burt Lancaster" in the 1956 movie, *The Rainmaker.* "In those days, I'd go to scientific meetings and feel I should sit in the back of the room. Now I get to sit up front and see the slides."

While "weather modification" is no longer a hot topic for scientific research, cloud seeding offers steady business for a few companies like Henderson's Atmospherics, Inc. in Fresno, Calif. It has more than 30 employees, owns 11 airplanes, nine radar systems, and computers that run simple weather-prediction models. A half-dozen or so projects — many in California — bring in $2 million to $4 million a year.

A key part of the business is helping augment snowfall in California and other parts of the West so power companies, irrigation districts and municipalities will have more water when the snow melts. At the Reno/Tahoe International Airport in Nevada, Henderson's company drops dry ice on "cold fog," a dense fog that forms when the air is colder than 32ºF. The dry ice turns the fog's water droplets into ice crystals that fall. The fog goes away and planes can take off and land again.

Henderson's first ventures into cloud seeding came when the California Electric Power Co. in Bishop, where he was chief hydrographer, decided to experiment after hearing of Vincent Schaefer's and Bernard Vonnegut's cloud-seeding work in 1946 and 1947.

The company bought a World War II P-38 Lightning fighter, rigged a cabin in the

Thomas Henderson, professional cloud seeder

nose (where a machine gun had been), fitted it with tiny windows, a seat and a three-inch square hole in the floor. Henderson sat in the little cabin tossing finely ground dry ice into clouds above the Sierra Nevada.

After 10 years of these experiments, Henderson "was really convinced there was something to it" and left the power company to go into the cloud-seeding business. "I had no idea where it was going, but it was a fun thing to do and I was fascinated with it."

By the mid-1990s Henderson's company and most other cloud seeders were still dispensing silver iodide using techniques that were little different from those developed by Vonnegut and others in the late 1940s and early 1950s.

"You'd think by now we'd have a lot of slick devices," Henderson says, "but we're still using the tools Vonnegut gave us.

"What has really changed is our ability to measure things with laser-based instruments that can measure droplet sizes [and] water concentrations while flying through clouds.

"Computer models have begun to emerge, too, largely for targeting. If you put

the material at Point A, does the precipitation really fall at Point B? They're not super complicated because we just don't understand enough about cloud behavior to make them too complicated."

Henderson says 50 years after the first cloud-seeding experiments, his company and others like it can increase the amount of precipitation from clouds by 10 to 15 percent. But the parts of the clouds they seed have to be below 32ºF. The next big breakthrough is to squeeze more water from warmer clouds. That could bring needed rain to dry areas such as those around the Mediterranean Sea.

Experiments in the 1990s seemed to show that new materials could squeeze rain from clouds above 32°F. But advances in cloud seeding come slowly.

"I thought it was going to be easy when I first started," Henderson says. But "the big, fundamental problem is you can't find two clouds that are alike. Most tests require similar kinds of things to test.

"We've looked at 65,000 to 70,000 clouds and we have never seen two exactly alike. They are like fingerprints. In clouds, the temperature is a little different, or one has more condensation nuclei than another. When there are so many parameters, there are bound to be pretty big differences."

Success depends on more cloud research, Henderson says. "I don't see any breakthroughs until we understand a lot more about precipitation. We need bigger computers and better [computer] models [of the atmosphere]. We need some dedicated people with good eyesight and good insight to make meaningful discoveries."

Scientific curiosity prompted Vonnegut's cloud-seeding experiments

Bernard Vonnegut says he "was not trying to learn how to modify clouds or anything like that" on Nov. 14, 1946, when he discovered that silver iodide can encourage clouds to produce precipitation. "I was just curious."

He was working with Vincent Schaefer, who had recently discovered dry-ice cloud seeding.

Of course, the experiment wasn't as casual as Vonnegut's comment might make it sound. He had good scientific reasons for trying silver iodide.

He had a Ph.D. in physical chemistry from the Massachusetts Institute of Technology. He had published his first scientific paper 10 years before. He had conducted research on topics as varied as the strength of glass rods, the effectiveness of smoke filters, and aircraft icing.

Vonnegut also brought to the laboratory "a natural inquisitiveness that is contagious. He has a bit of wonder about everything,"

says Bob Ryan, who completed his master's degree thesis at the State University of New York at Albany under Vonnegut. Ryan, chief meteorologist at WRC-TV in Washington, D.C., and president of the American Meteorological Society in 1993, describes Vonnegut's philosophy as "let's go in here and have some fun with science and see what we find out."

Vonnegut's continued cloud-seeding work at the General Electric Research Laboratory in Schenectady, N.Y., involved both theory and practical hands-on work, including inventing a thermometer to accurately measure the temperature of air from a fast-moving aircraft and a device to measure the number of condensation nuclei in the air.

In the introduction to his 1976 novel *Slapstick*, Bernard Vonnegut's younger brother, Kurt, describes Bernard's lab at GE as "a sensational mess ... where a clumsy stranger could die in a thousand different ways, depending on where he stumbled. The com-

pany had a safety officer who nearly swooned when he saw this jungle of deadfalls and snares and hair-trigger booby traps. He bawled out my brother. My brother said this to him, tapping his own forehead with his fingertips: 'If you think this laboratory is bad, you should see what it's like in here.'"

Fifty years after the experiment that assured his place in the history of meteorology, Vonnegut's mind is still churning out questions, creating theories and proposing experiments. In a couple of hours of conversation, topics he touched on included:

• How water drops act in an electrical field.

• Why tornado reports no longer note a sulfurous odor as the tornado hits. "The classic theory was that Lucifer himself was in the axis of the thing, you can see the fire, smell the brimstone." Reports since the 1920s don't note the odor. He speculates it might just be because we no longer have outhouses for tornadoes to smash.

---

study and forecast weather. Their research showed that plenty of tropical rain falls from clouds that never contain below-freezing temperatures. Ice crystals couldn't account for this rain.

This forced scientists back to the question: How can cloud droplets grow big enough to fall? One answer is "accretion," droplets bumping into each other and sticking together. This is how the planets were formed, when chunks of matter in space came together. In a cloud, drops that grow this way would start falling faster than others in the cloud. As they fell they would sweep up any smaller droplets they run into.

But questions remain. The air flowing around a falling drop would push many smaller drops out of the way, reducing the number that collide with and join the bigger drops. Scientists are still trying to discover just how efficient larger drops are at sweeping up smaller ones. They're also looking at how turbulence could mix

up drops, causing them to collide and grow.

Clouds without ice are called "warm clouds." If part of a cloud is below freezing, it's a "cold cloud." Many questions remain about exactly what goes on in warm and cold clouds. The study of these questions constitutes an entire branch of meteorology called cloud physics.

The elusive goal to control weather

One of the most persistent human hopes is to exercise some control over the weather, especially to create rain during droughts. This explains why many primitive cultures had rainmaking priests or wizards.

As the scientific world view took hold in Western societies, explaining and maybe even controlling the weather became a scientific goal. By the 19th century in the United States, rainmaking techniques were a mixture of ancient beliefs, scientific theories — many of

• What role atmospheric electricity might play in the world's weather other than causing lightning and experiments to find out what causes lightning.

• Possible connections between a thunderstorm's electricity and tornadoes. "Bob Ryan and I showed that with electrical discharges you could make a small whirlwind. But I don't see how this could scale up" to a full-sized tornado.

• A seeding experiment he'd like to perform on clouds that form over the Atlantic when frigid air moves over the ocean. "I can see these beautiful clouds, like a blackboard just ready to be written on. It would be fun to go out there and seed, watch it from a satellite."

• Finally, the story of an unsung hero of meteorology, Benjamin Thompson, who was born in Woburn, Mass., and became Count Rumford of Bavaria. "He was an American, not much of an American, who made a great contribution to meteorology." In 1776, he sided with the British and fled to England and later Germany, where among other things he was chief of police in Munich and was given his title. As a scientist "he was the

Bernard Vonnegut's lab was "a sensational mess," says his brother Kurt Vonnegut, the writer.

first to propose the theory that heat is not a substance, but the vibration of molecules."

Back in the 1950s, the ideas also kept coming when Vonnegut left GE for Arthur B. Little, Inc. "with no intention of working with

weather." But the Army was looking for a way to dissipate fog and Vonnegut wrote a proposal to look into electrical ways of doing it.

He didn't discover a way to clear fog, but the project led him into studies of atmospheric electricity, including lightning. In 1955, he proposed a new theory of how electrical charges build up in thunderstorms.

Vonnegut's theory holds that up and down air currents help build some of the charge. The prevailing theory in 1955 and now is that collisions among raindrops and ice particles are responsible. We look at this in more detail in Chapter 8.

In 1968, Vonnegut went to the State University of New York at Albany where he continued research in cloud physics and lightning. After retiring from the university, Vonnegut continued asking questions, teaching a few classes, going to scientific conferences, and writing scientific journal articles, including several noting problems with precipitation theories of lightning.

"People talk to me about proving your theory. I'm not trying to prove a theory. I'm trying to find out what's going on."

which have since been rejected — and sheer quackery. The same people who traveled the country peddling snake oils and miracle ointments sometimes sold the promise of rain as well.

It's often hard to separate the proposals of the outright quacks from honest scientific misconceptions of how the weather works. In 1893, the *Boston Globe* ran an article stating that in the course of 12 hours of fighting, one soldier would give off six gallons of perspiration, which, when multiplied by another thousand soldiers, would charge the atmosphere with 6,000 gallons of moisture and bring on rain.

That might sound reasonable until you start doing the math. In a 1952 study of small thunderstorms in Ohio and Florida, Roscoe R. Braham found that even a small thunderstorm takes in about 33 million gallons of water vapor; about 19 million gallons of it condenses into clouds, but only 4 million gallons fall as rain. Even

50,000 soldiers fighting for 12 hours would add only 300,000 gallons of water vapor to the air — far from enough to make even a small thunderstorm. And water vapor isn't enough to create a shower or thunderstorm. It needs a force to lift unstable air.

Other theories held that loud noises, such as cannon fire, cause rain. Versions of such theories held that clanging of church bells would dissipate dangerous storms and divert crop-damaging hail. But climbing into a church steeple — the highest place in a town or village — to ring bells when a thunderstorm approaches isn't wise. In the late 1700s, Great Britain banned ringing church bells during thunderstorms because lightning was killing too many bell ringers.

Concussion, or noise theories, were a mainstay of 19th-century rainmaking experiments. In 1892, a Washington, D.C. patent attorney, Robert Dyrenforth, used a $10,000 Congressional appropriation for rainmaking experiments in Texas. Dyrenforth sent up 12 balloons with 175 shells and 1,200 charges of explosives with no discernible results — except the noise.

Others wanted to move moist air around by mechanical means.

In 1880, New Yorker G.H. Bell described a 1,500-foot tower that would blow air up into the heavens to cause updrafts and create rain clouds. It would also draw clouds down to Earth and hold them until needed.

In 1918, Texas oilman Ira N. Terrill began fund raising for "Cyclerains," 200-foot concrete towers from the Rio Grande River to Wyoming. Terrill said the towers would divert "enough of the humid air [from the Gulf] westward and northward to give abundance of rain over the Great Plains country."

Such plans, like proposals after World War II to use nuclear weapons to control the weather, lacked an appreciation of the tremendous amounts of energy and moisture involved in even small thunderstorms. And, like the nuclear-bomb weather-control proposals, other environmental consequences were brushed aside.

Most of the pre-World War II rainmaking attempts were brute force efforts. They could be compared to pushing a car down the street. Today's rainmaking — the preferred term now is "cloud seeding" — efforts are more like making a car run better by giving it a tuneup.

Modern-day cloud seeding was born on Nov. 13, 14 and 15, 1946. On Nov. 13, Vincent Schaefer of the General Electric Research Laboratory in Schenectady, N.Y., released about three pounds of finely ground dry ice — frozen carbon dioxide — from an airplane into a cloud over Pittsfield, Mass.

As the airplane turned away from the cloud, "I looked toward the rear and was thrilled to see long streamers of snow falling from the base of the cloud through which we had just passed," Schaefer wrote in his laboratory notebook. "I shouted to [the pilot] to swing around, and as we did so we passed through a mass of glistening snow crystals." After another pass to release more dry ice, which was followed by more snow, Schaefer turned to the pilot and exclaimed, "We did it!"

The following day, Nov. 14, Bernard Vonnegut, one of Schaefer's colleagues at General Electric, discovered that silver iodide could be used to turn supercooled water droplets into ice; that is, it could be used for cloud seeding.

Then, on Nov. 15, *Science Magazine* appeared with Schaefer's paper on his July 1946 laboratory discovery of dry ice seeding. That lab work was the basis of the Nov. 13 flight.

These three November days set off an "explosion of cloud physics work around the world," Duncan Blanchard, also a General Electric scientist at the time, wrote in 1996.

These experiments led to the government-funded Project Cirrus, which used military aircraft for a variety of cloud-seeding and cloud physics experiments from 1947 into 1952. Irving Langmuir, the 1932 Nobel Prize winner in chemistry and head of the General Electric lab, was in charge.

Experimental results and Langmuir's scientific standing made cloud seeding one of the newest hopes for a better life. Respected scientists began talking about "weather modification," meaning not only squeezing more rain or snow from clouds but also doing such feats as turning hurricanes away from land or weakening them.

Beginning with the $790,116 the U.S. government spent on Project Cirrus from 1947 to 1952 — big research projects were much cheaper then — federal investment built to a peak in 1972 when about $18.6 million was spent on weather modification research.

But results were not clear-cut and scientists grew more aware of the complexity of clouds and storms. That, combined with pressures to cut government spending, led to federal weather modification spending being reduced to $3.1 million by 1995 and to zero for 1996.

Weather modification isn't dead. But today's cloud seeders don't talk of creating rain on clear days like their 19th-century forerunners. Instead, they talk about helping a storm produce 11 or 12 inches of snow instead of the 10 it would have dropped without help. In the U.S., cloud seeding is most used in California where every extra inch of rain or snow is vital.

USA's lowest normal annual precipitation (in inches)

Yuma, Ariz.	2.65
Las Vegas, Nev.	4.19
Barrow, Alaska	4.75
Bishop, Calif.	5.61
Bakersfield, Calif.	5.72
Barter Is., Alaska	6.49
Phoenix, Ariz.	7.11
Alamosa, Colo.	7.13
Reno, Nev.	7.49
El Paso, Texas	7.82

Rivers and streams have been flooding as long as water has flowed on the Earth. For thousands of years, humans have settled and farmed on flood plains, where crops thrive in the rich alluvial soils left by receding water.

Along with the advantages, living on rich soil left by past floods brings dangers. Places that have flooded before are likely to flood again.

Today we often visit places subject to flooding, without thinking of them as such. Many of our finest recreational treasures are next to rivers and streams that sometimes flood. Big Thompson Canyon, about 50 miles northwest of Denver, is a tragic example of what can happen when the waters rise.

Two-lane U.S. Highway 34 runs about 25 miles through the canyon, climbing about 2,500 feet from near Loveland, Colo., to the edge of the town of Estes Park. In some places, such as "The Narrows" at the canyon's east end, the walls jut almost straight up. In others, the canyon's sides slope gently enough to support a forest of ponderosa pine and Douglas fir. Houses, a few restaurants and other businesses are spread along the road and along the Big Thompson River, which is about as wide as two highway lanes.

Saturday, July 31, 1976, looked like a perfect day to enjoy being outdoors in Big Thompson Canyon. Colorado was celebrating 100 years of statehood with a long weekend. The weather was pleasant with only a threat of scattered afternoon and evening thunderstorms.

By late afternoon an estimated 2,500 to 3,500 people were staying in or driving through Big Thompson Canyon.

At 6 p.m. puffy clouds that had been growing over the Rockies had built into a thunderstorm. Most thunderstorms born above Colorado's Rockies drift eastward over the Plains. However, the 1976 storm stayed in

The Big Thompson Canyon Flash Flood was one of the most destructive in U.S. history, killing 139 people. A wall of water 19 feet high squeezed between the canyon walls. Even the river gauge was washed away; official readings were based on observation and high-water marks. In the flood's aftermath, the U.S. Army Corps of Engineers removed 197 cars and more than 300,000 cubic yards of debris — much of it wreckage from houses and other buildings.

place as it grew into an unusually efficient rainmaking machine. As the sun was setting in a partly cloudy sky in most of the rest of Colorado, the thunderstorm was sitting atop the slopes that define Big Thompson Canyon, pouring out rain.

At Drake, about halfway down the canyon, the river was carrying 137 cubic feet of water a second before the rain started at around 6 p.m. By 9 p.m. it was carrying 31,200 cubic feet of water a second. The National Weather Service estimated that the river's headwaters received 10 to 12 inches of rain, around 8 inches fell in less than two hours.

The canyon's steep, rocky slopes absorbed little of the rain. Tiny streams and rivulets grew, adding more water to the river. It quickly became deep enough and fast enough to tumble boulders up to 10 feet across. The flood washed away much of U.S. 34 and many of the houses and other buildings in the canyon.

Victims had little warning. The first alert came about 8:30 p.m. when Colorado State Patrolman Bob Miller, sent to investigate reports of rock and mud slides on the highway, radioed before escaping: "Advise them we have a flood. I'm up to my doors in water."

Many people who stayed in their cars died. Others, trapped by the rising water and washed-out

> Large quantities of water are measured in "acre-feet," the amount needed to cover one acre of land with one foot of water. An acre is slightly smaller than a football field.

road, scrambled up the steep canyon walls to shiver in light rain until rescued the next day. Those who stayed in their cars had little chance of survival. The roaring water smashed vehicles against rocks and ripped them to pieces. State Patrol Sgt. Hugh Purdy was killed in the flood. His car was identified a few days later only by a key ring labeled "Colorado State Patrol."

Survivors described the roar of the river as sounding like jet engines. In The Narrows, at the canyon's east end, the water piled up into a 19-foot wall filled with parts of cars, buildings and other debris, including hissing propane tanks. The flood-borne debris smashed into supports that held up a 227,000-pound water pipe where it crossed above the highway and river at the mouth of The Narrows. The flood pulled the pipe, filled with an estimated 873,000 pounds of water, from the canyon's walls and carried it a quarter of a mile.

The Big Thompson flood killed 139 people; six others were never found. It destroyed 418 houses and damaged 138 more. Fifty-two businesses were wiped out. Damage was estimated at $35.5 million.

Today, only someone familiar with the canyon before the flood can see any signs that it ever happened. Campgrounds, stores, restaurants and motels welcome visitors. Permanent and vacation homes still offer a wilderness escape. But now the river has more room to spread out than in 1976; people weren't allowed to rebuild in areas that floods are likely to scour.

Highway signs in Big Thompson and other Colorado canyons are the flood's most apparent legacy. They read: "Climb to safety! in case of a flash flood."

Floods are weather's big killer

The signs give good advice. Floods, especially flash floods, kill more people each year than hur-

What caused the Big Thompson flood

The causes of the July 31, 1976, Big Thompson Canyon flood — about 50 miles northwest of Denver — were similar to those of other flash floods, especially a flood that hit the South Dakota Black Hills June 10, 1972, killing 237 people.

4 Humid high-altitude air combining with weak winds meant the storm pulled in little dry air to weaken its rainfall.

3 Winds were less than 20 mph above 10,000 feet, too weak to move the storm away.

2 The unstable air continued rising as its water vapor condensed.

5 Hardly any of the rain soaked into the steep-sided canyon.

1 Winds from the east pushed very humid air up the mountains.

6 The river quickly went over its banks, filled with debris that acted like battering rams against downstream buildings, cars.

8 Flood water speeds up as it squeezes through narrow places, or is freed when a dam formed by debris bursts loose.

7 Water backed up in the canyon's narrow mouth. In floods, water often backs up as debris piles against bridges.

Normal river size

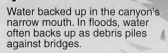

ricanes, tornadoes, wind storms or lightning. In the 1970s floods replaced lightning as weather's big killer. During the 1980s, floods killed an average of 110 people a year in the United States; about 60 percent of them died in their vehicles or while trying to flee from cars or trucks stalled in rising water.

The experts' advice: First, don't drive into flood waters, even the most innocent-looking ones. A street a few blocks from home that appears to be under only a foot or two of water could become a death trap. The placid-looking water over the road might have washed out part of the pavement. You could end up in two or three feet of water and that could be enough to wash you away. Second, if you're ever blocked by a flooded or washed-out road in a canyon or valley, get out of the car and head for higher ground. You're better off being cold and wet on a hillside watching a flood carry away your car than being in the car as it smashes against rocks.

Some Big Thompson victims were caught in houses, motels or other buildings. Many refused to heed warnings from police and others minutes before the flood hit. Experts have found that people balk at acting on warnings, or require confirmation before acting in almost all disasters. In fact, slowness to act is more common than panic when natural disasters hit.

A key home defense against floods is to find out how high the water can rise. Instead of relying on neighbors' memories of past floods, check with the local emergency management agency or the nearest National Weather Service hydrologist. Flood victims commonly say they never thought the water could climb as high or as fast as it did.

More and more places subject to floods are being protected by automated systems that sound an alarm when stream levels, amount of rainfall or both exceed dangerous levels. Defense against floods includes knowing whether your community has such a system and if so, how warnings are passed along.

Flash floods, East and West

Floods like the one that hit Big Thompson are called flash floods because the water rises quickly, maybe an hour or less after heavy rain begins. Forecasters often can tell when conditions are ripe for

World record rainfalls

1 minute
1.50 inches of rain, Nov. 26, 1970, at Barot, Guadeloupe, West Indies.

42 minutes
12 inches, June 22, 1947, Holt, Mo.

2 hours, 10 minutes
19 inches, July 18, 1889, Rockport, W.Va.

2 hours, 45 minutes
22 inches, May 31, 1935, D'Hanis, Texas.

9 hours
42.79 inches, Feb. 28, 1964, Belouve, La Reunion, Indian Ocean.

10 hours
55.12 inches, Aug. 1, 1977, Muduocaidang, Nei Monggol, China.

Why flood waters are so dangerous

Fresh water moving at only 4 mph, a brisk walking pace, exerts a force of about 66 pounds on each square foot of anything it encounters.

Look out for

A good safety rule: If you can't see the bottom of the water in a flooded area don't try to wade in it.

Double the water's speed to 8 mph and the force zooms to about 264 pounds per square foot.

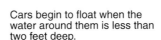

Cars begin to float when the water around them is less than two feet deep.

Once a car floats just a little, the moving water often carries it along into deeper water, or overturns it in a ditch.

More than 2,000 people died in flash floods at Johnstown, Pa., in 1889. At left, survivors at the ruins of the Sisters of Charity building. In 1990, Liberty County, Texas, had a one week warning of flooding to come. There were no deaths. The home above escaped floods because the owner had enough time to build a dike with plywood, tin and sandbags.

flash floods. They issue a "watch" saying, in effect, "Watch out, be alert for warnings." But they usually have little time to issue a "warning," which means a flash flood is actually occurring. Floods on larger rivers, in contrast, can often be accurately predicted days, or even a week or two, ahead.

While the Big Thompson flood was an extreme event, its causes were typical of Western floods. The weather had been dry. Hard, dry soil absorbs little water. What rain does fall almost immediately begins collecting in streams and rivers. In mountains, the water quickly drains downhill and can sweep streams that have been dry or nearly dry for months. Flood waters can rush miles down streams, to places where the sun has been shining all day. Sometimes victims have no reason to think a flood is headed their way.

In the East, where the ground generally absorbs more water, flash floods are more likely after the soil has been saturated by long rainy periods or when a quick thaw, maybe combined with heavy rain, fills streams with rain and melted snow. Hurricanes and tropical storms, which push ocean water inland, can cause flash floods as they dump heavy rain inland.

Dam failures are one of the most-feared causes of flash floods, with good reason. Two cases:

• America's deadliest flash flood hit Johnstown, Pa., May 31, 1889, when a 72-foot-high, 930-foot-long dam broke in the hills about 24 miles upriver from Johnstown. A wall of water, which was 76 feet high at one point and 23 feet high when it hit Johnstown, roared down the valley into the city. The official death toll was 2,209. The dam, completed in 1852, had fallen into disrepair. Repair work in 1879 was done poorly.

• Three minutes before midnight March 12, 1928, the 208-foot-high, 700-foot-long St. Francis Dam across San Francisquito Canyon, Calif., broke, spilling more than 138,000 acre-feet of water. The flood killed more than 400 people as it washed down the Santa Clara Valley to the Pacific Ocean south of Ventura, about 54 miles from the dam. The dam, owned by the Los Angeles Department of Water and Power, was only 22 months old when it failed. Investigators said the rocks of the canyon's walls weren't strong enough to support it.

Since then inspection and regulation of dams, especially large ones, have improved. But smaller, uninspected dams can still present a flash flood hazard.

The greatest floods

Rain falling over a wide area or snow melting over a large area is needed to cause floods that can cover hundreds of square miles and last for weeks. It's not unusual for such floods to be fed in part by individual flash floods in smaller streams feeding into rivers.

The Great Mississippi River Flood of 1927 was one of the most damaging ever to hit the United States and led to vast improvements in flood control. It began with above-normal rain. The precipitation for the entire Mississippi River Basin had been above normal since the fall of 1926.

Flooding began that fall at Memphis, and the Mississippi stayed high there until the spring floods began. By late March the Mississippi and most of the rivers that flow into it — such as the Ohio, the Tennessee and the Cumberland — were at flood stage. Major flooding began during the first three weeks of April when more than nine inches of rain fell on southern Missouri and most of Arkansas. This rain was added to water from melted snow coming down the Mississippi from the north.

The river broke through levees along the Mississippi and its tributaries in more than 120 places, flooding more than 26,000 square miles of land in seven states. In some places, the river ran 80 miles wide. The flood forced more than 600,000 people from their homes. Since officials could forecast when the flood would arrive — except for unexpected levee breaks — most people were warned in time to escape. Still, the flood killed 246 people.

As a result of the flood, Congress adopted the Flood Control Act of 1928, which gave the U.S. Army Corps of Engineers the job of Mississippi River flood control. It led to today's elaborate system of reservoirs and better-designed levees. Project engineers also straightened the river in places to allow water to flow faster to the Gulf of Mexico. A flood way was added on the river's west side near Cairo, Ill., to divert floodwaters from the main channel. And the Bonnet Carre Spillway was built north of New Orleans to divert some Mississippi water to Lake Pontchartrain instead of allowing it to press against the levees and flood walls in New Orleans. The work was finished in time for the 1937 flood, which exceeded the flows in 1927, but did relatively little damage.

The Great Mississippi Flood of 1993

The Great Flood of 1993 on the upper Mississippi River, the Missouri River and scores of rivers flowing into them covered parts of nine states and lasted three months. For the first time in history, major floods came down both the Mississippi and Missouri rivers at the same time.

It was unusual not only because of its size and duration but also because it occurred in summer, not spring like most Midwest floods.

Even though the worst flooding occurred from June

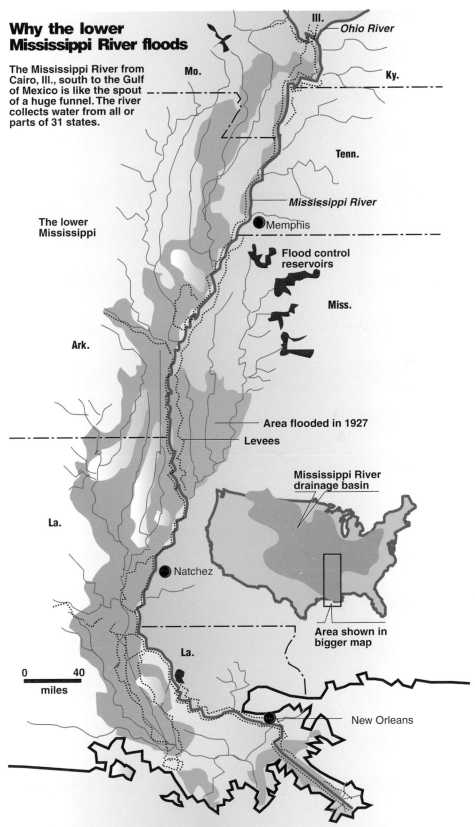

Why the lower Mississippi River floods

The Mississippi River from Cairo, Ill., south to the Gulf of Mexico is like the spout of a huge funnel. The river collects water from all or parts of 31 states.

The lower Mississippi

Ohio River

Ill.

Mo.

Ky.

Tenn.

Mississippi River

Memphis

Flood control reservoirs

Miss.

Ark.

Area flooded in 1927

Levees

La.

Natchez

Mississippi River drainage basin

Area shown in bigger map

La.

0   40
miles

New Orleans

into August 1993, we could say the flood really began in November 1992 when, after a slightly dry summer and early fall, more rain than normal began falling on the north-central USA.

87

HOW WEATHER AND GRAVITY CYCLE THE EARTH'S WATER

Without water life as we know it could not be possible. No plant or animal can live without it. If the atmosphere didn't carry water, life would have never left the sea for land. More than 97 percent of the Earth's water is salt water in the ocean. Evaporation of just a tiny part of the ocean's water and the atmosphere's movement of that water vapor inland to fall as rain and snow make life possible more than a few feet from the ocean's edge.

HOW THE HYDROLOGICAL CYCLE OPERATES

On an average day, 40,000 billion gallons of water are in the atmosphere over the United States as water vapor or liquid water and ice in clouds. Imagine this as the water in an average-sized, 40-gallon bathtub.

1 About 4,200 billion gallons fall on the land as precipitation on an average day. Think of this as draining 4.2 gallons from our 40-gallon bathtub.

2 An average of 2,800 billion gallons evaporate from the land into the air each day. This is like putting 2.8 gallons back in the bathtub.

3 This leaves about 1,400 billion gallons to soak into the ground or to run down rivers back to the oceans. If nothing else happened, the atmosphere would be missing this 1,400 billion gallons at the end of the day. Our 40-gallon tub would be 1.4 gallons low. The atmosphere — or our tub — never runs dry because water evaporating from the oceans and carried by winds over the land makes up the difference.

Fresh ground water

WHERE THE WORLD'S WATER IS FOUND

Here is an experiment to show what share of the Earth's water is available for drinking and irrigation.

All of the Earth's water

Let's use a one liter container to represent all of the Earth's water.

Salt water in oceans

972 milliliters would be contained in the oceans as salt water ...

All Earth's fresh water

... leaving only 28 milliliters as fresh water

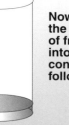

Now let's divide the 28 milliliters of fresh water into smaller containers as follows:

HOW STREAMS RISE AND FALL

While the cycle of water evaporating and falling as rain and snow balances over large areas and long times, precipitation at particular places can be high enough to cause floods or low enough to cause droughts from time to time.

DRY WEATHER

Water is always evaporating from streams, lakes and the soil. Growing plants release water vapor into the air. This is called transpiration.

WET WEATHER

Water from falling rain and melting snow soak into the ground, raising the level of the water in the ground — the water table. After the rain stops, underground water continues flowing into streams, sometimes helping keep streams at flood levels after the rain stops.

Water table

During dry spells, water in the ground flows into streams and lakes, replacing water flowing away or evaporating.

In long dry spells, the top of the underground water drops below stream's bottom; the stream dries up.

HOW WE DEFEND AGAINST FLOODS AND DROUGHTS

Dams are a major defense against floods and droughts. During wet periods water is stored in reservoirs behind dams instead of flowing downstream to cause floods. It can then be used during dry weather. During unusually wet weather, dam keepers open floodgates, letting water flow downstream. That ensures water doesn't wash over the dam's top, which could weaken it. Earthen embankments, called levees, and flood walls protect many places from floods.

Flood wall

Dam

Reservoir

Levee

Satline ground water

Ocean

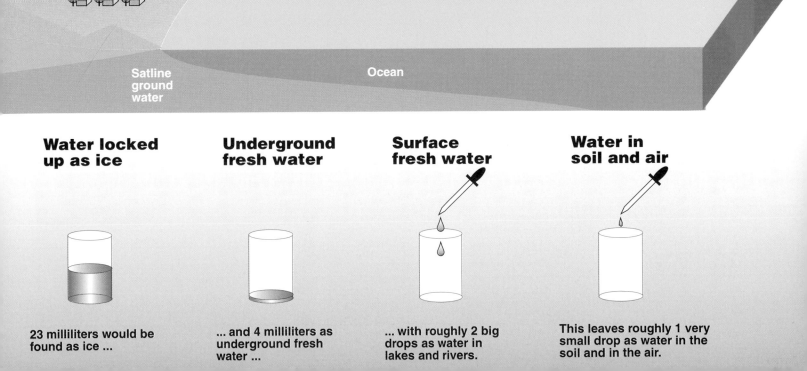

Water locked up as ice	Underground fresh water	Surface fresh water	Water in soil and air
23 milliliters would be found as ice ...	... and 4 milliliters as underground fresh water ...	... with roughly 2 big drops as water in lakes and rivers.	This leaves roughly 1 very small drop as water in the soil and in the air.

Farmers in the 1930s learned a bitter lesson — drought is a normal part of the Plains' climate — in what became known as the Dust Bowl (below). At far right, Forest Lee Graham of Pomarja, S.C., surveys the damage to his farm in the summer of 1986. And on the Mississippi River at Memphis, Tenn., a drought in 1988 dropped water levels and left boats aground.

Greater-than-normal precipitation continued through January 1993. February and March were on the slightly dry side, but not dry enough. As other parts of the nation began drying out from a wet winter, in the Upper Mississippi Valley an average of 16.13 inches of rain fell from April through June, the wettest spring since records began in 1895.

Wet fields delayed or prevented spring planting. Streams and rivers were beginning to fill up. By mid-June, widespread minor to moderate flooding was occurring across the Upper Mississippi Basin, but the water was flowing south with no signs of major flooding.

Then, the skies opened up on June 19, 20 and 21. That started serious flooding that lasted into August. The graphic on Page 95 shows the major events.

During the summer, flood water covered more than 17 million acres in the upper Mississippi and Missouri valleys. The flood damaged or destroyed more than 22,000 homes and forced more than 85,000 residents to evacuate. Water completely flooded more than 75 small towns. More than 6,500 National Guard members and thousands of volunteers worked to shore up levees.

At least the flood moved slowly enough for forecasters to warn those in its path. The death toll among the millions threatened was 48.

Even with all of the damage it did, the Great Flood of 1993 could have been worse. The U.S. Army Corps of Engineers says reservoirs held back water that would have made flooding even higher in many places. In addition, even though many levees failed, most held. For example, the 11-mile-long, 52-foot-high St. Louis flood wall kept the river out of the heart of the city.

The Corps estimated that without flood-control measures, the Great Flood of 1993 could have cost twice as much as the $20 billion in damage it's estimated to have done.

Unlike in 1927, the 1993 flood did not lead to a call for more flood control measures. Instead it prompted a debate about the best use of flood plains. Should many of the flooded areas have been developed or should they have been left as nature's safety valve during unusually wet spells?

The Great Flood of 1993 followed other costly natural disasters such as Hurricane Andrew in 1992, prompting the Administration and Congress to look for ways to cut costs of disaster relief. Before the end of 1993, new laws enabled the federal government to cover up to 75% of the cost of relocating homes and businesses from flood plains. By the fall of 1994, funding had been approved to completely move four communities to higher ground.

Over the coming years, lawmakers are likely to come up with more ways to discourage people from living on flood plains, and to require those who do live or farm there to buy more private insurance.

How average precipitation varies from year to year

Less than 16 inches
16 to 32 inches
32 to 48 inches
More than 48 inches

Annual USA precipitation

(in inches)

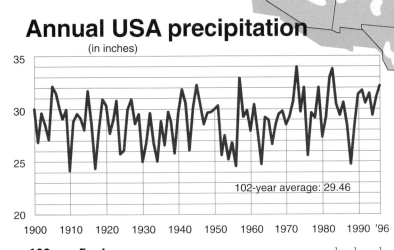

102-year average: 29.46

1900 1910 1920 1930 1940 1950 1960 1970 1980 1990 '96

100-year floods

In planning flood protection, hydrologists and engineers talk about "10-year floods" or "100-year floods" as a way of assessing the odds of water reaching certain levels. It's all based on averages. A 100-year flood describes water levels that, on the average, should occur once a century. But having 100-year floods in back-to-back years doesn't mean nature is up to something strange. A place could also go a couple of hundred or more years without a 100-year flood. The best way to think of a 100-year flood is one that has a one percent (one in 100) chance of occurring in any given year.

The great droughts

Droughts are a big part of American history, especially in the West. Yet each time a major drought arrives, it's almost as though something brand new is happening. For instance, the drought that peaked in 1988 encouraged talk about the greenhouse effect and global warming.

While it is possible that the Earth's climate is warming up and that a warmer Earth would have more serious and prolonged droughts, you don't need to invoke greenhouse warming to explain the drought of 1988. A quick look at history shows that nature managed to create droughts as bad as, or worse than, the 1988 drought long before a greenhouse effect could have started warming the Earth.

The Homestead Act of 1862 — which gave settlers free title to 160 acres of land if they lived on and cultivated that land for five years — helped set the stage for one of the United States' most socially disruptive droughts. Thousands of settlers who moved west to the Great Plains in the late 1860s and 1870s didn't realize the region was enjoying an unusually wet period.

Railroads promoted settlement with the slogan "Rain follows the plow," which was a respectable enough scientific opinion in the middle of the 19th century. There's no doubt that growing plants add tremendous amounts of water vapor to the air. And there's good reason to believe that denuding the ground of vegetation, as has happened in parts of Africa, can help turn the climate drier. But, as settlers on the Plains learned during the drought that lasted from

Record dry spells: The longest rainless period in the U.S. was 767 days — Oct. 3, 1912 to Nov. 8, 1914 — at Bagdad, Calif.

The longest rainless period in the world was 14 consecutive years — October 1903 to January 1918 — at Arica, Chile.

Red Cross: Volunteers help flood victims sort out their lives

When flood waters recede, those who've fled their homes often return to a demoralizing scene. A dark line of mud on the walls traces the flood's depth. The odor of all the things that have been swirling around in the water and the mold that's already started growing are overwhelming. Everything left behind is soaking wet, covered by silt or destroyed.

Flood victims feel they've "lost everything. They come in and they don't know where they're going to go, or where they're going to sleep that night," says Norma Schoenhoeft, a Red Cross volunteer from Columbus, Ind., who's been helping flood victims since the early 1960s. "That just totally devastates them, and the children, too.

"Personal things are what people always miss the most. It's usually photographs and personal mementos. We get a lot of people who say things like 'I lost everything that my mother had given me.' It's hard missing those treasured items. They're just gone forever."

As with other disasters, Red Cross disaster workers and professionals rush to areas inundated by flood water to help ensure that victims are safe from the flood and then have shelter, food, clothing and medical aid. After the flood recedes, workers such as Schoenhoeft set up service centers to help victims pull their lives back together.

"After the water starts receding, we're still real busy," she says. "We're busy setting them up in apartments, or getting them settled in with relatives. Even after the flood water recedes, they still don't want to go back to their house, or they just can't because the water has rotted out the floors and the furniture. So we just concentrate on finding them an apartment or a motel room

Norma Schoenhoeft at a flood site in Shreveport, La.

where they can be safe . . . a lot of them are just so tired and worn down, we give them medical help."

Volunteers from the Red Cross and other community and religious groups pitch in to help victims make some of the dozens of phone calls needed to start getting their lives back together.

Often, more intangible help is needed, Schoenhoeft says. "You know, we sit and cry with them. We aren't supposed to, but sometimes it gets pretty bad. A lot of the times, when they come in and they're upset, we'll just go over and hold their hand and say, 'It's OK. You just go ahead and cry,' and let them talk about anything until they feel they are ready. We don't push these people like they're on an assembly line here, because we're very concerned. If we weren't, we wouldn't be volunteering for this."

Schoenhoeft is one of 3 million Red Cross volunteers nationwide; about three

volunteers for each paid staff member. About 3,000 are trained and on-call for disaster assignments around the country.

Her advice for anyone threatened by flood is to heed the warnings and head for a safe place. If you have time, gather up insurance policies, birth certificates, and any papers you might need. Schoenhoeft notes that such papers usually can be replaced, but having them makes recovery easier. You should also think about taking things that can't be replaced, such as family scrapbooks or photos.

"To me the most important thing is for these people to walk out of here with their pride," Schoenhoeft says. "I've always felt that way. And I think that's more important than clothing or housing or anything else. When we leave a place we leave a part of ourselves here and we take a part of them home with us."

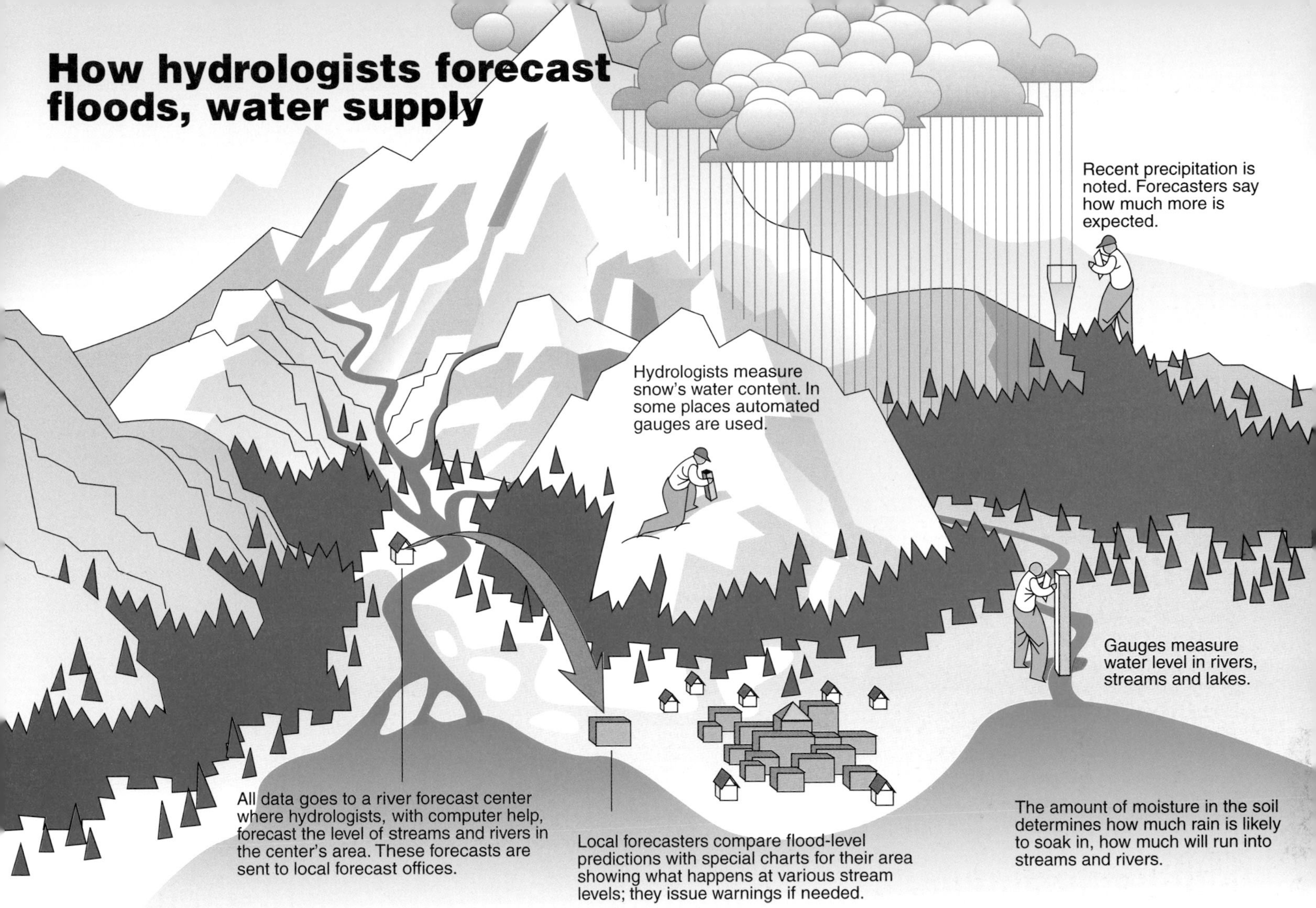

How hydrologists forecast floods, water supply

Recent precipitation is noted. Forecasters say how much more is expected.

Hydrologists measure snow's water content. In some places automated gauges are used.

Gauges measure water level in rivers, streams and lakes.

All data goes to a river forecast center where hydrologists, with computer help, forecast the level of streams and rivers in the center's area. These forecasts are sent to local forecast offices.

Local forecasters compare flood-level predictions with special charts for their area showing what happens at various stream levels; they issue warnings if needed.

The amount of moisture in the soil determines how much rain is likely to soak in, how much will run into streams and rivers.

1887 to 1896, covering the Plains with crops isn't enough to overcome the global climatic patterns that bring droughts.

By 1890, thousands who had set out to find a new life were fleeing back East or to the West Coast. The population of many counties in Kansas and Nebraska dropped by one quarter to one half. About 400,000 of the more than a million families who had tried to homestead the Plains managed to keep their land.

The next big drought, which affected much of the United States but which hit the Plains the hardest, began in 1934 and didn't end until 1941. Two years during this period, 1934 and 1936, are the country's driest on record. Thousands of acres of land on the Plains dried out and turned to dust that at times blew all the way to the East Coast. This "Dust Bowl," like the one in the 1890s, displaced thousands. This time, however, many found jobs working on the huge dams that were being built partly in response to the drought.

The drought of 1987-89, which was most severe in the summer of 1988, was in some places more severe than the Dust Bowl of the 1930s. But there's no evidence that the drought alone forced vast numbers of families off their farms. Loans and other aid enabled farmers to hang on even though the drought cut grain production by 31 percent. A Plains drought in the 1950s had shown that the aid programs enacted after the Dust Bowl meant Plains farmers no longer were forced to abandon their land in large numbers as farmers did in the 1890s and 1930s.

This isn't to say the 1987-1989 drought wasn't expensive and painful. The official government estimate put the cost at $39 billion and many experts argue that this estimate is low. Grain surpluses built up in previous years kept people from going hungry and prices didn't soar despite the crop losses. But farmers and rural communities were hurt. And an estimated 5.1 million acres of forest burned, including highly publicized fires in and around Yellowstone National Park.

What drought means

Information collected and analyzed by the National

Climatic Data Center in Asheville, N.C., helps put the 1987-1989 drought, and other weather swings, in perspective. Reliable national records go back to 1895, just as the 1890s drought was ending. The map and graph from the Asheville Center on Page 91 show how normal precipitation varies in different parts of the country and how precipitation has varied from year to year for the entire United States.

The map shows why the term "drought" is relative. If only 30 inches of rain fell during a year in South Florida, the region would be considered to be suffering from a serious drought; 30 inches is only a little more than half a year's normal rainfall. But 30 inches of precipitation just about anywhere in the West, except for the mountains of California and the Northwest, would be double the normal amount.

Drought is "abnormal dryness" for a particular region; exact definitions are complicated. Hydrologists have formulas that take into account such things as the amount of precipitation that fell during particular times and the amount of water in the soil, lakes and rivers. From this they determine whether a region is suffering a mild, severe or extreme drought, or whether it's abnormally wet.

The chart of annual rainfall on Page 91 helps explain why the 1987-1989 drought seemed so bad. It ended one of the wettest periods in 100 years. That period began in the late 1960s and included the two wettest years on record, 1973 and 1983.

Making sense of drought

Global weather patterns establish the conditions that dry out particular areas. Once a drought starts, it tends to feed upon itself until eventually a change in the global pattern ends it. While 30-day and 90-day forecasts have some success at predicting when large areas of the United States will be drier or wetter than average, they have little success at predicting when droughts will begin or end.

After the 1988 drought some researchers said patterns of sea surface temperatures in the tropical Pacific could have been a major trigger. Thunderstorms, which pump huge amounts of moisture and heat into the upper atmosphere, are most numerous over the warmest parts of tropical oceans. Heat carried high into the air by such thunderstorms can change long-wave patterns of high-altitude winds. Shifts in locations of warmest water temperatures can change global patterns.

But, other researchers advanced strong arguments that other aspects of global weather patterns could have started the 1988 drought; the ocean temperature pattern then could have made it persist. The lesson is, don't expect to be able to forecast droughts based only on one aspect of weather patterns, such as ocean temperatures.

Another lesson that today's climatologists stress is that droughts are completely natural. The climate experts say governments ought to prepare for a 100-year drought because no matter what else the climate does, we'll face droughts in the future.

Working to contain water

People were building dams to manage floods and droughts before written history began. Efforts to forecast American floods go back to the 19th-century beginnings of the National Weather Service. The United States undertook gigantic efforts to control floods and assure water supplies beginning in the 1930s. Huge water projects not only promised to bring rivers under control but also provided needed jobs during the Depression. The work speeded up after World War II and continued through the 1960s, until budget and environmental concerns began slowing the pace in the 1970s.

Army Corps of Engineers figures show that at the end of 1932 the United States had 133 million acre-feet of water stored behind dams. By 1944 that figure had increased to 279 million acre-feet and by 1969 to 753 million acre feet.

Even at the height of water project building, few experts seriously thought we would be able to build enough dams, levees and other structures to take care of all our potential water problems. People still need to forecast when and where floods are likely and how high the floods will be.

In the United States, hydrologists at 13 National Weather Service River Forecast Centers have the responsibility for forecasting floods. Some centers also make water-supply forecasts. Their work is a kind of accounting, but instead of keeping track of money, they keep track of water. Their ledgers include how much water is in lakes, rivers and streams; how much rain has fallen recently and how much is expected; how much water melting snow should produce; and how much more water the soil can absorb. This watery bookkeeping enables them to forecast when floods are likely and how the water supply should hold out.

The old saying "Still waters run deep" might apply to human nature, but not to rivers.

Slow-moving streams can be shallow or deep, but in general as moving water becomes deeper, it tends to run faster.

Why?

Gravity causes water to run downhill, while friction against a stream's bed and banks tends to hold it back. The water's speed de-pends on a balance bet-ween gravity and friction.

As a stream becomes wider or deeper, a smaller share of the water is rubbing against the banks or bed. The gravity-friction balance then tilts more toward gravity and the water speeds up.

Highlights of the 1993 Midwest flood

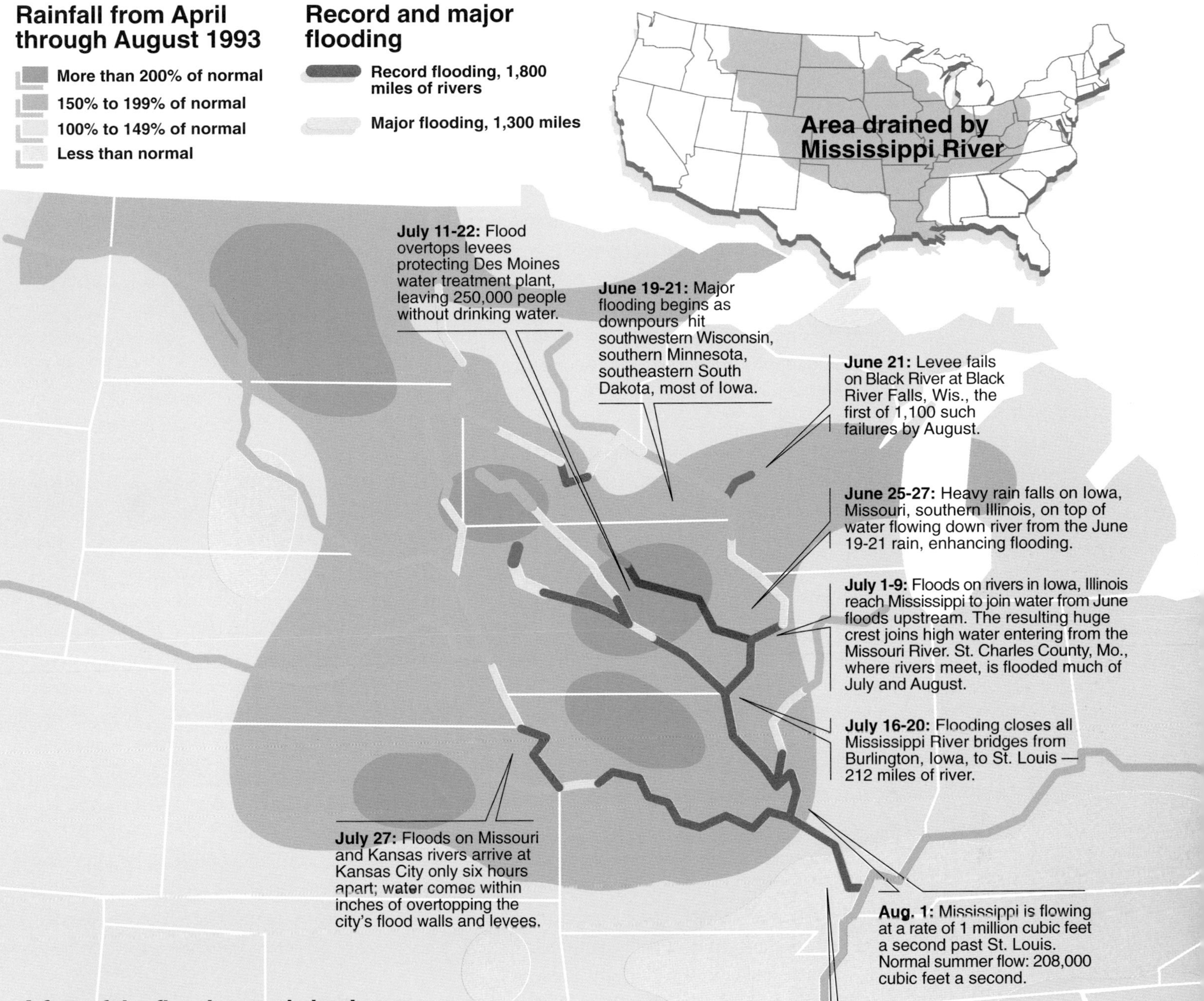

Rainfall from April through August 1993

- More than 200% of normal
- 150% to 199% of normal
- 100% to 149% of normal
- Less than normal

Record and major flooding

- Record flooding, 1,800 miles of rivers
- Major flooding, 1,300 miles

Area drained by Mississippi River

July 11-22: Flood overtops levees protecting Des Moines water treatment plant, leaving 250,000 people without drinking water.

June 19-21: Major flooding begins as downpours hit southwestern Wisconsin, southern Minnesota, southeastern South Dakota, most of Iowa.

June 21: Levee fails on Black River at Black River Falls, Wis., the first of 1,100 such failures by August.

June 25-27: Heavy rain falls on Iowa, Missouri, southern Illinois, on top of water flowing down river from the June 19-21 rain, enhancing flooding.

July 1-9: Floods on rivers in Iowa, Illinois reach Mississippi to join water from June floods upstream. The resulting huge crest joins high water entering from the Missouri River. St. Charles County, Mo., where rivers meet, is flooded much of July and August.

July 16-20: Flooding closes all Mississippi River bridges from Burlington, Iowa, to St. Louis — 212 miles of river.

July 27: Floods on Missouri and Kansas rivers arrive at Kansas City only six hours apart; water comes within inches of overtopping the city's flood walls and levees.

Aug. 1: Mississippi is flowing at a rate of 1 million cubic feet a second past St. Louis. Normal summer flow: 208,000 cubic feet a second.

Aug. 7: Main crest reaches Cairo, Ill., where the Ohio River joins the Mississippi. From here to the Gulf of Mexico the Mississippi is much wider and deeper; the flood water causes only a small rise in levels.

A few of the flood records broken

	Flood stage in feet	1993 record in feet	Previous record in feet and date
Mississippi River			
Quad Cities, Ill.	15	22.6	22.5 on April 28, 1965
Burlington, Iowa	15	25.1	21.5 on April 25, 1973
Quincy, Ill.	17	32.2	28.9 on April 23, 1973
St. Louis	30	49.6	43.2 on April 28, 1973
Chester, Ill.	27	49.7	43.3 on April 30, 1973
Des Moines River			
Des Moines, Iowa	23	34.3	29.8 on April 11, 1965
Missouri River			
St. Joseph, Mo.	17	32.7	26.8 on April 22, 1952
Kansas City, Mo.	32	48.9	46.2 on July 14, 1951
Jefferson City, Mo.	23	38.6	34.2 on July 18, 1951
St. Charles, Mo.	25	39.5	37.5 on Oct. 7, 1986

While the basic laws of physics determine how water acts, keeping track of it for even a relatively small area is dauntingly complicated. The amount of water evaporating or transpiring from plants changes with the wind, temperature and humidity. How much water soaks into the ground depends on the types of soil — which can vary across a region — and how wet the ground already is.

Even determining exactly how much rain is falling is harder than it might look. Rainfall is measured at weather stations or by volunteer observers, miles apart. The amount of rain falling at one place isn't a good guide to how much is falling just a mile away.

As anyone who's ever driven through thunderstorms knows, a downpour can be making visibility almost zero at one place, while the sun is shining a mile down the road. A rain gauge in the downpour or the sun won't give a good picture of what's going on over an area of several square miles, such as the area drained by a stream subject to flash flooding.

Weather radar gives the big picture of where rain is falling, but in the past didn't do a good job of depicting how much rain was coming down. The Next Generation Weather Radars, which the National Weather Service completed installing around the U.S. in 1996, do a much better job of measuring rainfall amounts. The computers that are part of the new radars keep track of how much rain has fallen in various locations. These figures can alert forecasters to flash flood dangers and give a better picture of how much water will be flowing down streams.

Even with improved radar, forecasters can never expect to detect every flood on every tiny stream in a country as large as the United States.

This is one reason why since the 1960s cities, towns, counties and other agencies have been installing flood warning systems on streams where flash floods can be especially deadly. These systems consist of automated rain gauges or stream level gauges, sometimes both, connected to a central computer by radio or telephone. Relatively inexpensive personal computers can be programmed to monitor the reports from the gauges and sound an alarm when rainfall or water height reach dangerous levels.

Trying to predict floods and droughts

When water is flowing down streams, keeping track of it gets tricky. Fast-moving flood waters can reshape stream bottoms, changing the amounts of water being carried and upsetting hydrologists' calculations. The Great Flood of 1993 was a hydrologist's nightmare, not only for these reasons but also because flooding water breached hundreds of levees.

When water begins roaring through a hole in a levee, that water isn't going to raise the flood level as high as expected downstream. Forecast flood levels turn out to be too high. But water that breaks through a levee has to go somewhere. In many cases it flows "downstream" on the normally dry side of a levee until it reaches another break and rejoins the river. Now the water can make the river higher than forecast downstream.

At times during the 1993 Midwest floods, river forecasters took to the air to scout levee breaks and then tried to use the information for forecasts.

The 1993 flood showed hydrologists that they needed to greatly improve their computer models. Fortunately, more powerful computers are giving hydrologists the tools they need to keep up with the swirl of flood waters.

Forecasting floods is only part of the hydrologist's job. Managing water resources, especially reservoirs, is also vital. Imagine being in charge of a large reservoir that's becoming full. If you knew that the next few weeks, or even month or two were going to be dry, you would fill the reservoir nearly to the brim to supply water during the dry spell. But if you knew the next few weeks were going to be wet, you'd let water out now so that when flood water began flowing into the reservoir it would have room. You would avoid having to release large amounts of water all at once, causing floods downstream.

Better forecasts for precipitation a month ahead or even longer in the future could help water managers do a better job with these kinds of decisions.

Hydrologists know people are going to continue to live on flood plains and that the risks of floods are more likely to increase than decrease over the next several years as the population increases. They also know droughts will continue to arrive from time to time. Their hope is to reduce the danger from floods and to help manage scarce water resources during droughts with better forecasts, both of what a flooding river is likely to do in the next few hours and couple of days, and of how much precipitation is likely over the next month or two or longer.

Meteorologists like to tell about the forecaster who gets a call from an irate man: "Hey, buddy, how about coming over and shoveling out the six inches of partly cloudy in my driveway."

Anyone who has done much forecasting of winter storms is almost sure to have predicted "partly cloudy skies" for the following day, only to wake up the next morning to find a half-foot of snow on the ground. That same forecaster at one time or another has probably called for a big snowstorm only to wake up to partly cloudy skies.

Such "busted" forecasts are an annoyance to the person whose car is covered by six inches of "partly cloudy" or to children who dreamed of a school "snow day" only to wake and find the ground bare and school open. But they are more than an annoyance.

Our schedules today don't allow for weather delays. An unforecast snowstorm traps travelers and forces thousands of people to make last minute changes. A forecast storm that fizzles closes schools and businesses unnecessarily and costs highway departments thousands of dollars in pay for crews that may have been called in at overtime rates.

As forecasters move into the era of improved technology and computers, they are seeing improvements in their winter storm predictions. Scientists are using new technologies, especially better radar, to capture more detailed pictures of storms and to untangle the complications of snow and ice.

Big snowstorms are almost always a product of an extratropical cyclone. These storms have numerous ways of making forecasters look bad.

When an extratropical cyclone is on the move in winter, forecasters' problems begin with determining the storm's exact track and timing. The path is vital to the forecast because rain is likely on the storm's warmer

Rime ice covers a Mount Washington Observatory building atop the 6,288-foot New Hampshire mountain.

Eskimos have many names for snow. Here are some and a guide to pronouncing them:

<u>Aniu</u> (pronounce *an-e-YOU*): falling snow
<u>Apun</u> (*a-PYUN*): snow on the ground
<u>Mauja</u> (*mow-YA*): deep, soft snow
<u>Pukak</u> (*pu-CACK*): snow that can cause avalanches
<u>Qaliq</u> (*ka-LEEK*): snow that collects on trees
<u>Qannik</u> (*con-EEK*): snowflake
<u>Upsik</u> (*OOP-sick*): compacted snow

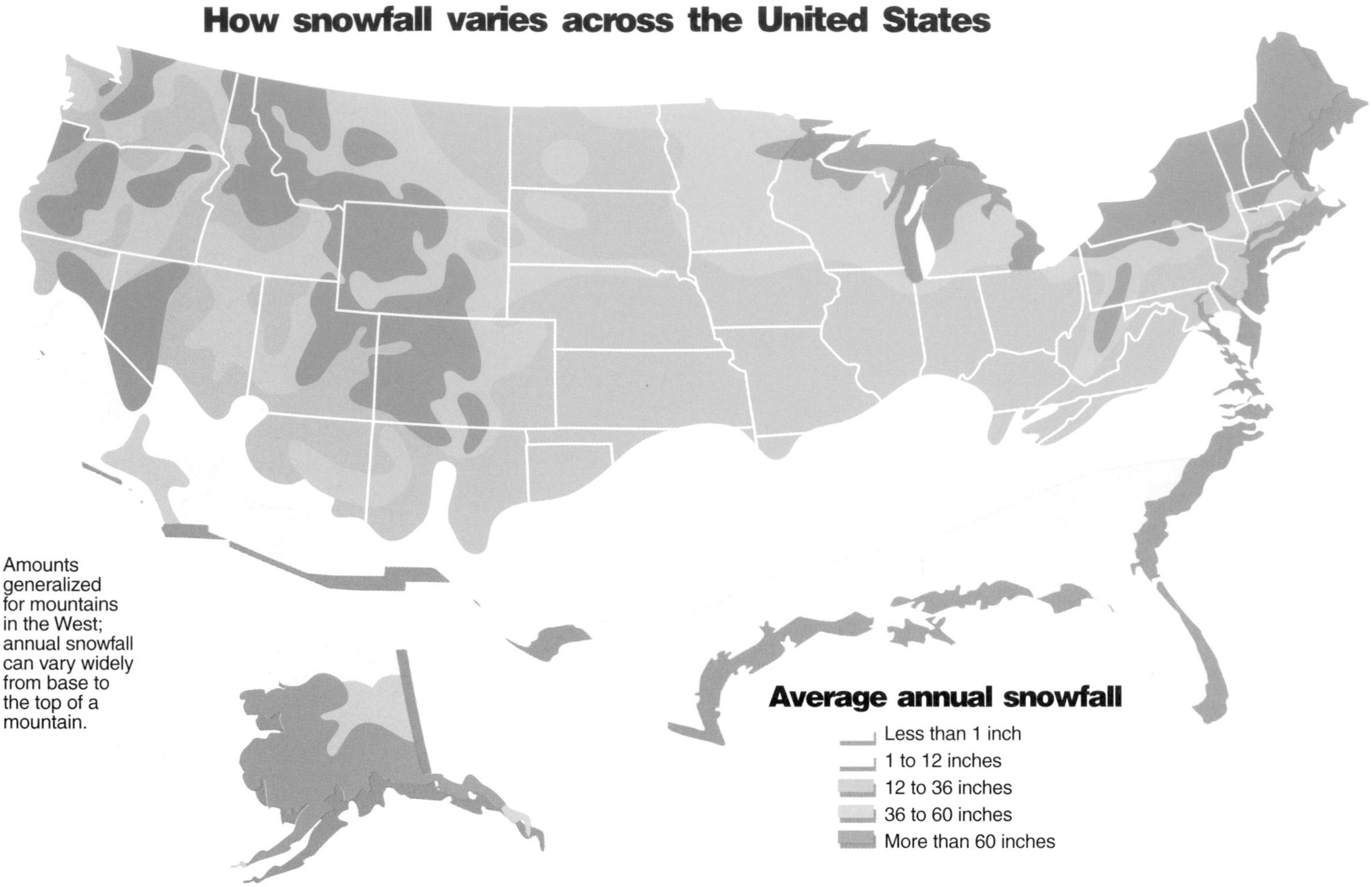

How snowfall varies across the United States

Amounts generalized for mountains in the West; annual snowfall can vary widely from base to the top of a mountain.

Average annual snowfall

- Less than 1 inch
- 1 to 12 inches
- 12 to 36 inches
- 36 to 60 inches
- More than 60 inches

southern and eastern sides, while snow is probable on the northern and western sides. Often a mixture of rain, freezing rain, sleet and snow will fall somewhere in the middle. While one city is being buried in snow, another an hour away may be having rain.

A successful winter storm forecast should say what kind of precipitation — rain, freezing rain, or snow — will fall, when it will start, how long it will last, how hard it will be, and exactly where it will fall. But storms' small-scale antics sometimes wreck such forecasts.

Storms can slow down or speed up. And even if the forecaster is right on the money with the storm's track and timing, the prediction can still go wrong. As researchers look at winter storms in more detail, using technology such as Doppler radar that detects wind speed and direction and even the air's up and down movements, they're finding that the details of a winter storm are incredibly complicated.

For three years, from 1988 through 1990, scientists closely examined winter storms in a 190-mile circle centered on Champaign, Ill. They found that small-scale changes from rain to snow to ice and back were happening in as little as 30 minutes.

If forecasters knew when such sudden changes were coming, they could keep their forecasts updated. But they don't always know what causes the changes, which means they can't predict them.

Mohan K. Ramamurthy of the University of Illinois at Champaign-Urbana, says the Illinois project has helped lead to new ideas about small-scale disturbances embedded in storms. But, he says, "it's still not a science that's been perfected." In some cases an event can be "like a big footprint you see on the ground and just don't know where it came from."

Winter's wide variety

In the three years of the Illinois study, 26 storms moved across the region. Their variety illustrates the range of Midwestern winter weather. Many of these storms looked similar in the early stages. In one, precipitation fell for only 30 minutes; in another, for 28 hours. One storm spawned an extremely damaging tornado. In some storms thunder and lightning accompanied heavy snow. Some brought heavy snow to Champaign; others, temperatures in the 50s with rain. One storm caused extensive floods east of the research area in Kentucky

Small-scale atmospheric waves complicate storm forecasting

When scientists look closely at storms, they sometimes turn up evidence of "solitary waves" in the air that can't be seen by the conventional National Weather Service network. But these waves have definite effects on the weather. For forecasters, the waves are wild cards that bring unexpected weather changes, such as bursts of heavy snow or wind shifts in parts of a storm but not in the whole storm. Scientists have theories about what causes such waves, but they don't know enough about them to forecast when and where they'll form, how strong they'll be, and where they'll go.

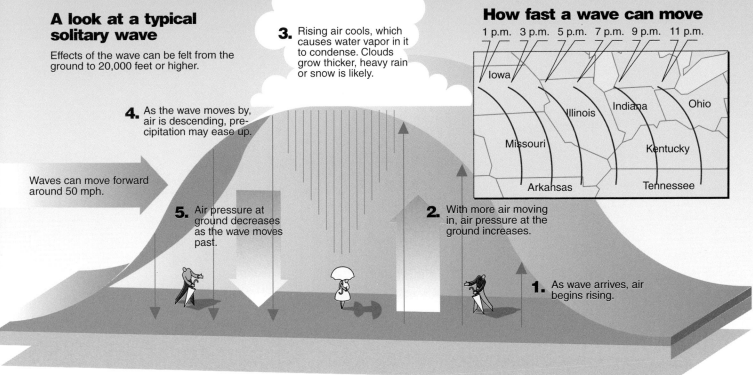

A look at a typical solitary wave

Effects of the wave can be felt from the ground to 20,000 feet or higher.

3. Rising air cools, which causes water vapor in it to condense. Clouds grow thicker, heavy rain or snow is likely.

4. As the wave moves by, air is descending, precipitation may ease up.

Waves can move forward around 50 mph.

5. Air pressure at ground decreases as the wave moves past.

2. With more air moving in, air pressure at the ground increases.

1. As wave arrives, air begins rising.

How fast a wave can move

1 p.m. 3 p.m. 5 p.m. 7 p.m. 9 p.m. 11 p.m.

Iowa · Illinois · Indiana · Ohio · Missouri · Kentucky · Arkansas · Tennessee

---

and West Virginia. While forecasters could distinguish the 30-minute snowfall from the 28-hour snowstorm ahead of time, most storms produced at least a few surprises.

The worst storm hit Champaign on Valentine's Day 1990. Freezing rain coated the area with 1½ inches of ice that did more than $9 million in damage and left the Champaign area without power for a week. To the north, 9 inches of snow fell on Chicago while rain was falling on southern Illinois.

Ramamurthy says that storm illustrates just one of the practical pitfalls of studying winter storms. "After two hours of freezing rain, all our instruments lost power. We couldn't launch [weather] balloons. We couldn't turn on the radar." With power out for five days, the temperature inside his house fell to 35°F. Weather study took second place. Researchers were busy trying to keep their water pipes from freezing.

The giant storms

A major United States winter storm often begins with a storm from the Pacific Ocean hitting the West Coast. Here the dividing line between rain and snow depends mostly on elevation because the ocean air is mild. The same storm will drench San Francisco with rain, while burying higher parts of the Sierra Nevada, such as Donner Pass, under 10 or 15 feet of snow. Storms normally have little moisture left after they cross the Sierra into the Rockies, but they often cover the Rockies with a few feet of dry, powdery snow.

After weakening in the mountains, storms normally begin to strengthen on the Plains east of the Rockies. A storm can follow a variety of paths as it moves eastward. But no matter which path it follows, the storm begins drawing in warm, humid air from the Gulf of Mexico and cold, dry air from Canada as it crosses the Plains and heads east. The contrasting air masses supply both the energy and moisture needed for blizzards.

From the Midwest, storms follow various tracks. Some move into Canada across the Great Lakes with little effect on the Southeast. Sometimes a storm will seemingly die west of the Appalachians, but the storm's cold air and counterclockwise motion in the upper atmosphere will move across the mountains to stir up a "secondary" storm along the Atlantic Coast. These secondary storms account for some of the heaviest snowfalls in the East. Storms moving up the East Coast from the Gulf of Mexico also can explode into major snow-

Snowiest U.S. places outside Alaska (in inches per year, based on 1961-1990 averages from National Weather Service reporting stations):

Paradise Ranger Station, Wash.	683
Crater Lake, Ore.	489
Wolf Creek Pass, Colo.	442
Silver Lake, Utah	407
Berthoud Pass, Colo.	390
Government Camp, Ore.	287
Climax, Colo.	264
Bowman Dam, Calif.	249
Holden Village, Wash.	243
Bozeman, Mont.	236

How a snowflake is born and grows

The shape of snow crystals depends on the temperatures and to some extent on the amount of water vapor in the air. Crystals often take on complex forms because they spend time in areas with different conditions.

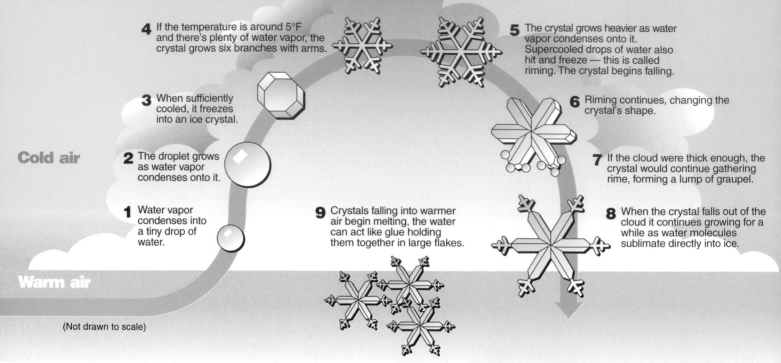

4 If the temperature is around 5°F and there's plenty of water vapor, the crystal grows six branches with arms.

5 The crystal grows heavier as water vapor condenses onto it. Supercooled drops of water also hit and freeze — this is called riming. The crystal begins falling.

3 When sufficiently cooled, it freezes into an ice crystal.

6 Riming continues, changing the crystal's shape.

Cold air

2 The droplet grows as water vapor condenses onto it.

7 If the cloud were thick enough, the crystal would continue gathering rime, forming a lump of graupel.

1 Water vapor condenses into a tiny drop of water.

9 Crystals falling into warmer air begin melting, the water can act like glue holding them together in large flakes.

8 When the crystal falls out of the cloud it continues growing for a while as water molecules sublimate directly into ice.

Warm air

(Not drawn to scale)

---

storms in the mid-Atlantic and Northeastern states.

About 10 inches of wet snow will melt down to the equivalent of an inch of rain.

Twenty to 40 inches — sometimes more than 50 inches — of dry, powdery snow can melt to just an inch of precipitation.

The beauty of snowflakes

When you're in no hurry to travel and a power failure isn't turning your home into a walk-in freezer, you can enjoy the beauty of a big storm's snow and ice.

Wilson A. Bentley, who spent almost 50 years photographing snow crystals, helped create the picture most of us have of "snowflakes." When children fold paper and cut out six-sided "snowflakes" they are creating figures much like those Bentley photographed. Artists and designers are still using Bentley's photographs as patterns. Such snowflakes are really snow crystals. Flakes are collections of crystals.

Like his parents before him, Bentley spent his life as a farmer in Vermont. He began observing snow when his mother gave him a microscope for his 15th birthday on Feb. 9, 1880. After putting a few snow crystals on a slide and looking at them through the microscope, Bentley was hooked for life. For more than a year, Bentley sketched what he saw, but he realized the sketches weren't capturing what he was seeing.

After obtaining the necessary camera and equipment, Bentley worked for three years, teaching himself photomicrography. Finally, on Jan. 15, 1885, he took the first photomicrographs anyone had ever taken of an ice crystal. "The day that I developed the first negative made by this method, and found it good, I felt almost like

This snowflake is actually a casting. The photographer "caught" a snowflake on a slide and put a drop of a clear mounting liquid on it. The snowflake melted, leaving behind its image in the mounting medium. The photo was taken through a 20x microscope lit from below through colored filters.

falling on my knees beside that apparatus and worshipping it," he said. "It was the greatest moment of my life."

This began his lifetime of photographing the shapes of snow. In addition to photographing snow crystals and other atmospheric phenomena, Bentley was ahead of his time in speculating about the different conditions that caused different kinds of crystals. In some of his articles in the *Monthly Weather Review* and other journals, Bentley discussed the idea that different forms in the same crystal resulted from falling through different air temperatures. No scientists of Bentley's time, the early years of the 20th century, looked into his theories.

His articles were anything but dry scientific reports. In an 1898 article, Bentley wrote of an ice crystal: "A careful study of this internal structure not only reveals new and far greater elegance of form than the simple outlines exhibit, but by means of these wonderfully delicate and exquisite figures much may be learned of the history of each crystal and the changes through which it has passed in its journey through cloudland. Was ever life history written in more dainty hieroglyphics!"

Bentley's book, *Snow Crystals,* which he published in 1931 with W. J. Humphreys, contains more than 2,300 photos of snow crystals along with other photos of frost patterns. Most of the snow crystals are intricate, six-sided plates or stars.

If you spend enough time closely examining snow crystals, you'll find plenty of the beautiful, symmetrical crystals like the ones in Bentley's book. But such crystals are far from the most common. The most widespread are small and irregular with no easily identifiable form.

Scientists as far back as Johannes Kepler (1571-1630) had wondered about the hexagonal form of snow crystals. Today's scientists know that snow crystal shapes reflect the hexagonal organization that water molecules assume when they freeze. This organization is dictated by the way electrical attractions bond water molecules.

Like Bentley, Ukichiro Nakaya, who worked at Hokkaido University in Japan from 1930 until his death in 1962, photographed snow crystals. As a scientist he was interested in discovering which conditions led to what kinds of crystals. Beginning more than 30 years after Bentley had first suggested how temperatures shape ice crystals, Nakaya began several years of catching snow crystals and making detailed notes of the weather conditions. He also created snow crystals in the laboratory, precisely controlling the temperature.

Nakaya's key discovery was a confirmation of what Bentley had suspected: temperature largely determines whether a crystal will grow into a flat, platelike form or into a long, columnlike or prismlike form.

Many natural crystals are combinations of the two because they spend time in areas with different temperatures as they're carried higher in updrafts or as they fall toward the ground.

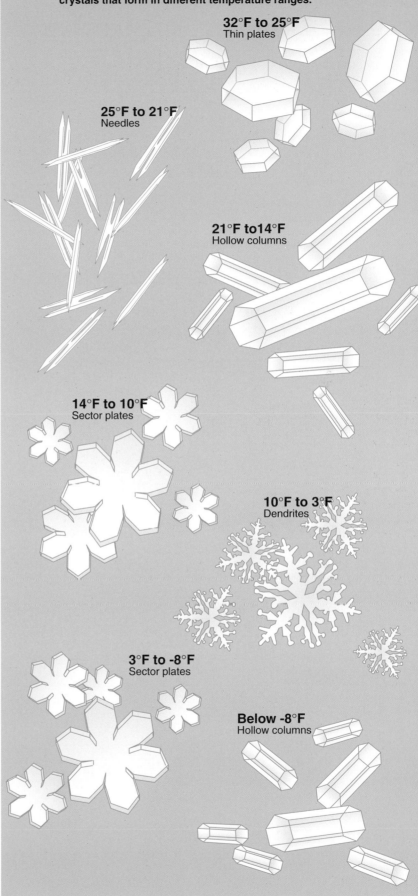

Temperature determines crystal shape

The temperature, and to some extent the amount of water vapor available, determine the shape of snow crystals. Here are the kinds of crystals that form in different temperature ranges.

32°F to 25°F
Thin plates

25°F to 21°F
Needles

21°F to 14°F
Hollow columns

14°F to 10°F
Sector plates

10°F to 3°F
Dendrites

3°F to -8°F
Sector plates

Below -8°F
Hollow columns

WINTER PRECIPITATION

In most of the United States, winter storms bring more than snow. Most of the country also has rain and sleet in the winter. Except during the coldest months in the northernmost states, most winter storms can glaze roads with freezing rain, which can cause more accidents than snow, or they can bring rain to turn the snow that's fallen into a mess. This graphic illustrates how the flow of relatively warm air over below-freezing air can cause rain, freezing rain, sleet and snow. Such overrunning is common ahead of a warm front. As a storm moves, its fronts are moving, often changing the kind of precipitation falling on a particular location. Three things complicate the picture even more: Smaller scale air flows; the cooling of air as falling precipitation evaporates; and local geography. It's easy to see why forecasters can't always say what kind of precipitation will fall in winter.

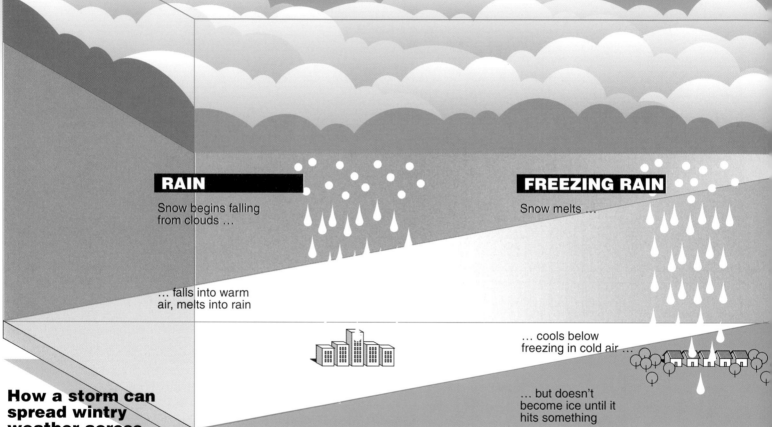

RAIN

Snow begins falling from clouds …

… falls into warm air, melts into rain

FREEZING RAIN

Snow melts …

… cools below freezing in cold air …

… but doesn't become ice until it hits something

How a storm can spread wintry weather across the United States

Sometimes a storm can cross the United States from the Pacific to the Atlantic bringing snow, freezing rain, sleet, rain and cold to most of the nation. It's more common, however, for winter storms to affect only parts of the country because they form east of the Rockies or head northward into Canada before reaching the East Coast.

SUNDAY

Low pressure center of storm

A Pacific storm hits the West Coast. Rain falls along the coast with showers as far south as Los Angeles. Snow falls on elevations as low as 3,000 feet in the Cascades of Washington and Oregon and above 5,000 feet in California's Sierras.

MONDAY

Cold, dry air

Warm, moist air

The storm is weakening as the West's mountains disturb its wind flow. Snow is falling from the Sierras across the Great Basin and in the Rockies. Warm, humid air from the Gulf of Mexico is beginning to flow into the storm. As this humid air moves upward over the Rockies' east slopes the air cools and its moisture creates "upslope" snow. Cold air is flowing in from the north.

TUESDAY

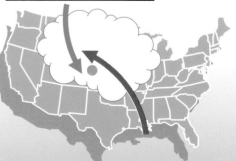

The storm, now free of the mountains, is growing stronger on the Plains, pulling in both more warm air and cold air. Snow, freezing rain and sleet are falling across the central and northern Plains into the Midwest. Thunderstorms are breaking out across the southern Plains. The Rockies are turning colder as snow trails off to flurries.

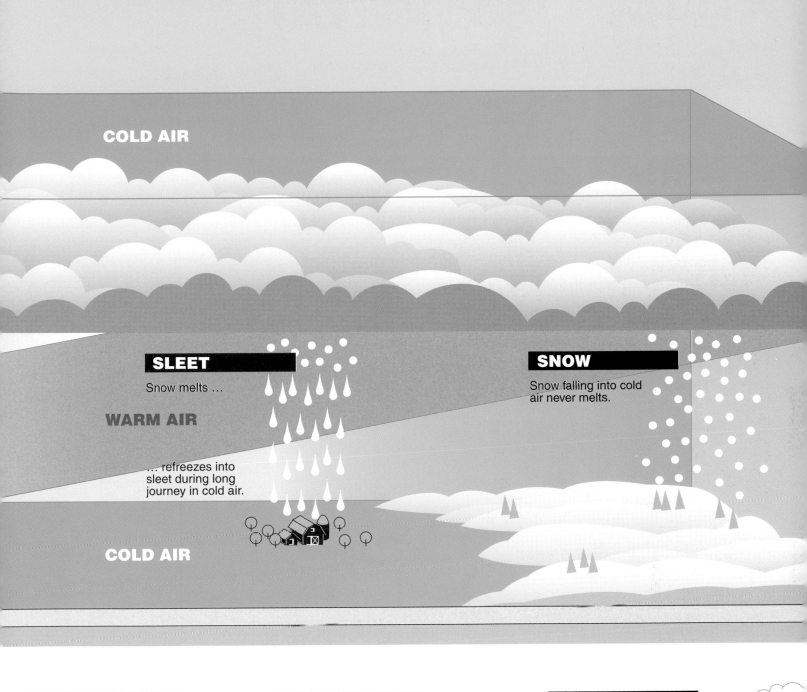

COLD AIR

SLEET

Snow melts …

WARM AIR

… refreezes into sleet during long journey in cold air.

COLD AIR

SNOW

Snow falling into cold air never melts.

WEDNESDAY

The storm's center is beginning to weaken as it moves into the western Appalachians. But heavy snow continues to fall across the Midwest as bitter cold air flows across the Plains and into the south-central states. Thunderstorms are moving across the Southeast and freezing rain and sleet are icing roads in the Northeast.

THURSDAY

As the storm's old center weakens in the Appalachians, a new center forms off the North Carolina coast and begins moving northeastward. Heavy snow falls in the Appalachians while the big cities along the East Coast see a mixture of rain, freezing rain, ice pellets and snow.

FRIDAY

The storm is rapidly moving northeastward across eastern Canada and the Atlantic. Snow continues falling on New England as cold air brings clearing skies to the East Coast.

Forecasting dangerous and deadly airplane icing is improving

For pilots, winter's dangers go far beyond airplanes skidding off icy runways. Ice that forms on aircraft can be deadly because it erodes the lifting force that keeps a plane or helicopter in the air. Ice can even snatch control of the plane from a pilot.

The National Transportation Safety Board (NTSB) said that an ice-caused loss of control probably caused the Oct. 31, 1994 crash of a commuter airliner near Roselawn, Ind., killing all 68 aboard. That crash encouraged everyone — pilots, commuter airlines, aircraft makers and regulators, weather forecasters and researchers — to focus more attention on icing.

Pilots have feared ice ever since planes began flying in clouds. Freezing drizzle in freezing rain or supercooled water in clouds turns to ice almost immediately on contact with an aircraft. Even a thin coating of ice can change a wing's shape, disrupting the flow of air around it and reducing lift. And as the Roselawn crash showed, ice can cause an airplane to roll nearly upside down when it disrupts the flow over the ailerons at the ends of the wings.

All pilots have to worry about ice coating their planes as they sit on the runway waiting for takeoff. But pilots of large, high-flying jets know that once they take off, they can climb quickly above the danger area to altitudes where the clouds are thin, made of ice crystals that don't stick to airplanes. When these long-range jets land, their descent through clouds is usually quick enough to avoid dangerous icing

However, pilots of smaller airplanes, such as commuter planes that fly at lower altitudes, have to be worried about flying inside the ice-clouds for long periods. For these planes, accurate icing forecasts are essential.

But accurate icing forecasts are difficult. Tiny changes in the amounts of water in clouds or in air temperatures can make one cloud produce ice on an airplane and a neighboring cloud produce no ice at all.

Marcia Politovich heads the icing research program at the National Center for Atmospheric Research (NCAR) in Boulder, Colo. After the Roselawn crash, her group was asked to find out what happened. This was no theoretical challenge: If one plane went down, others could too. No one, including weather forecasters, had any reason to expect unusually heavy icing the day of the crash. What happened?

The National Weather Service had saved that night's radar readings and other detailed data. NCAR's Politovich and research meteorologist Ben Bernstein, and scientists from the University of Wyoming, the National Oceanic and Atmospheric Administration, and the University of Washington analyzed the data and came up with an answer.

They concluded that relatively unusual conditions in the clouds, where the airplane had been circling in a holding pattern, created supercooled cloud droplets that were much larger than normal. Cloud droplets can be as big as 50 micrometers across. These were around 200 to 500 micrometers across. Aircraft icing protection systems are designed for the smaller droplets.

Pilots have long avoided landings and takeoffs in freezing drizzle or freezing rain. But the existence of drizzle-sized drops high in clouds was something new. "We don't know for sure that these large drizzle drops existed at the time the plane went down," says Politovich. But the weather conditions and the way the airplane acted make supercooled large droplets the most likely suspect.

Such icing is "a relatively rare event," she adds. "We just don't know very much about freezing drizzle, the size of the drops, how much water is present. We don't have engineering specifications to characterize the atmosphere" for designers of ice protection systems. "Computer models don't handle large droplet icing well. It's hard to generate the large droplets in wind tunnels."

While researchers worked to understand more about how drizzle-sized drops form in clouds, pilots needed better forecasts immediately. Fortunately, pilots didn't have to wait until all of the scientific questions had been answered. Bernstein had been working on an algorithm (a mathematical problem-solving procedure) to predict airplane

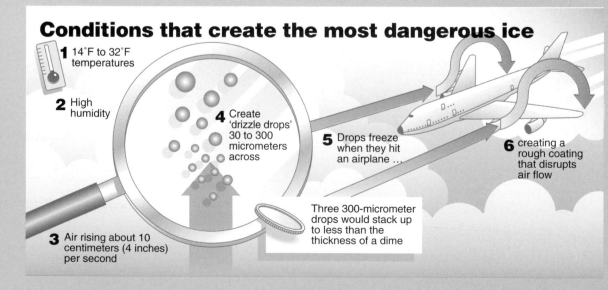

Conditions that create the most dangerous ice

1 14°F to 32°F temperatures

2 High humidity

3 Air rising about 10 centimeters (4 inches) per second

4 Create 'drizzle drops' 30 to 300 micrometers across

Three 300-micrometer drops would stack up to less than the thickness of a dime

5 Drops freeze when they hit an airplane …

6 creating a rough coating that disrupts air flow

Marcia Politovich and Ben Bernstein study conditions that can lead to deadly airplane icing.

icing conditions over small areas that looked like it would help forecasters right away.

Dave Sankey, head of weather research for the Federal Aviation Administration (FAA), says "it would have taken forever" to jump through the normal bureaucratic hoops to get Bernstein's work into the hands of forecasters. The FAA, which contracts with NCAR for icing research, and the Weather Service are in different federal departments.

Instead, Sankey arranged for Bernstein to go directly to the Weather Service's Aviation Weather Center in Kansas City, Mo. "I had a huge chunk of computer code," Bernstein says. "They plugged it in and made sure it worked. There was satisfaction all the way around. The Aviation Weather Center was able to put out a really well-defined product that said, 'Here's a potentially serious hazard.' I

don't want to overstate the benefits, but it is a new tool."

The Center, using Bernstein's technique and others', issued its first forecast for freezing drizzle-sized droplets on Jan. 9, 1996. Reports from two pilots, who encountered icing and fled in time, confirmed that the forecasting techniques work.

Icing forecasts are not new, but accurate, pinpoint ones are. For years pilots had complained that icing forecasts has covered vast areas, causing unnecessary detours or even canceled flights.

Politovich says scientists are reducing the size of those danger areas. And she expects computer models to continue to improve the forecasts. She also expects help from the more detailed pictures of clouds that new weather satellites are sending back. Satellites can

detect supercooled water droplets in clouds. In fact, they can "see" about 1,500 feet down into clouds. But clouds are often composed of layers, and satellites see only the top of the top layer of clouds.

In addition to learning more about in-flight icing, the NCAR researchers also are looking for ways to keep ice off airplanes when they're on the ground. During the winter of 1995-96, NCAR researchers at Chicago's O'Hare International Airport set up a network of snow gauges that weighed the snow to check for water content.

Data from those gauges plus Doppler weather radar data help airlines determine the danger of ground icing and do a better job of de-icing planes before takeoff. Airport officials found the extra data also helped them make better decisions about plowing runways, ramps, parking lots and airport roads.

Undertaking research directed at solving particular problems isn't for all scientists, Politovich says. "Some say, 'I'm in pure research. You can't schedule breakthroughs; enlightenment doesn't just come when you want it to.' We have a mix of people in the icing program. Some are very applied types. They enjoy getting out, talking to people outside of meteorology. Some don't do this so well."

She says, as head of the program, "a part of the whole game is figuring out how people can do well at what they're best. I like to go out in the world, come back and say, 'Hey the real world wants this, can you apply your skill to it?'"

How graupel, ice pellets and hail form

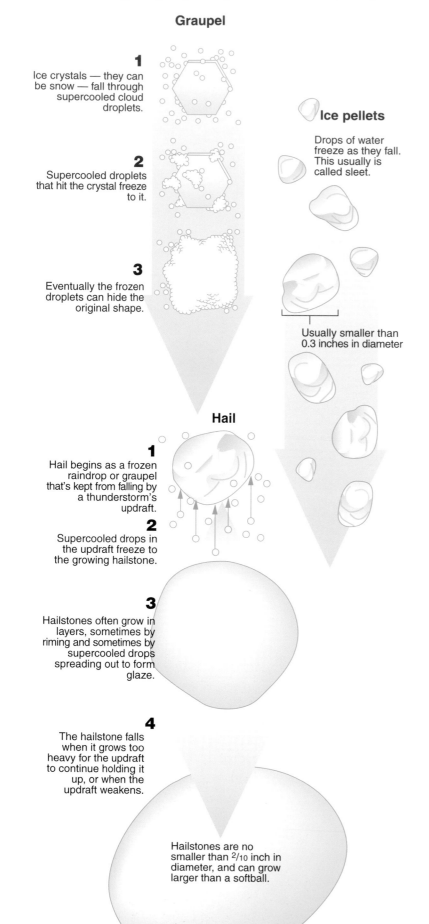

Graupel

1 Ice crystals — they can be snow — fall through supercooled cloud droplets.

2 Supercooled droplets that hit the crystal freeze to it.

3 Eventually the frozen droplets can hide the original shape.

Ice pellets

Drops of water freeze as they fall. This usually is called sleet.

Usually smaller than 0.3 inches in diameter

Hail

1 Hail begins as a frozen raindrop or graupel that's kept from falling by a thunderstorm's updraft.

2 Supercooled drops in the updraft freeze to the growing hailstone.

3 Hailstones often grow in layers, sometimes by riming and sometimes by supercooled drops spreading out to form glaze.

4 The hailstone falls when it grows too heavy for the updraft to continue holding it up, or when the updraft weakens.

Hailstones are no smaller than 2/10 inch in diameter, and can grow larger than a softball.

Are all snow crystals different?

Just about everyone has heard "no two snowflakes are ever alike." The closer you look at that statement, says Charles Knight of the National Center for Atmospheric Research, the less it means.

To begin with, no two crystals ever will be alike if alike means every molecule of one crystal is in exactly the same place as every molecule of another. Even a single crystal isn't the same from second to second because some water molecules constantly are breaking free of the ice to become vapor, while other molecules are joining the ice.

Most people think of "alike" as meaning that the crystals have to look the same. By this looser standard, the statement that no two crystals are alike isn't true, Knight says. Many tiny snow crystals are simple, hexagonal plates with no obvious difference in shape.

Even more complicated crystals can be alike, as Knight's wife, Nancy, also a cloud physicist, discovered in 1989. She photographed twin snow crystals snared by a research aircraft on a cloud-study flight about 20,000 feet over Wausau, Wis. The crystals were thick, hollow columns 250 micrometers long and 170 micrometers wide. (A human hair is about 100 micrometers in diameter.)

These twin crystals, however, weren't the elaborate, six-sided figures most people think of as typical snow crystals. Considering how many water molecules go into such crystals and how many ways these molecules can arrange themselves, the odds are against anyone finding two starlike crystals that look the same. So while the "no two snow crystals are alike" isn't literally true, Knight says, "it's a nice way of saying snow crystals can be very complex."

Other kinds of ice

Snow makes up only seven of the 10 kinds of frozen precipitation in the international snow classification system. The others are graupel (pronounced GROU-pel), ice pellets and hail.

When we start naming the different kinds of ice, confusion begins. Scientists don't always agree on the names. Here we'll talk about snow, graupel, ice pellets and hail as the kinds of ice that fall from the sky. Each is described in the way commonly used by most American scientists. But the official guide for United States weather observers has different descriptions. And neither set of definitions uses the common word "sleet," which is defined differently in various parts of the English-speaking world. In the United States, "sleet" refers to frozen raindrops, or ice pellets.

To understand why ice takes so many forms, start with this fact: Water can cool below 32°F — often called the freezing point of water — without ice crystals forming. Supercooled drops of water range in size from tiny cloud droplets to drizzle to raindrop-size. These drops turn into ice when they come in contact with an ice nucleus or something solid. Supercooled raindrops leave a glaze of ice on sidewalks, roads, trees, power lines, your car's windshield and your front steps. This is called freezing rain. Smaller supercooled drops are freezing drizzle.

Why don't the drops freeze when they cool below 32°F? Charles Knight puts it this way: "Water below freezing wants to be ice. It just has to learn how." Supercooled water "learns" how to become ice by coming in contact with ice crystals or ice nuclei. An individual cloud drop or raindrop isn't likely to freeze until it meets an ice nucleus or ice that's already formed. When supercooled rain or drizzle is falling, a few ice crystals or ice nuclei are likely to be mixed in or on the surface being hit. These crystals are the seeds that turn the supercooled water into a sheet of ice.

If the drops are tiny, they freeze instantly on contact, creating tiny balls of ice with air between them. This is "rime." The trapped air gives rime a milky white appearance. When larger supercooled drops hit, they spread out before freezing, creating smooth ice or "glaze." Small drops also can create glaze if the temperature is warm enough to slow their freezing. Rime and glaze can be mixed together.

Glaze from freezing rain often is destructive. Since the water spreads out to freeze, the ice is smooth. Cars skid on roads and people fall on sidewalks. The ice's weight pulls down power lines and tree limbs. Since tiny drops freeze on contact to make rime, they keep their shape; the ice isn't slick. Clouds of supercooled droplets create icy mountain wonderlands when they deposit thick coats of rime on trees and buildings.

The struggle against ice and snow

For centuries people who lived in places covered by winter snow looked forward to the season's first big snowstorm. Horse-drawn sleds zipped along roads that wagons plodded over in the summer — or couldn't manage at all when rain turned roads into bogs.

With the advent of railroads, snow and ice presented a new kind of travel barrier in the middle of the 19th century. Deep snow could block the rails. Usually a plow blade mounted on a locomotive, maybe with another locomotive or two pushing, could clear the rails. Keeping the tracks open also meant hiring hundreds of workers to shovel snow from around switches or other places where plows didn't do the job.

At times, however, blizzards on the Plains and snow in mountains overwhelmed simple plows and crews of shovelers. A dozen locomotives working together sometimes were unable to push a plow through the snow blocking mountain railroads. The big breakthrough in railroad snow-clearing came with the development of rotary plows in the mid-1880s. The rotor, often 10 to 12 feet high, consisted of blades that hurled snow off to the side. Similar plows still are used to cut through deep snow both on railroads and highways.

Until the mid-1950s most public works depart-

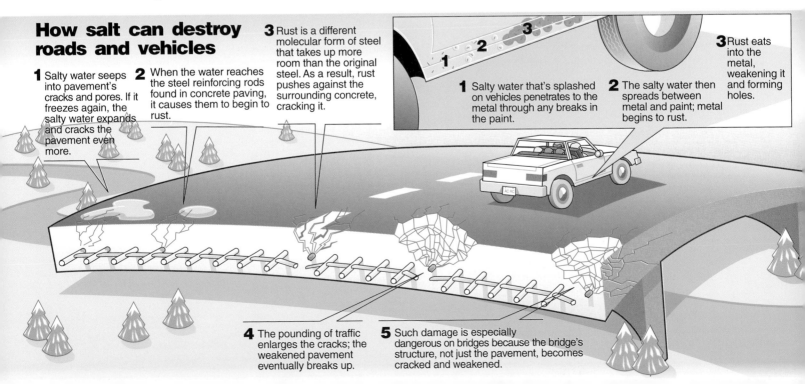

How salt can destroy roads and vehicles

1 Salty water seeps into pavement's cracks and pores. If it freezes again, the salty water expands and cracks the pavement even more.

2 When the water reaches the steel reinforcing rods found in concrete paving, it causes them to begin to rust.

3 Rust is a different molecular form of steel that takes up more room than the original steel. As a result, rust pushes against the surrounding concrete, cracking it.

1 Salty water that's splashed on vehicles penetrates to the metal through any breaks in the paint.

2 The salty water then spreads between metal and paint; metal begins to rust.

3 Rust eats into the metal, weakening it and forming holes.

4 The pounding of traffic enlarges the cracks; the weakened pavement eventually breaks up.

5 Such damage is especially dangerous on bridges because the bridge's structure, not just the pavement, becomes cracked and weakened.

ments fought ice by spreading sand or cinders on roads to improve traction. Plows pushed the snow away, and drivers often used tire chains to drive on snowy streets. By the 1960s, however, a "dry pavement" snow policy was the rule in many places, especially in the East. Drivers wanted snow removed completely and quickly without leaving icy roads behind. Snow tires were fine. Tire chains were a bother. Citizens voted out politicians who couldn't get ice and snow quickly removed.

Salt was the answer. It's relatively cheap, it melts ice unless the temperature is extremely low, and it's easy to spread on roads. Its downside wasn't immediately evident.

But by the 1970s the damage from salt was becoming obvious. Not only were cars and trucks rusting out, corrosion was destroying bridges and streets. People saw the environmental damage signaled by dying trees along roads. Despite the costs of repairing and rebuilding bridges and roads, a return to "natural" snow and ice removal — the spring thaw — was unlikely.

Today in the United States, keeping roads and highways clear of ice and snow costs an estimated $2 billion a year. In addition, various studies say damage, mostly from the salt used to melt ice, costs individuals and governments between $800 million and $2.5 billion a year. Much of this is corrosion damage to roads, bridges and vehicles. But it also includes environmental problems such as salt damage to trees, streams and even wells, which can't be used when underground water becomes too salty to drink.

John Qualls, the California Department of Transportation regional manager responsible for Interstate 80 across Donner Pass, sums up the salt dilemma: "I am concerned about salt use. I don't like to see vehicles rusting, I don't like to see trees dying, and I don't like to see streams polluted. But neither do I like to see the broken and twisted bodies of accident victims, or massive traffic tie-ups due to vehicles' inability to traverse

the highway. At this time, considering the level of service the public demands and the budgetary constraints we work under, we have no choice but to judiciously use salt or some other effective de-icer to keep our highways open."

Judicious use of salt is being made easier by technological improvements and "sensible salting" educational campaigns. The idea is to use no more salt than needed to keep a road from turning dangerously slick. Since the 1970s, salt spreaders have been changed to key the application rate to the salt truck's speed. Old spreaders ran at the same speed whether the truck was going 40 mph, 10 mph or was stopped. Calcium chloride is sometimes mixed with salt. And while it is corrosive, too, calcium chloride makes the salt more effective, cutting down the amount needed.

Improvements both in vehicles and tires also have helped. Today's cars and trucks don't rust out as easily as those of a decade or two ago. Tire companies have developed rubber that has a better grip on ice or wet pavement. Some states allow winter use of metal tire studs that make skidding more unlikely.

Bridges and roads are being designed to withstand corrosion. Since the big problems arise from corrosion of the reinforcing bars inside concrete, bars are being coated with epoxy. Salt also can be kept away from the reinforcing bars by adding something that blocks water, such as a wax, in the concrete between the road surface and the bars.

New ice fighters

Other materials melt ice or keep it from bonding to roads, but all are more expensive than salt and many pose environmental problems of their own. Many airports and some highway maintenance departments use the liquid ethylene glycol, but it's considered a hazard to fish and other aquatic life, and it is expensive.

Calcium magnesium acetate (CMA) is being used

Snow blindness is a temporary loss of sight caused by bright sunlight reflected from snow. It usually lasts from several days to a week. Occasionally a person will have trouble distinguishing between colors after snow blindness and sees everything colored red for a time. In most cases, snow blindness disappears when a person rests the eyes and remains indoors. Wearing sunglasses or goggles usually prevents snow blindness.

How snow fences protect roads

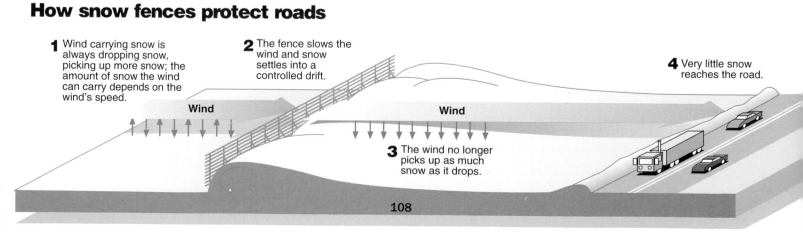

1 Wind carrying snow is always dropping snow, picking up more snow; the amount of snow the wind can carry depends on the wind's speed.

2 The fence slows the wind and snow settles into a controlled drift.

3 The wind no longer picks up as much snow as it drops.

4 Very little snow reaches the road.

Wind

Wind

High-tech road patrol

Sensors embedded in pavement and weather instruments near roads help highway officials get the jump on ice and snow.

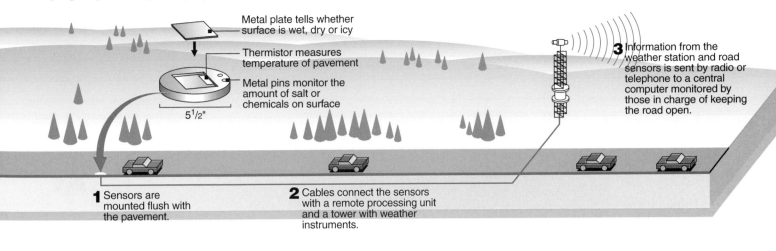

Metal plate tells whether surface is wet, dry or icy

Thermistor measures temperature of pavement

Metal pins monitor the amount of salt or chemicals on surface

$5^{1}/_{2}$"

3 Information from the weather station and road sensors is sent by radio or telephone to a central computer monitored by those in charge of keeping the road open.

1 Sensors are mounted flush with the pavement.

2 Cables connect the sensors with a remote processing unit and a tower with weather instruments.

in environmentally sensitive areas because it doesn't seem to harm the environment and is far less corrosive than salt. But it costs around $650 a ton compared to $25 to $50 a ton for ordinary rock salt. Some states require its use on new bridges and in areas that are sensitive to environmental damage. The calcium and magnesium come from dolomitic lime, a common material. The acetate, which is what makes CMA so expensive, is made from natural gas. Scientists are looking for ways to cut CMA's price by producing acetate at a lower cost. Making it by fermentation from plant material, such as corn, might be the answer.

Researchers have looked into materials that could be mixed into the pavement of new roads to keep ice from sticking. Others are looking for new ways, not involving chemicals, to break ice loose from roads. None has shown much promise. Research that does show promise is the development of more efficient plows, which could cut the cost of snow removal.

Finally, keeping so much snow from piling up on the road can be more effective than many highway engineers suspect, says Ronald Tabler, who specializes in snow, dust and wind engineering. "Usually snow is ignored in the design of highways and other facilities," he says. "The added expense is thought not to be worthwhile by engineers in warm offices miles from the project. If you're not out there in the blowing snow, it doesn't seem to be as important a factor as cost."

Tabler explains that large, well-designed snow fences greatly reduce the amount of snow drifting onto highways. "We can say, in general, it costs about 100 times more to plow snow than to store it with fences." Studies of road-clearing costs and accidents in Wyoming

on Interstate 80 between Laramie and Walcott Junction show that the stretch's $1.9 million worth of fences have saved at least that much in property damage from accidents in 15 years.

Wyoming believes in highway snow fences. "There's probably more wood in snow fences in Wyoming than there is wood in houses," Tabler jokes. Montana and snowy parts of Arizona also use 12-foot-high snow fences. Higher fences are required in the West where the wind generally blows harder. In the East, eight-foot fences have done well in tests in the Buffalo, N.Y., area and are likely to become more common.

The best hope for snow control

Improvements in weather forecasting and automated systems that keep track of what's happening to roads offer the best hope for cutting the use of salt. Sensors and computer hookups that feed up-to-the-minute road and weather information to maintenance supervisors are being set up to supply the needed data.

"This information is invaluable in determining when they should take some action," says David Minsk of the National Research Council. "They have the knowledge to make the proper decisions and this offers the biggest potential of cost savings. They can avoid the useless call-up of crews; false alarms can be extremely costly. On the other hand, if they try to save money and not call the crews out in time, then there are extra costs in removing the snow and ice."

While such systems tell a highway superintendent what's going on now, good forecasts of what's going to happen in the next hour or two also are vital. If salt or other chemicals can be spread just as freezing rain or

Winter still wins some battles at the infamous Donner Pass

The 150-mile trip from Sacramento, Calif., to Reno, Nev., by car on Interstate 80 or by train on Amtrak's California Zephyr, cuts across one of America's major snow battlegrounds. While both routes stay open with only a few delays most winters, California Department of Transportation (Caltrans) and Southern Pacific Railroad officials know they never can have total victory over the snow.

The highway and railroad climb from a few feet above sea level in the Sacramento Valley, where even a trace of snow is rare, to more than 7,000 feet as they cross the Sierra Nevada at Donner Pass, where a normal winter's snow is easier to measure in feet than inches.

"The snow is so heavy with moisture that it makes an ice pack. We can plow down to the ice pack, but it's often too cold to melt" with salt or other de-icers, says John Qualls, Caltrans' regional maintenance manager whose responsibilities include the Donner Pass highway.

"This combined with the grades and curves means at times it's impossible to keep roads open. They just get so slippery you can't steer a vehicle."

California's wide climate variation gives Qualls and his crews headaches that their counterparts in other snowy places don't have. If, for example, you spend a winter in Wisconsin, you're bound to learn the basics of winter driving.

But Donner Pass and its surrounding ski areas are only a two- or three-hour drive for millions of motorists from the Central Valley or the San Francisco Bay Area who never see snow at home.

"I have to say well over 90 percent are inexperienced snow drivers," Qualls says of Donner Pass motorists. Even those who make regular ski trips to the Sierra are likely to find the road clear most of the time. "If they've made several trips over the Pass they've found the road sanded and salted and they expect to go 55 miles per hour. They'll try to drive 55 with chains on icy roadways." The resulting accidents cause more delays than the weather.

The Donner Pass is named for the Donner Party, a group of inexperienced travelers. Old movies about settlers heading westward featured experienced guides leading the way. That certainly wasn't the case when the party led by George and Jacob Donner and James Reed left Independence, Mo., in May 1846 for Sutter's Fort in Sacramento.

From May until late July 1846, the party traveled from Independence, Mo., to Fort Bridger, Wyo., with few problems. But at Fort Bridger, instead of taking a proven route, the party took a "shortcut" to Utah's Great Salt Lake. Rugged mountains turned the trip into a month-long ordeal. It took until Oct. 20 to reach the east side of the Sierra Nevada, at a point close to where Reno, Nev., is located today.

They knew winter was coming

Snow falls at Donner Summit Lodge during a March 3, 1991 storm. Efforts to dig out were slowed because of the continual snowfall that left a total of more than 62 inches.

and had been told that it arrives early in the Sierra. Yet, instead of hurrying across the mountains before snow began, the party stopped for a few days to let their cattle regain strength.

The party's lack of experience probably accounts for the decision to continue into the mountains even though the weather was threatening. On Oct. 28, 1846, a storm unloaded about five feet of snow as the party was reaching the pass. It trapped the settlers, their wagons and cattle. Only 45 of the 79 men, women and children survived the winter.

The story of starvation, death — and eventually cannibalism — in what came to be called Donner Pass became part of the folklore of America's westward expansion. But such dangers didn't stop people from crossing the Pass on the way to California, especially after gold was discovered in 1848. In 1849, an estimated 200 people a day were flowing westward through Donner Pass.

Despite its height and ruggedness, Donner Pass was the best route for North America's first transcontinental railroad. The Central Pacific Railroad was built eastward from Sacramento, beginning in 1863, across Donner Pass and Nevada to Promontory, Utah, where the symbolic golden spike linked it with the Union Pacific Railroad on May 10, 1869.

In 1861, the Central Pacific's chief engineer Theodore D. Judah, who was an experienced Sierras explorer, wrote a naive explanation of how to keep the railroad open in winter: "It is only necessary then to start an engine with snow plows from the summit each way after the commencement of a storm, clearing the snow as it falls. A similar course of procedure at each successive storm will keep the track open during the entire winter."

The nearly 40 feet of snow that fell during the 1866-67 winter, when trains were running on the completed part of the line, proved him wrong. Snow closed the railroad more than half the winter despite the work of an army of shovelers. Small plows on the front of locomotives, which worked well in the East, were useless in heavy Sierra snow. After that winter the railroad began building sheds over the tracks in the areas where snow drifted the deepest. By 1873 more than 30 miles of wooden sheds covered tracks through the mountains.

In places not covered by the sheds, "bucker" plows, weighing up to 19 tons and pushed by as many

as a dozen locomotives, tried to keep the railroad open with the help of hundreds of shovelers. Rotary plows that toss snow far off to one side were first used during the winter of 1889-90, when nearly 65 feet of snow fell on Donner Summit. Even with the new plows and, at one time, more than 1,200 shovelers — who were paid $2 a day — the railroad was closed for 16 days in January 1890.

Most of the railroad's old wooden sheds have been torn down, but from Interstate 80 you can see the concrete snowsheds that cover about a mile of track.

Snow still closes the railroad from time to time during the worst winters. The 20th century's most dramatic snow battle began on Jan. 13, 1952, when snowslides trapped the *City of San Francisco* with 226 passengers and crew aboard. The slides happened during a six-day storm that dumped about 16 feet of snow atop the eight feet already on the ground.

Even rotary plows were unable to free the stranded train. After a couple of days the locomotive, which had been supplying heat, ran out of fuel and water pipes in the passenger cars were freezing. But rescuers reached the train after three days and none of the passengers was seriously hurt.

Even with today's version of the rotary plows that worked so well on the railroad beginning in 1890 — today's highway, rotary snow blowers can toss aside 3,000 tons of snow an hour — Qualls and his crews can't keep Interstate 80 open all of the time. Sometimes blowing snow cuts visibility so much the road must be closed. But, Qualls says, "We feel that just considering the snow itself, if we didn't have traffic to put up with, there hasn't been a snowstorm in recent history [through which] we could not have kept the road open."

During a major storm, Qualls can call on about 56 people each shift to operate 12 rotary plows, 20 motor graders, 17 combination sanders and plows, three trucks to apply liquid de-icers and the smaller vehicles needed for enforcement and supervision. Two private tow trucks strong enough to push stranded tractor-trailer trucks out of the way also are available.

Qualls, who started working on Sierra highways in 1958, says over the years he's seen increased efforts to keep roads open in storms. People in California expect to be able to get around by car, no matter how bad the weather, he says. "We have an obligation to do all in our power to make that happen."

The last time snow closed the Donner Pass train route for a significant time was from May 31 to April 7, 1982, after a storm brought the season total snowfall to 796 inches. That was second only to the record 819 inches — over 68 feet — that fell in the winter of 1931-32.

Charles and Nancy Knight: Chasing clouds to catch their secrets

Charles Knight in his refrigerated laboratory.

What the Knights learned about hailstones:

- Some hailstones have onion-skin layers from traveling up and down several times in a storm.

- Some stones grow while balanced in an updraft and have little layering.

- Some hailstones form around raindrops that are carried high into the storm and freeze.

- Some hailstones form around ice crystals.

Studying clouds would be easier if scientists could bring them into the laboratory.

Since that's not possible, cloud physics demands ingenuity.

Physicists Charles and Nancy Knight at the National Center for Atmospheric Research have spent more than 20 years in the occasionally wild — and sometimes painful — study of what makes clouds work as they form ice crystals, water droplets and precipitation.

In the early days, donning rain slickers and hard hats, the Knights would head out into Plains thunderstorms, hand-gathering hailstones for study back in the lab.

"It's actually kind of exciting," Charles Knight says. "There's lots of lightning, a few tornadoes. It certainly keeps your attention. We never had any harrowing escapes, but we got some bruises from hailstones."

Later — to eliminate the bruises and increase the number of specimens — the Knights rigged a funnel in the roof of a van. Hailstones falling into the funnel were put into a container and cooled to -70°F to be preserved for the ride back to the lab.

"Ideally we'd plan to park in a storm's path," he says. "But storms didn't often cooperate and we'd find ourselves driving 90 miles an hour down highways trying to find where the hail was."

The Knights cut thin slices from hailstones and studied their crystal structure under a microscope to learn how the stones grew. They found several patterns of growth, which showed that older and relatively simple ideas of how hail forms were wrong.

Now they've moved on to other areas of cloud study where simple-looking questions have complicated answers. Consider: How soon after a cloud begins forming does rain begin to fall? At first glance, the answer should be simple. Couldn't a scientist with a stopwatch go outside and time a cloud from the moment it starts forming until rain begins?

But there's a little problem in getting a cloud to stay together. "There are a lot of reports of people who can make clouds go away by watching then," Charles

Knight jokes. "One cannot outsmart clouds. If you watch a little cloud, chances are it will grow a little and disappear." Most clouds aren't rainmakers.

Charles Knight is getting around that problem by combining cameras and radar when rain clouds are likely to form. The radar record and photos trace the history of clouds that are successful in producing rain.

Like most of today's atmospheric scientists, the Knights combine observations of the phenomena they're studying and computer models based on a combination of theory and observation.

In atmospheric science, physics and chemistry theory are translated into computer models.

"You come up with the theories," Charles Knight explains, "then you model them. But without the observations you don't know what you have to model. Without observations you don't know if there's a problem. Field observations are where I get my ideas. If I had to punch keys on a computer all day, I'd go stale pretty fast."

Questions about exactly how clouds form rain go back decades, but have received less attention recently because fewer scientists are focusing on clouds now than in the 1970s, when interest in weather modification ran high.

"A major problem can sit around without progress and most people lose interest," says Charles Knight. "A few stubborn people like myself are still interested."

snow begins, the ice won't freeze on the road and less salt will be needed. Knowing whether snow is likely to end in the next half-hour, or continue for another six, can help highway managers use resources effectively.

Highway superintendents aren't the only ones who need better forecasts of winter weather. Any transportation, school, government, or business official who has to decide whether to shut down when snow threatens needs better forecasts. As far as that goes, ordinary people trying to decide whether to stock up for a few days snowbound at home need help in making that decision.

The 161 new Doppler radars that National Weather Service, Defense Department and Federal Aviation Administration installed across the USA from 1988 to 1996 are beginning to help forecasters do a better job of predicting what snowstorms will do. The Doppler radars "see" rain and snow as the old radars did, but in finer detail. They also show wind speeds and directions. "We're seeing things we've never seen before," says Paul Kocin, a National Weather Service research meteorologist. "On the old radars, a snowstorm was a blob. Now we're seeing the banded structure."

These better pictures of snowstorms are helping forecasters do a much better job of saying where a storm will dump its heaviest snow and when the snow should begin and end in each part of a region. But Doppler radar doesn't help until the storm is arriving. Forecasts made a day or two before a storm arrives are based on computer models that use current data to calculate what the weather should do in the future.

New radars, better research, and improved computer models will bring more accurate winter storm forecasts for children looking forward to a "snow day," pilots flying into wintry clouds, maintenance supervisors worried about deploying snow plows, or those who don't like being surprised by six inches of "partly cloudy" in their driveways.

A snow fence blocks drifting snow along I-80 in Wyoming.

Effective snow fences have to store a winter's worth of blowing snow. Scientists determine how much snow is likely, enabling engineers to design good fences.

A 4-foot-high fence will store 4 tons of snow for each foot of fence length.

An 8-foot-high fence will store 18 tons per foot of fence length.

By early morning April 2, 1974, meteorologists at the National Weather Service knew big trouble was brewing for April 3.

Cold air was spreading snow in the Rockies, while to the southeast, air over the 75°F water of the Gulf of Mexico was growing more humid. Computer forecasts showed a low-pressure center would form just east of the Rockies and strengthen as it moved eastward.

The computer forecasts also showed that on April 3, jet stream winds would be blowing faster than 100 mph from Texas to New England. And forecasts showed winds around the growing low-pressure area would be doing two things: pushing dry air eastward across the Mississippi River and pulling warm, extremely humid air northward from the Gulf to the Ohio Valley.

It was a perfect setup for treacherous thunderstorms, the kind that spin out the strongest tornadoes.

Thunderstorms are dangerous in several ways. Tornadoes are their most feared product. Some thunderstorms blast the Earth with damaging straight-line winds or spew chunks of ice — hail — that can beat a field of corn into the ground. Thunderstorms are responsible for many flash floods, weather's biggest killer in the United States. And all thunderstorms generate lightning, the USA's second biggest weather killer.

The thunderstorms of April 3 and 4, 1974, hit the Plains with what became known as the Super Outbreak, 127 tornadoes — the largest, most damaging tornado outbreak in history. Winds in at least six of the tornadoes were faster than 261 miles per hour, making them Category 5 storms on the 1 to 5 Fujita Tornado Scale. That's about a normal decade's worth of Category 5 storms. Some of the April 3, 1974, tornadoes are among the strongest ever recorded.

Three hundred fifteen people in 11 states were

Lightning strikes northwest Tucson, Ariz. On the average, around 40 million lightning strokes hit the ground each year in the United States.

The most dangerous place to be when a tornado hits is in a mobile home. In the United States, tornadoes killed 286 people from 1990 through 1996. Of these, 112 people, more than a third, were killed in mobile homes.

A perfect setup for tornadoes

A strong extratropical cyclone brought warm, humid air into conflict with dry air and cool, humid air over the central United States the afternoon of April 3, 1974, triggering the strongest tornado outbreak ever recorded. The map shows the situation at 3 p.m. EST on April 3 as the worst part of the outbreak was beginning.

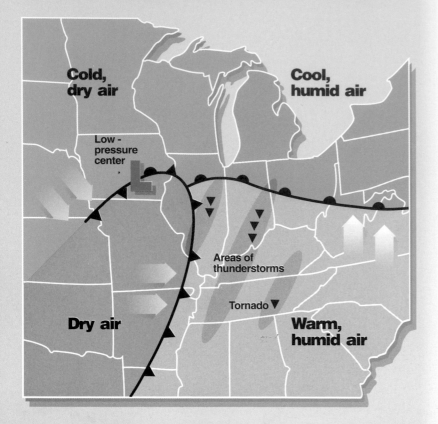

Cold, dry air

Cool, humid air

Low-pressure center

Areas of thunderstorms

Tornado ▼

Dry air

Warm, humid air

Paths of tornadoes April 3-4, 1974

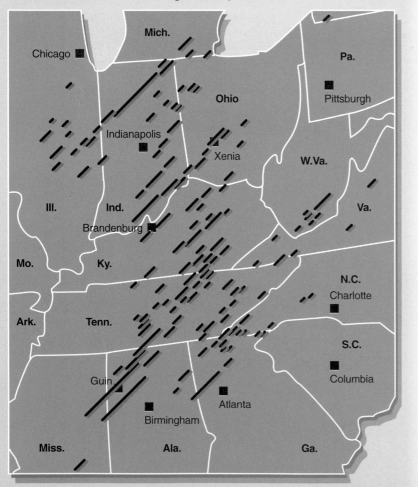

Mich.

Chicago

Pa.

Ohio

Pittsburgh

Indianapolis

Xenia

W.Va.

Ill.

Ind.

Va.

Brandenburg

Mo.

Ky.

N.C.

Charlotte

Ark.

Tenn.

S.C.

Columbia

Guin

Atlanta

Birmingham

Miss.

Ala.

Ga.

---

killed; 6,142 were injured. Without warnings, many more would have died. Total damage was more than $600 million. The Red Cross estimated that 27,590 families suffered some kind of loss.

The tornadoes began in Indiana around 9:30 on the morning of April 3, reaching a peak that afternoon and evening along lines from Alabama into Ohio. The Super Outbreak finally ended with pre-dawn twisters April 4 in West Virginia and Virginia.

Hardest hit was Xenia, Ohio, 16 miles east of Dayton, where a half-mile wide tornado smashed homes, businesses and buildings in the town and at the nearby campuses of Central State and Wilberforce universities.

In Xenia the tornado lifted freight cars of a passing train and threw them across streets. It demolished Central High School an hour after most students had left for the day. A teacher and students rehearsing for a play fled the auditorium seconds before the twister tossed two school buses through the wall and onto the stage.

Tornadoes, severe storms are local

While tornadoes like those of the Super Outbreak are extremely fierce, they are also small-scale events compared to extratropical cyclones and hurricanes.

The area directly threatened by the Super Outbreak tornadoes covered around 490,000 square miles. Yet tornadoes and other violent winds damaged only about 600 square miles inside that area. But, as Frederick Ostby, former director of the National Severe Storms Forecast Center, points out, everyone in the 490,000-square-mile area had an equal chance of being hit.

While conditions were right for fierce thunderstorms across the entire area, small-scale atmospheric events have to come together in exactly the right way to create a tornado. Even with the best technology and highest skills, forecasters have no way of knowing more than about 20 minutes ahead of time exactly where a tornado will form and where it will go.

This doesn't mean we're completely at the mercy of tornadoes.

First, only a little more than one percent of the tornadoes that hit the United States have winds stronger than 200 mph, putting them in Categories 4 and 5 on the Fujita scale (graphic on Page 127). Yet these twisters account for more than 70 percent of all tornado deaths. These large, most dangerous tornadoes are the ones that new forecasting technology can detect up to 20 minutes before they hit.

Second, people who know what to do have a good chance of staying safe, if they stay alert to warnings and then act quickly and correctly.

Basic building blocks of thunderstorms

To see how thunderstorms create tornadoes and other hazards, it's important to know how they work.

In Chapter 5, we saw how water takes up energy when it evaporates and then releases this energy — called latent heat — when it condenses into liquid. We also saw how this heat slows the cooling of rising air, making it more likely to continue rising. In such a case, we end up with warm air rising in some places and cool air descending in others. This process is called "convection." Thunderstorms are "convective storms."

Thunderstorms need unstable air, a temperature profile with warm air near the ground and cold air aloft. This explains why thunderstorms are more likely in the spring and summer than in the fall and winter. In spring and summer the sun warms the ground, which warms the air near the ground. Air near the ground also is warmed in the fall, but there's an important difference. In the spring the air aloft retains its winter cold; the air will be more unstable than in the fall when the air aloft retains its summer warmth.

The growth of thunderstorm knowledge

In his 1841 book, *The Philosophy of Storms*, J. P. Espy, a leading 19th century American scientist, presented the basic idea of how thunderstorms are powered by latent heat released as water vapor condenses.

The first attempts at scientific forecasting of tornadoes were begun in the 1880s by John P. Finley of the U.S. Army Signal Service — later the Signal Corps — which handled the nation's weather forecasting at the time. In 1884 and 1885, based on his weather observations, Finley produced generalized forecasts that stated tornadoes would be possible on certain days. Finley realized that in the central United States days much like those that brought the Super Outbreak were the most dangerous. In 1886, the Army discontinued the experiment because, it said, "the harm done by a [tornado] prediction would eventually be greater than that which results from the tornado itself." The authorities envisioned deaths from panic if people heard tornadoes were possible on a certain day. Tornado forecasts for the public didn't begin again until 1952.

As Finley's forecasts showed, scientists understood

A tornado touches down near Enid, Okla., in 1960.

what general conditions favor thunderstorms. But a detailed picture of thunderstorms didn't begin emerging until after World War II.

In 1946 and 1947, Horace R. Byers and R. R. Braham led the federally funded Thunderstorm Project that studied storms in Florida and Ohio. Their 1949 book, *The Thunderstorm*, outlined the basic picture of convective storms that researchers have been building on since. The drawings on the next page show the basic model of thunderstorms that Byers and Braham worked out.

By the 1950s, researchers understood that thunderstorms need a supply of warm, humid, unstable air and something to give that air an initial upward shove. Scientists understood that shove could come from heating of air near the ground, lifting of warm air by undercutting cold air — such as along a cold front — or wind pushing air up hills or mountains. But it was also known that thunderstorms sometimes occurred when none of the three known conditions supplying the upward shove was present.

Until recent years scientists puzzled over scattered thunderstorms that seemed to pop up at random. Forecasters still sometimes refer to "pop up" thunder-

Tornado: A strong, rotating column of air extending from the base of a cumulonimbus cloud to the ground.

Funnel cloud: A rotating column of air extending from a cloud but not reaching the ground.

Severe thunderstorm: A thunderstorm with winds 58 mph or faster or hailstones three-quarters of an inch or larger in diameter.

The birth, life and death of a thunderstorm

A single-cell thunderstorm, shown here, is the simplest kind of thunderstorm. The life cycle of other kinds are essentially the same. The cycle shown here would take approximately one hour.

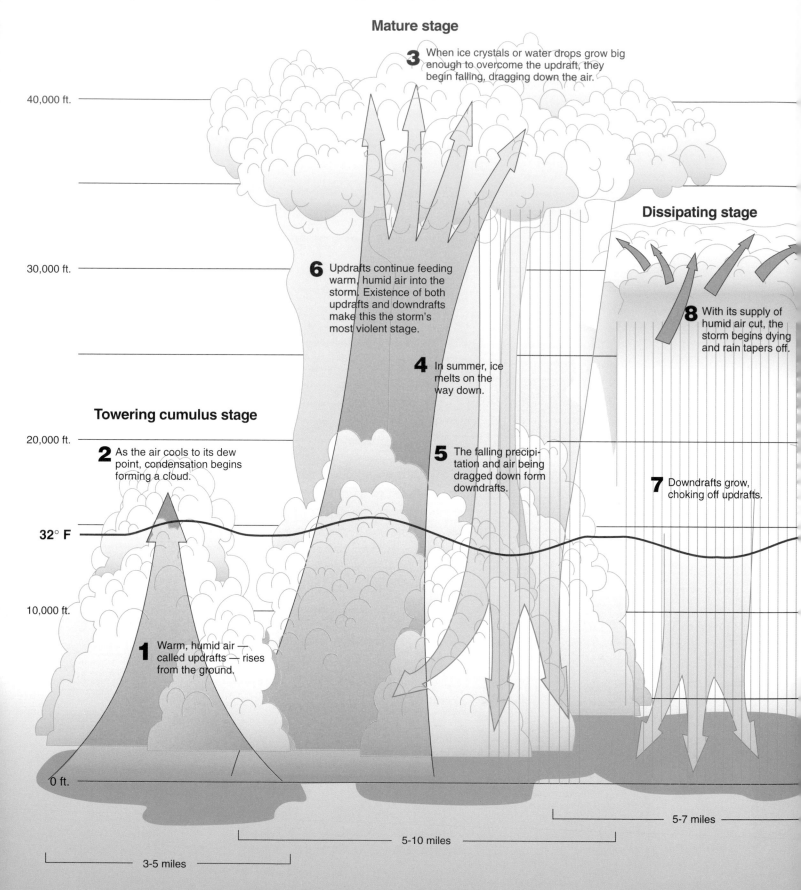

Mature stage

3 When ice crystals or water drops grow big enough to overcome the updraft, they begin falling, dragging down the air.

40,000 ft.

30,000 ft.

Dissipating stage

6 Updrafts continue feeding warm, humid air into the storm. Existence of both updrafts and downdrafts make this the storm's most violent stage.

8 With its supply of humid air cut, the storm begins dying and rain tapers off.

4 In summer, ice melts on the way down.

Towering cumulus stage

20,000 ft.

2 As the air cools to its dew point, condensation begins forming a cloud.

5 The falling precipitation and air being dragged down form downdrafts.

7 Downdrafts grow, choking off updrafts.

32° F

10,000 ft.

1 Warm, humid air — called updrafts — rises from the ground.

0 ft.

5-7 miles

5-10 miles

3-5 miles

storms. Observations during the 1980s took most of the mystery out of these storms. Doppler radar and other technologies showed that earlier, less-detailed observations had simply missed some faint air boundaries that can get thunderstorms started.

These boundaries can occur where air flowing from one thunderstorm meets air from another, maybe more than 100 miles away from either original storm. If an upper-level disturbance moves over the boundary, it can help trigger thunderstorms.

In Texas, and sometimes farther to the north on the Plains, thunderstorms often are triggered on the "dryline." This is the boundary between warm but dry air moving east, and downhill, from the Southwest deserts, and warm, humid Gulf of Mexico air moving northwest.

The humid-dry air boundary triggers thunderstorms, often severe storms. Researchers are still trying to discover exactly how the dryline causes severe thunderstorms.

Our understanding of thunderstorms began growing during the 1950s and 1960s as radar became a regular tool for National Weather Service forecasters. Researchers also looked into how the dynamics of the upper air could affect small-scale events such as thunderstorms.

Another key to understanding thunderstorms was the establishment in 1964 of the National Severe Storms Laboratory in Norman, Okla. The lab became a center for Doppler radar observations of thunderstorms. In 1972, the lab and the nearby University of Oklahoma began sending teams of meteorologists and students out to "intercept" storms. These storm chasers gathered an impressive collection of instrument observations as well as still photos and films in and around severe thunderstorms.

On-the-scene photos and data collected by the

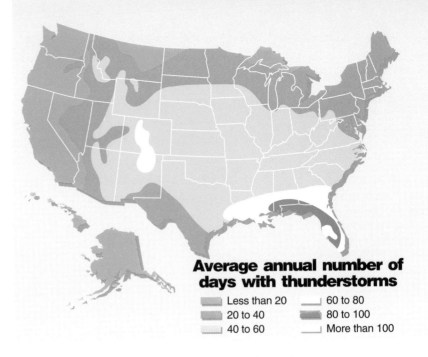

Average annual number of days with thunderstorms

- Less than 20
- 20 to 40
- 40 to 60
- 60 to 80
- 80 to 100
- More than 100

chasers were vital for the scientists trying to understand the complex patterns they were seeing on their Doppler radars. Over the years scientists put other sophisticated instruments to work observing thunderstorms from both the ground and airplanes. The resulting flood of data, combined with computer modeling, has led to today's picture of thunderstorms and tornadoes.

More complicated kinds of storms

The single-cell thunderstorm shown in the drawing on the opposite page is relatively uncommon, but it is the basic building block of more complex storms.

Multicell storms, which are clusters of single-cell storms, are the most common kind of thunderstorms. Air flowing outward from one storm can supply the upward push of warm, humid air needed to trigger other cells.

Often we hear of "air mass" thunderstorms. These

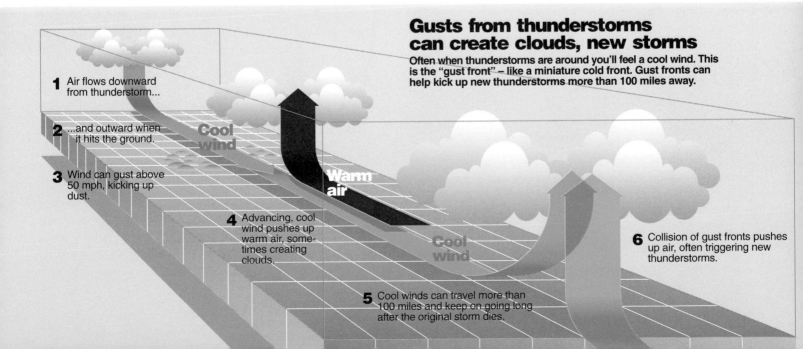

Gusts from thunderstorms can create clouds, new storms

Often when thunderstorms are around you'll feel a cool wind. This is the "gust front" – like a miniature cold front. Gust fronts can help kick up new thunderstorms more than 100 miles away.

1 Air flows downward from thunderstorm...

2 ...and outward when it hits the ground.

3 Wind can gust above 50 mph, kicking up dust.

4 Advancing, cool wind pushes up warm air, sometimes creating clouds.

5 Cool winds can travel more than 100 miles and keep on going long after the original storm dies.

6 Collision of gust fronts pushes up air, often triggering new thunderstorms.

Cool wind

Warm air

Cool wind

are storms that grow during the afternoon on warm, humid days when no fronts or large-scale storm systems are around. Such storms can be single-cell or multicell.

While single-cell or scattered multicell storms aren't the parents of strong tornadoes, they have other dangers. Lightning is always a thunderstorm hazard; small storms that produce only one or two lightning bolts have killed people. As we'll see later, small thunderstorms also can produce deadly winds called "downbursts."

Conditions needed for killer storms

Fierce thunderstorms, ones that cause the strongest winds and largest hail, require special conditions. This is why such storms are relatively rare and also why they occur more often in the United States than anywhere else.

These conditions include:

• Unstable air that will continue rising when it's forced upward.

• A disturbance in the upper-air flow, such as a short wave as described in Chapter 4.

• A layer of stable air, especially an inversion, atop warm, humid air near the ground. This holds back convection as air near the ground becomes hotter during the day. Otherwise, fair weather clouds or even thunderstorms could grow early in the day, using the available energy. An inversion is like a lid on a boiling pot. Eventually, when enough steam builds up, the lid blows off. When the hot, humid air below an inversion finally breaks through, large thunderstorms explode upward, growing to 50,000 feet in minutes.

• A fresh supply of warm, humid air flowing in near the ground. High-altitude cold air flowing in also helps.

• Wind speeds that increase with altitude. Fast-moving, high-altitude winds interacting with the updraft promote rotation in the storm, which helps keep the storm going. These winds also make thunderstorms tilt, rather than go straight up. This adds to their strength. A tilted thunderstorm means precipitation from high in the storm doesn't fall back into the updraft, choking it. Instead, precipitation falls into dry air, which adds to a storm's energy.

Precipitation falling into cool, dry air is a feature of the strongest kinds of thunderstorms: squall-line storms and supercells.

As we saw in Chapter 5, when water evaporates it cools the surroundings. Precipitation evaporating in dry

The National Weather Service credits improved technology, better warnings and increased awareness with the reduction in USA tornado deaths. Number of tornado deaths by decade, 1930-96.

1930-39:	1,945
1940-49:	1,786
1950-59:	1,419
1960-69:	945
1970-79:	998*
1980-89:	522
1990-96:	286

*Decade of the Super Outbreak

Storm chasers deliberately put themselves near a twister's path

At first glance, deliberately seeking out tornadoes seems like a foolish thing to do.

Howard Bluestein disagrees: "I wouldn't do it, if it were dangerous. We believe we understand the structure of storms and we do not put ourselves in any danger of being hit by a tornado. The biggest danger is from driving on narrow country roads. The other danger is lightning."

Since 1977, Bluestein has led University of Oklahoma storm chasers trying to help answer the question: Why do some thunderstorms produce tornadoes while others do not?

It has been a safe enterprise. Scientists and their students who seek out thunderstorms likely to produce tornadoes "look at storms as a doctor looks at a patient. Things that are not obvious to the average person are clear to those who know what to look for."

Howard Bluestein, tornado researcher

Such knowledge tells the chasers what a storm is doing and where it's likely to travel. They can approach closely from safe directions.

Storm chasers' roles have evolved since the early 1970s. Then the job was to correlate what a storm was doing with what researchers were seeing on Doppler radar at the National Severe Storms Laboratory across town from the university.

From 1981 through 1983, the storm chasers deployed the Totable Tornado Observatory, TOTO, in a twister's path. TOTO, which not coincidentally is the name of Dorothy's dog in *The Wizard of Oz*, was a 400-pound package of instruments the chasers hoped would stand up to a twister while measuring what was going on.

Chasers managed to place TOTO close to twisters and obtained some useful information, Bluestein says, but never saw how it would work in the heart of a twister. "We gave up on it; we decided it was much too difficult to get in the path."

Cross-section of a squall line

A squall line is made of several thunderstorms. It can be more than 100 miles long. Squall lines are often 50 to 150 miles ahead of an advancing cold front.

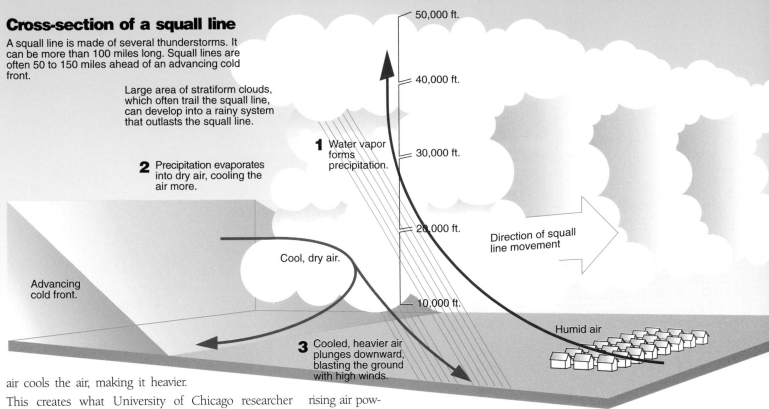

Large area of stratiform clouds, which often trail the squall line, can develop into a rainy system that outlasts the squall line.

2 Precipitation evaporates into dry air, cooling the air more.

50,000 ft.

40,000 ft.

30,000 ft.

1 Water vapor forms precipitation.

20,000 ft.

Direction of squall line movement

Cool, dry air.

Advancing cold front.

10,000 ft.

Humid air

3 Cooled, heavier air plunges downward, blasting the ground with high winds.

air cools the air, making it heavier. This creates what University of Chicago researcher Theodore Fujita calls a "cold air balloon" that plunges downward. Heat released by condensing water vapor in rising air powers all thunderstorms. The heat taken up by evaporating liquid water adds to the power of squall-line and

---

Today the emphasis is on getting more precise meteorological readings in and around thunderstorms. In 1984, the chasers began releasing weather balloons around storms. Since 1987, Bluestein's team has been using a portable Doppler radar for precise measurements that can't be obtained by large radars miles away.

Measurements and photos by chasers over the years have helped forecasters and those trying to build computer models of storms. They've also helped improve the training of tornado spotters who look for signs of storms when a watch goes out.

Bluestein first became interested in tornadoes at age five when a tornado hit Worcester, Mass., about 40 miles from his home. He was also interested in astronomy and electronics, interested enough to build his own radios and become a licensed amateur radio operator when he was 12.

At Massachusetts Institute of Technology,

University of Oklahoma graduate students probe a tornado near Hodges, Okla., with a portable Doppler radar May 13, 1989.

he majored in electrical engineering, but began considering a career in meteorology in his senior year. Bluestein credits Professor Fred Sanders at MIT with his career choice. "He was a good role model. He was inspiring; he was extremely enthusiastic about what he was doing and he seemed to really enjoy it."

Bluestein joined the University of Oklahoma meteorology department in 1976. "I was reluctant to go to Norman (Oklahoma) and I intended to stay for a year," he says. But the pull of research was more powerful than the pull of home. So Bluestein is still in Norman, conducting research, teaching and leading teams of storm chasers.

Although "I never put notches in the bed post," Bluestein estimates he's seen "on the order of a hundred or more tornadoes. They've never been scary. I consider myself a naturalist," Bluestein says. "Nature is incredibly powerful. Tornadoes are incredibly powerful and beautiful."

Most thunderstorms are 'multicell clusters'

The most common kind of thunderstorms are called "multicell cluster" storms because they're made of several "cells" with some of the cells dying while others grow. The graphic here shows what's going on inside one of these storms, which would look like a single boiling cloud to someone watching it from miles away. As individual cells are born, grow and die, the storm produces bursts of rain, wind and maybe hail.

Inside a multicell thunderstorm

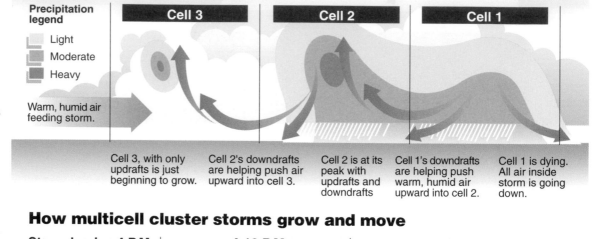

Precipitation legend
- Light
- Moderate
- Heavy

Warm, humid air feeding storm.

Cell 3 | **Cell 2** | **Cell 1**

Cell 3, with only updrafts is just beginning to grow.

Cell 2's downdrafts are helping push air upward into cell 3.

Cell 2 is at its peak with updrafts and downdrafts

Cell 1's downdrafts are helping push warm, humid air upward into cell 2.

Cell 1 is dying. All air inside storm is going down.

How multicell cluster storms grow and move

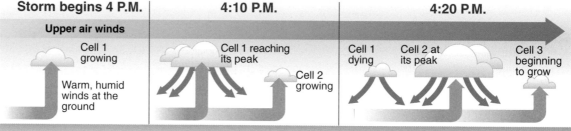

Storm begins 4 P.M. | **4:10 P.M.** | **4:20 P.M.**

Upper air winds

Cell 1 growing

Warm, humid winds at the ground

Cell 1 reaching its peak

Cell 2 growing

Cell 1 dying

Cell 2 at its peak

Cell 3 beginning to grow

Severe thunderstorms sometimes are reported as being green. Researchers have confirmed that some storms are green, but they haven't pinned down why. One theory is that hail causes the green color, but researchers have seen green storms that didn't produce hail. And many severe storms don't look green. To be safe, don't count on a thunderstorm turning green before you run for shelter.

---

supercell thunderstorms by sending cooled air hurtling downward.

Squall-line thunderstorms

In simple terms, squall lines are lines of thunderstorms. But they're more than that. They are an example of what meteorologists call a "mesoscale convective system."

"Mesoscale" refers to weather systems up to about 250 miles across. "Convective" means thunderstorms are involved, and "system" means the whole thing is organized. Squall lines are found in both the tropics and in the middle latitudes. We'll look at the middle-latitude squall lines such as those common across the USA east of the Rockies.

Mid-latitude squall lines produce strong, straight-line winds, hail and sometimes tornadoes. They can be especially disruptive to aviation because the individual thunderstorms in a squall line can top out above 50,000 feet, higher than the 30,000 to 40,000-foot cruising altitude of jet airliners. Since pilots don't want to fly through such storms, a squall line can force airplanes to take detours of 200 miles or more.

Squall lines can form along cold fronts, but the strongest and most common lines form as much as 100 or more miles ahead of an advancing cold front in the warm sector of an extratropical cyclone. Such lines are known as "prefrontal squall lines."

"We really don't know why prefrontal squall lines are the most prevalent," says David Jorgensen, head of NOAA's Mesoscale Research and Applications Division in Boulder, Colo. "This is an area of active research." Possibly they are triggered by solitary waves as shown on Page 99. Upper-air disturbances could also play a role.

In addition to the thunderstorms, squall lines often trail a large area of flat, stratus clouds that produce steady rain for hours after the thunderstorms and their heavier rain have moved on.

Tornadoes: concentrated violence

Large tornadoes stir up the fastest winds ever found at the surface of the Earth, believed to be as fast as 300 mph in rare cases. The smallest tornado might last only a few seconds, travel only a few yards and create winds no stronger than 50 mph. Rare, monster twisters have lasted for hours, have been a half mile wide, traveled more than 200 miles and created 300 mph winds.

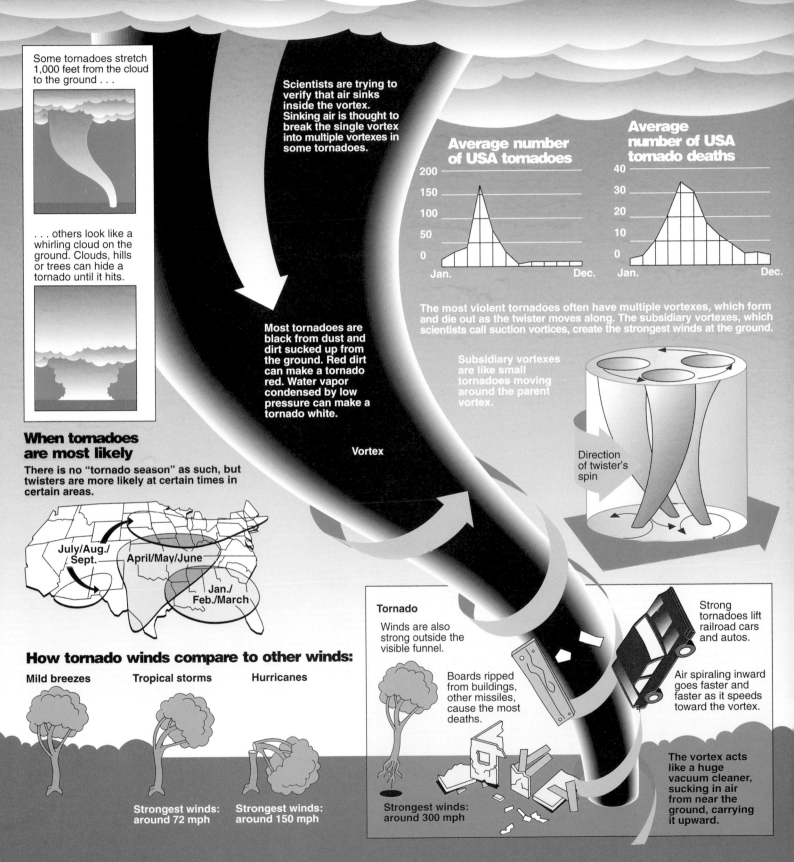

Some tornadoes stretch 1,000 feet from the cloud to the ground . . .

. . . others look like a whirling cloud on the ground. Clouds, hills or trees can hide a tornado until it hits.

Scientists are trying to verify that air sinks inside the vortex. Sinking air is thought to break the single vortex into multiple vortexes in some tornadoes.

Most tornadoes are black from dust and dirt sucked up from the ground. Red dirt can make a tornado red. Water vapor condensed by low pressure can make a tornado white.

Vortex

Average number of USA tornadoes

200
150
100
50
0
Jan.   Dec.

Average number of USA tornado deaths

40
30
20
10
0
Jan.   Dec.

The most violent tornadoes often have multiple vortexes, which form and die out as the twister moves along. The subsidiary vortexes, which scientists call suction vortexes, create the strongest winds at the ground.

Subsidiary vortexes are like small tornadoes moving around the parent vortex.

Direction of twister's spin

When tornadoes are most likely

There is no "tornado season" as such, but twisters are more likely at certain times in certain areas.

July/Aug./Sept.

April/May/June

Jan./Feb./March

How tornado winds compare to other winds:

Mild breezes

Tropical storms

Strongest winds: around 72 mph

Hurricanes

Strongest winds: around 150 mph

Tornado

Winds are also strong outside the visible funnel.

Boards ripped from buildings, other missiles, cause the most deaths.

Strongest winds: around 300 mph

Strong tornadoes lift railroad cars and autos.

Air spiraling inward goes faster and faster as it speeds toward the vortex.

The vortex acts like a huge vacuum cleaner, sucking in air from near the ground, carrying it upward.

123

A SUPERCELL: THE KING OF THUNDERSTORMS

Ordinary thunderstorms are born, reach their peaks and die within an hour or so. Supercell thunderstorms can last for hours. Supercells — sometimes called steady-state thunderstorms — are the most organized kind of thunderstorm and the most dangerous because they spawn the strongest tornadoes. During their lifetimes, supercells can travel more than 300 miles across the countryside, spawning one large tornado after another as well as bringing heavy rain and hail.

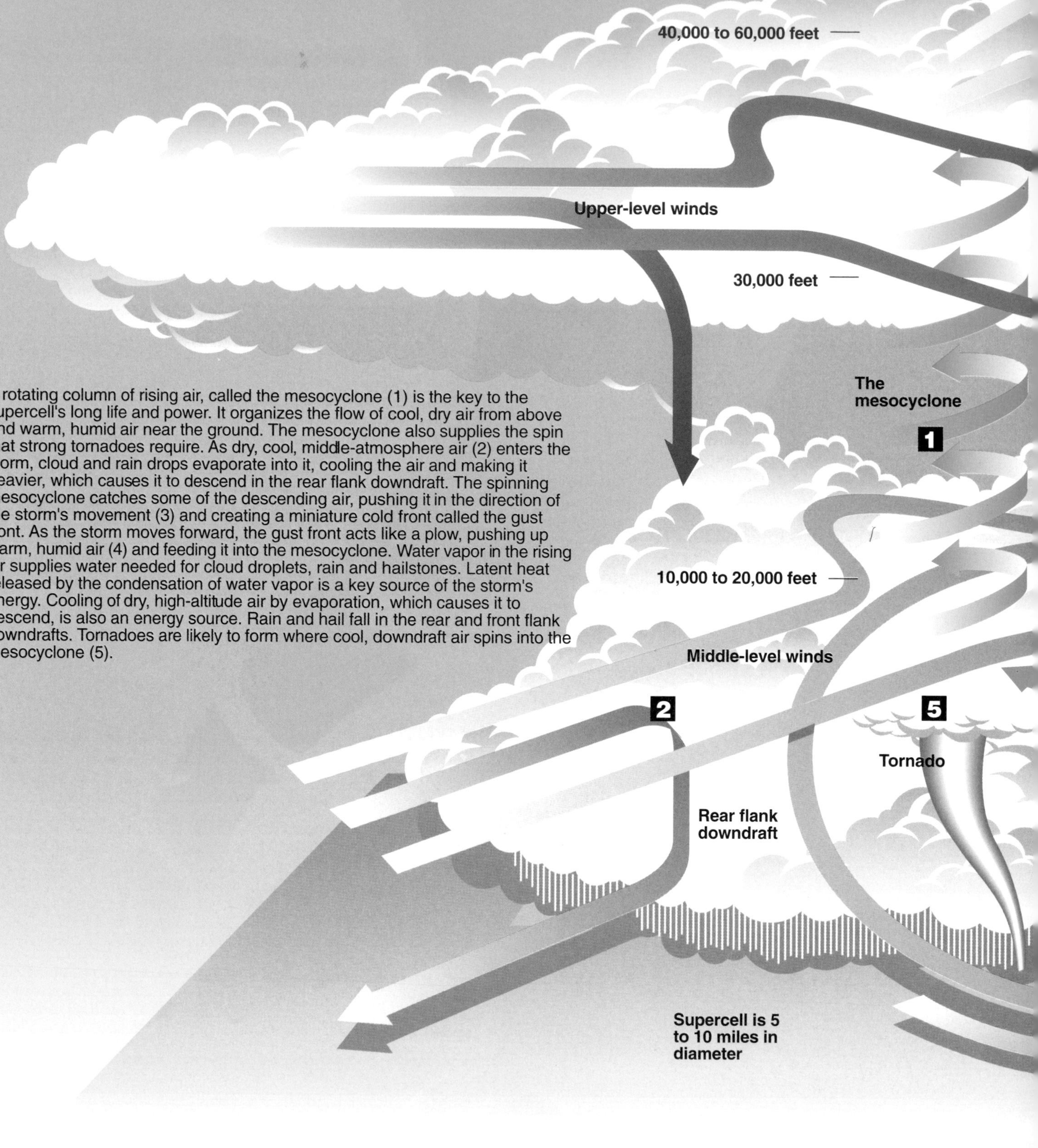

40,000 to 60,000 feet ——

Upper-level winds

30,000 feet ——

The mesocyclone

1

A rotating column of rising air, called the mesocyclone (1) is the key to the supercell's long life and power. It organizes the flow of cool, dry air from above and warm, humid air near the ground. The mesocyclone also supplies the spin that strong tornadoes require. As dry, cool, middle-atmosphere air (2) enters the storm, cloud and rain drops evaporate into it, cooling the air and making it heavier, which causes it to descend in the rear flank downdraft. The spinning mesocyclone catches some of the descending air, pushing it in the direction of the storm's movement (3) and creating a miniature cold front called the gust front. As the storm moves forward, the gust front acts like a plow, pushing up warm, humid air (4) and feeding it into the mesocyclone. Water vapor in the rising air supplies water needed for cloud droplets, rain and hailstones. Latent heat released by the condensation of water vapor is a key source of the storm's energy. Cooling of dry, high-altitude air by evaporation, which causes it to descend, is also an energy source. Rain and hail fall in the rear and front flank downdrafts. Tornadoes are likely to form where cool, downdraft air spins into the mesocyclone (5).

10,000 to 20,000 feet ——

Middle-level winds

2

5

Tornado

Rear flank downdraft

Supercell is 5 to 10 miles in diameter

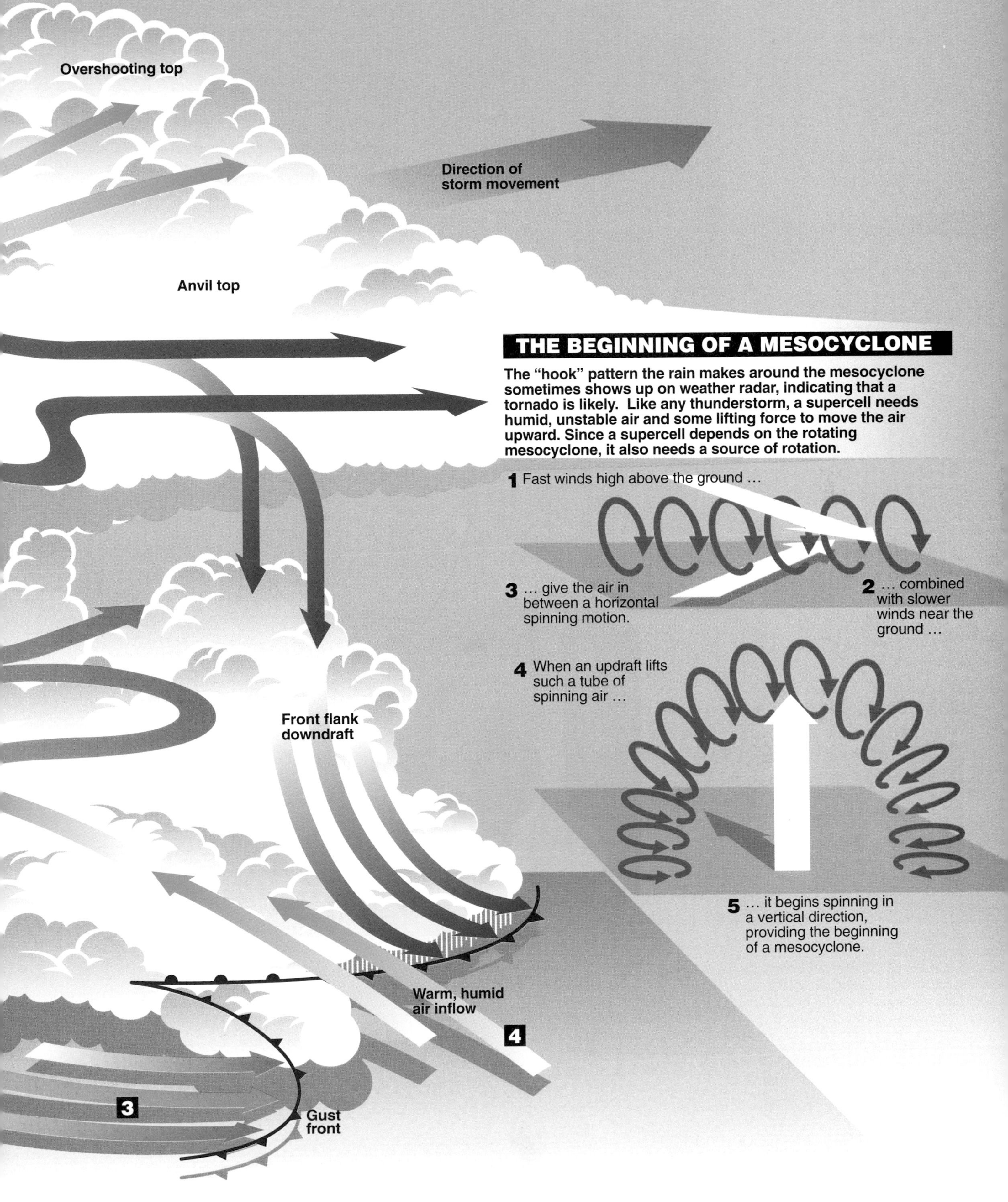

Overshooting top

Anvil top

**Direction of
storm movement**

THE BEGINNING OF A MESOCYCLONE

The "hook" pattern the rain makes around the mesocyclone sometimes shows up on weather radar, indicating that a tornado is likely. Like any thunderstorm, a supercell needs humid, unstable air and some lifting force to move the air upward. Since a supercell depends on the rotating mesocyclone, it also needs a source of rotation.

1 Fast winds high above the ground …

3 … give the air in between a horizontal spinning motion.

2 … combined with slower winds near the ground …

4 When an updraft lifts such a tube of spinning air …

5 … it begins spinning in a vertical direction, providing the beginning of a mesocyclone.

Front flank downdraft

Warm, humid air inflow

4

3

Gust front

"Heat lightning" is ordinary lightning too far away for its thunder to be heard. Instead of individual strokes, heat lightning often illuminates an entire cloud as cloud droplets diffuse the light.

Aerial view of a waterspout about 2,000 feet high spotted 14 miles north of Key West, Fla.

Since the early 1980s, researchers have been interested in the organized systems of clouds and rain that sometimes develop in the stratiform clouds formed by squall lines. These clusters of clouds take on a circulation of their own, which outlasts the parent squall line. These clusters, known as "mesoscale convective complexes" — a kind of mesoscale convective system — supply about 80 percent of the growing-season rain on the Plains and across the Midwestern Corn Belt.

Supercells: The kings of thunderstorms

When conditions are right, a large thunderstorm will form either by itself or on the southwestern end of a squall line. Unlike a normal multicell thunderstorm that might last 40 minutes, this storm will last several hours. It can travel 200 miles or more, often spinning out a series of strong tornadoes. This is a "supercell." It is an appropriate name.

As shown in the graphic on Pages 124-125, an extraordinarily complicated interplay of atmospheric forces creates supercells. The key to the storm's long life and power is a tilted, rotating column of air.

Middle-atmosphere winds flowing around this barrel-shaped "mesocyclone" send cool downdraft air into the forward and rear parts of the storm instead of into its

How a journey to Hiroshima led to the discovery of downbursts

While T. Theodore Fujita is best known as the inventor of the tornado damage scale that carries his name, his place in the history of science was assured by his discovery of downbursts.

His 1974 finding was based on things he first saw in 1945. Fujita, then a 24-year-old assistant professor of physics at Japan's Meiji College of Technology, visited Nagasaki and Hiroshima to study the nuclear blasts that had destroyed the two cities. He calculated how high above the ground the bombs had to explode to create their unique starburst damage patterns.

Twenty-nine years later, while flying over areas raked by the April 3 and 4, 1974 Super Tornado Outbreak, he saw damage patterns that sent his thoughts back to 1945.

"If something comes down from the sky and hits the ground it will spread out . . . it will produce the same kind of outburst effect

Theodore Fujita with a tornado generator

that was in the back of my mind from 1945 to 1974," he says.

At that time meteorologists knew thun-derstorms produced downdrafts, but they assumed these didn't hit the ground hard enough to cause damage. Fujita called Horace R. Byers — "my mentor" — to describe what he had seen and what conclusions he had drawn. Byers helped Fujita coin the term "downburst" and encouraged him to publish his theory.

Byers, who with R. R. Braham conducted the groundbreaking Thunderstorm Project in 1946 and 1947, is a key figure in Fujita's life. In 1949, Fujita came across an article by Byers on thunderstorms. He translated two of his own thunderstorm papers into English, sent them to Byers and began a correspondence that led to Byers inviting Fujita to the University of Chicago in 1953.

Initially, few scientists accepted Fujita's downburst theory. But with the aid of the National Center for Atmospheric Research, Fujita set up a research project near Chicago

center. The storm's organization keeps the downdrafts from choking off the updraft and ensures that the flow of warm, humid air into the storm — a power source — continues for hours. Evaporation of precipitation into dry air also adds to the storm's power as it does in squall-line thunderstorms.

A close look at tornadoes

While tornadoes come in several varieties, here we'll classify them into two general kinds:

The first kind of tornado forms in the outflow of air from a thunderstorm. It has a shallow, localized vortex resembling a narrow, rope-like tube. Appearing barely attached to the parent storm cloud, these tornadoes usually are weak, with winds rarely exceeding 150 mph, F-2 on the Fujita scale, and normally last less than five minutes. They spin up like swirls of water running past obstacles in a fast-moving stream.

"Gustnadoes," waterspouts, and "landspouts" are all this first kind of tornado. A gustnado is a vortex whipped up by intense, small-scale wind shear along the outflow boundary, or "gust front," of a thunderstorm;

Downbursts are a major thunderstorm hazard

Downbursts are any winds blasting down from a thunderstorm; the most dangerous kind is a microburst in which the damaging winds cover an area less than 2.5 miles across. At times, they can produce winds of 150 mph or faster.

What causes microbursts

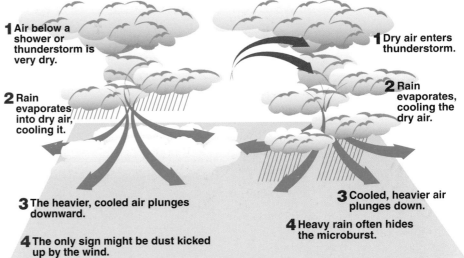

DRY MICROBURST

1 Air below a shower or thunderstorm is very dry.

2 Rain evaporates into dry air, cooling it.

3 The heavier, cooled air plunges downward.

4 The only sign might be dust kicked up by the wind.

WET MICROBURST

1 Dry air enters thunderstorm.

2 Rain evaporates, cooling the dry air.

3 Cooled, heavier air plunges down.

4 Heavy rain often hides the microburst.

thus the name "gust-nado."

At one time a waterspout was defined as a "tornado over water," a definition you'll find in some out-of-date

in 1978 that detected 50 downbursts in 42 days.

Today, other scientists talk about Fujita's observational genius and his ability to visualize quickly what the atmosphere is doing.

Fujita says doubts about communicating in English when he first came to the United States strengthened his ability to communicate with sketches and drawings.

He began his aerial surveys of tornado damage with flights after the 1965 Palm Sunday Outbreak and over the years has traveled more than 40,000 miles in small airplanes taking photos of wind storm damage.

Fujita says few of his graduate students enjoyed these studies, which involved peering through a camera viewfinder from a small plane in a steep bank, maybe bouncing around in turbulence. They didn't share one of his vital attributes: "I never get airsick."

Ground and aerial surveys of storm damage led Fujita to develop his tornado scale in the late 1960s. Until then records listed only

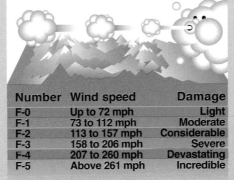

Fujita Tornado Scale

Number	Wind speed	Damage
F-0	Up to 72 mph	Light
F-1	73 to 112 mph	Moderate
F-2	113 to 157 mph	Considerable
F-3	158 to 206 mph	Severe
F-4	207 to 260 mph	Devastating
F-5	Above 261 mph	Incredible

total numbers of tornadoes with no objective way of listing their strength. Fujita's scale correlates damage with wind speeds.

Now those who survey thunderstorm damage try to determine whether it was caused by a tornado or a downburst as well as ranking it on the one to five scale. Damage up to F-3 on the scale is common from downbursts, Fujita says. He's also sure that much damage attributed to tornadoes in the past was really from downbursts. "After I pointed out the existence of downbursts," he says,

"the number of tornadoes [listed] in the United States decreased for a number of years."

Even though he began studying tornadoes soon after coming to the United States in 1953, and was called "Mr. Tornado" in a 1972 *National Geographic* magazine article, Fujita didn't see a tornado until June 12, 1982, "a date I'll remember after I forget my birthday."

This was during a National Center for Atmospheric Research project in the Denver area. He was at one of the project's three Doppler radar stations "when I noticed a tornado maybe was coming down. I told all the radars to scan that area. My first sighting of a tornado was one with the best tornado data ever collected."

Fujita has retired from teaching, but still has a laboratory at the University of Chicago. "When people ask me what my hobby is, I tell them it's my research," he says. "I want to spend the rest of my life in air safety and public safety, protecting people against the wind."

Open windows don't save houses

If a tornado ever threatens your house, don't run around opening windows. It wastes time you should be using to take shelter. Open windows aren't needed to keep unequal air pressure from making the house explode as once thought. Tornado winds, not unequal pressure, destroy buildings.

1 Air pressure inside tornado can be 10 percent lower than outside . . .

2 . . . but houses have openings other than windows that will relieve pressure differences.

3 Winds as low as 60 mph can lift roofs that aren't well attached . . .

4 Flying debris often breaks windows, allowing wind inside to push up on the roof and out on the walls.

5 If wind rips off the roof, the walls often fall outward, leading to the mistaken impression that air pressure had 'exploded' the building.

books. But as researchers studied waterspouts in places such as the Florida Keys, they realized that they weren't like the large tornadoes that do so much damage on the Plains.

For one thing, while the large tornadoes come from thunderstorms, waterspouts commonly come from rapidly-growing cumulus clouds that have not become thunderstorms and often never grow into thunderstorms.

Waterspouts grow by stretching an existing vortex, air that's whirling as it rises into the growing cloud. The American Meteorological Society's *Glossary of Weather and Climate* defines a waterspout as: "Usually a tornado-like rotating column of air (whirlwind) under a parent cumuliform cloud occurring over water; waterspouts are most common over tropical and subtropical waters and tend to dissipate upon reaching shore."

As scientists learned more about waterspouts, they realized that some tornadoes over land had more in common with them than with the strong tornadoes that do the most damage. These twisters were given the name "landspouts."

The second kind of tornado is the strongest. These storms are found where air is flowing into a thunderstorm, in the updraft area. Doppler radar measurements show that such tornadoes begin at the storm's middle levels and then grow both up into the storm and down toward the ground. Often appearing as a broad cylinder or cone sometimes a mile wide, their winds can reach 200 to 300 mph, F-4 to F-5 on the Fujita scale.

As with many weather phenomena, the dividing lines between the different kinds of tornadoes aren't always clear-cut. Even an expert might have a hard time using a particular twister's appearance alone to say whether it was a landspout or the second kind of tornado. The appearance of the cloud it's attached to might make it possible to tell which kind of twister it is, but a firm answer might require radar images of the storm.

Dust devils are yet another kind of whirlwind. Unlike waterspouts, landspouts, gustnadoes and large

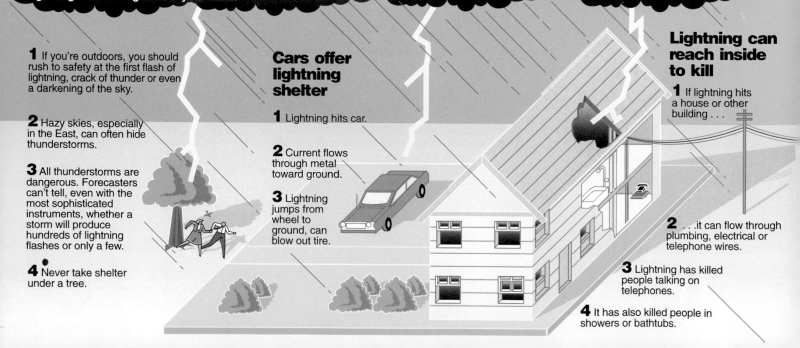

Ways to avoid becoming a victim of lightning

In the United States lightning is the second biggest weather killer, next to floods. Most victims are struck in the open, such as on beaches or golf courses, or when they take shelter from the rain under a tree. But lightning can be dangerous in your home.

1 If you're outdoors, you should rush to safety at the first flash of lightning, crack of thunder or even a darkening of the sky.

2 Hazy skies, especially in the East, can often hide thunderstorms.

3 All thunderstorms are dangerous. Forecasters can't tell, even with the most sophisticated instruments, whether a storm will produce hundreds of lightning flashes or only a few.

4 Never take shelter under a tree.

Cars offer lightning shelter

1 Lightning hits car.

2 Current flows through metal toward ground.

3 Lightning jumps from wheel to ground, can blow out tire.

Lightning can reach inside to kill

1 If lightning hits a house or other building . . .

2 . . .it can flow through plumbing, electrical or telephone wires.

3 Lightning has killed people talking on telephones.

4 It has also killed people in showers or bathtubs.

tornadoes, dust devils are not attached to a cloud. Instead, the whirling wind rises and disappears into clear air. They are most common in deserts, but they are seen in other locations. Wind speeds rarely are stronger than 72 mph and dust devils rarely cause any damage.

How tornadoes get their twist

The discovery of the large, spinning mesocylones in supercell thunderstorms was a breakthrough, but it didn't tell scientists what starts tornadoes spinning.

The onset of rotation under the supercell is called "tornadogenesis." Ask scientists what happens during tornadogenesis and you're likely to get several different responses.

In 1986, Richard Rotunno of the National Center for Atmospheric Research wrote, "No single theory explains all the commonly observed features" of tornadoes and supercells.

Even as 1990s technology gave scientists better data on tornadoes, answers about the source, or sources, of rotation remained unclear.

Scientific thinking suggests there is a connection between updraft air in the storm's core and downdraft air behind the storm. Temperature and windshift boundaries generated by the interaction of a warm updraft and cool downdraft seem to be important, but researchers aren't sure why or how.

Another question: Do tornadoes spin up from the ground or spin down from the rotating cloud? The physical evidence supports both possibilities, further complicating the picture. But two years of scientific storm chasing on the Plains may have given researchers their best clues yet.

While scientists are working out the details of how tornadoes acquire their spinning motion, they have rejected old ideas that it is created by the Coriolis Effect (explained in Chapter 3) acting on inflowing winds. In theory, a tornado could get its spin this way, but it would take more time than some tornadoes seem to have.

Also, if the Coriolis Effect were the only source of tornado rotation, all Northern Hemisphere tornadoes would spin counterclockwise.

Most Northern Hemisphere tornadoes do spin counterclockwise, but clockwise-spinning twisters have been photographed.

Chasing tornadoes for science

To understand how tornadoes work — the first step toward doing a better job forecasting them — scientists

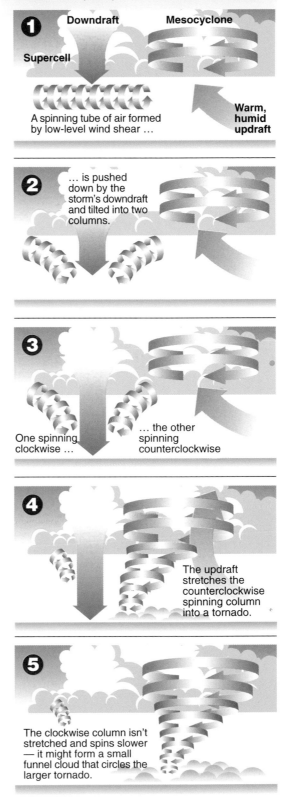

How tornadoes form

Tornado researchers think a supercell's warm updraft and cool downdraft are linked to the formation of the strongest tornadoes. In the same manner a tube of spinning air is lifted into the storm to form the mesocyclone, a small part of that tube is pushed and tilted into a vertical column under the storm. The exact method is still unclear, but when the two rotating columns of air connect, a tornado often spins up.

1 Downdraft   Mesocyclone

Supercell

A spinning tube of air formed by low-level wind shear …

Warm, humid updraft

2 … is pushed down by the storm's downdraft and tilted into two columns.

3 One spinning clockwise …   … the other spinning counterclockwise

4 The updraft stretches the counterclockwise spinning column into a tornado.

5 The clockwise column isn't stretched and spins slower — it might form a small funnel cloud that circles the larger tornado.

HOW CLOUDS CREATE LIGHTNING AND THUNDER

Lightning streaks within clouds, between clouds and even from clouds to clear air. But the flashes that concern us most are those that go from clouds to the ground because they kill people, start fires, splinter trees and knock out electrical power.

ELECTRICAL CHARGE IN A TYPICAL THUNDERSTORM

Lightning occurs when electricity travels between areas of opposite electrical charge within a cloud, between clouds or from a cloud to the ground. Scientists still are trying to discover exactly how positive and negative charges build in different parts of a cloud. This drawing is a simplified version of charges in a typical thunderstorm.

Thin area of negative charge at cloud top.

Areas of positive charge high in the cloud.

Strong area of negative charge in region of temperatures around 5°F; containing water vapor, liquid water and ice.

Thin area of positive charge at bottom of cloud.

Positive charge builds up on ground below the cloud. It stays under the cloud as the cloud moves.

WHY THUNDER RUMBLES

When a lightning bolt flashes through the sky we see it instantly. Thunder — the sound that lightning creates — takes a few seconds longer to reach us.

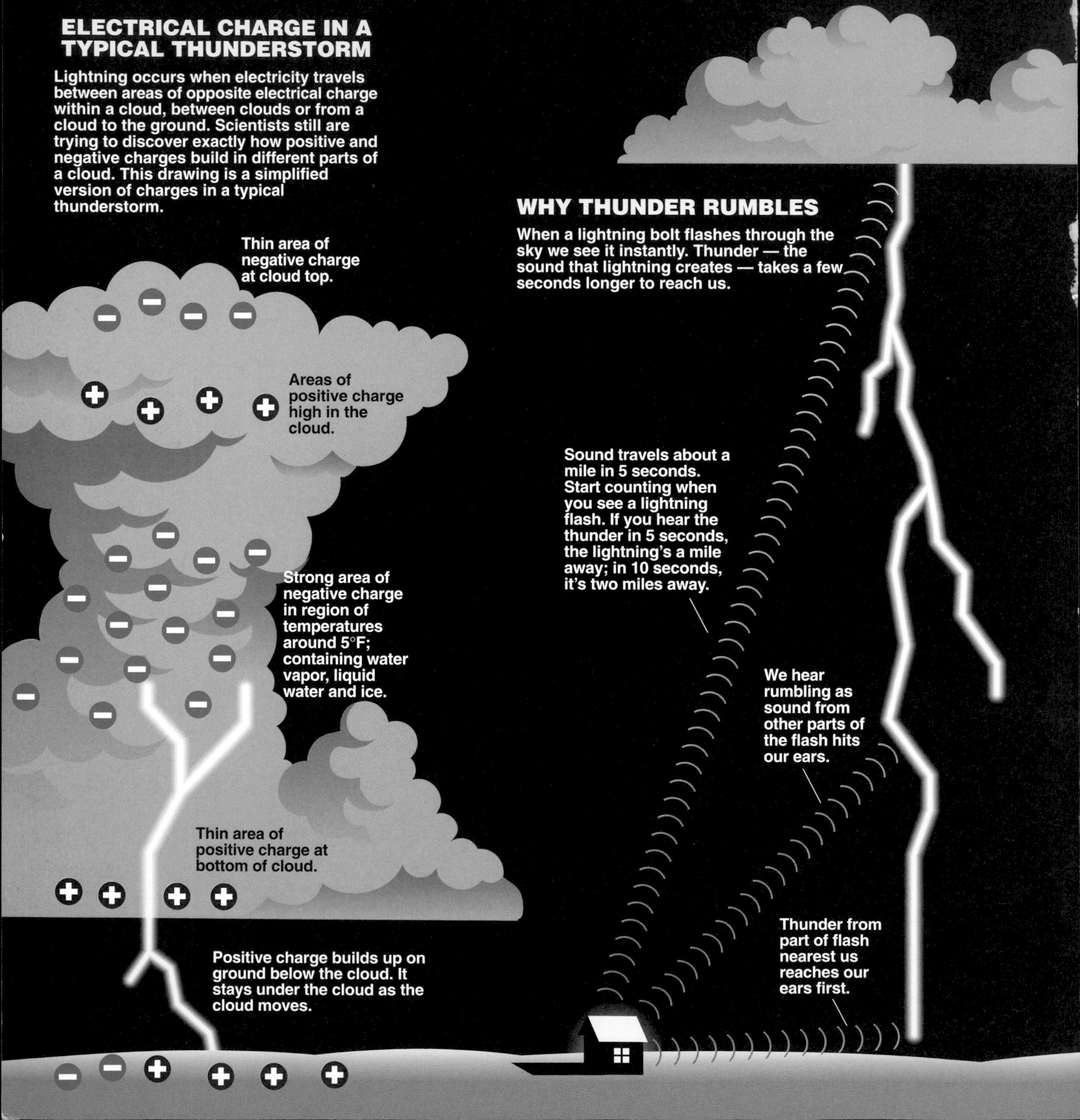

Sound travels about a mile in 5 seconds. Start counting when you see a lightning flash. If you hear the thunder in 5 seconds, the lightning's a mile away; in 10 seconds, it's two miles away.

We hear rumbling as sound from other parts of the flash hits our ears.

Thunder from part of flash nearest us reaches our ears first.

ANATOMY OF A LIGHTNING STROKE

Lightning flashes when the attraction between positive and negative charges becomes strong enough to overcome the air's high resistance to electrical flow. Here's how it happens in less than a second:

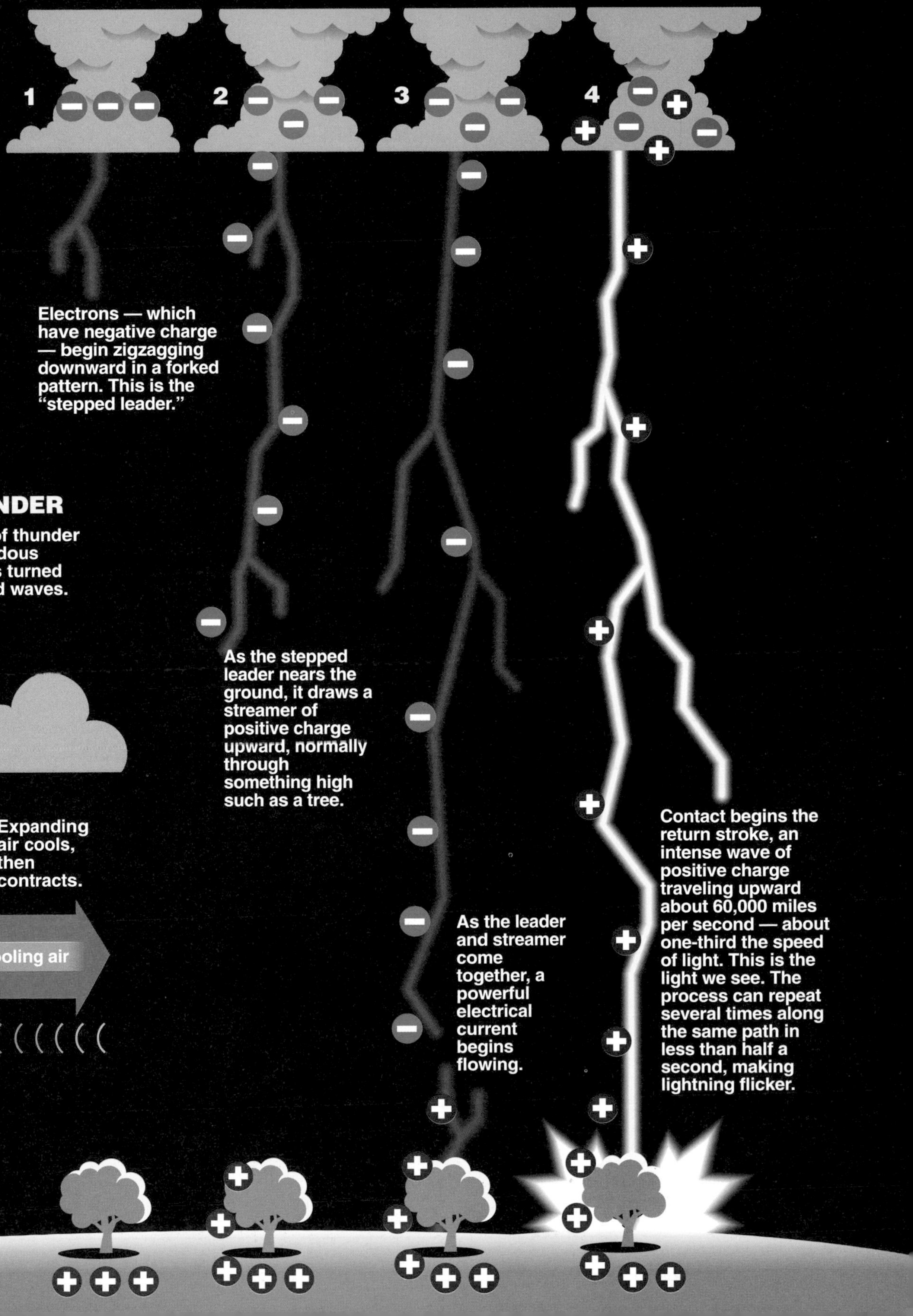

1

2

3

4

Electrons — which have negative charge — begin zigzagging downward in a forked pattern. This is the "stepped leader."

HEAT MAKES THUNDER

We hear claps and rumbles of thunder because some of the tremendous energy of lightning flashes is turned into heat and then into sound waves.

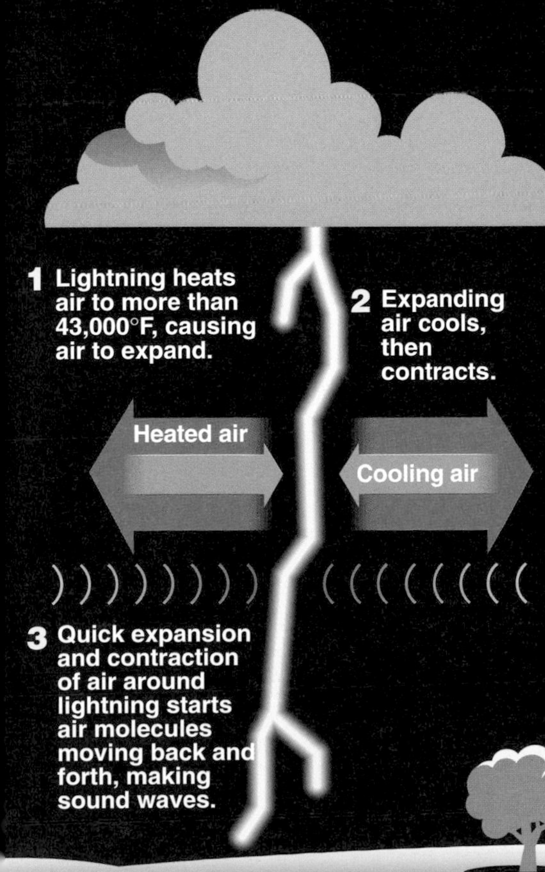

1 Lightning heats air to more than 43,000°F, causing air to expand.

2 Expanding air cools, then contracts.

Heated air

Cooling air

3 Quick expansion and contraction of air around lightning starts air molecules moving back and forth, making sound waves.

As the stepped leader nears the ground, it draws a streamer of positive charge upward, normally through something high such as a tree.

As the leader and streamer come together, a powerful electrical current begins flowing.

Contact begins the return stroke, an intense wave of positive charge traveling upward about 60,000 miles per second — about one-third the speed of light. This is the light we see. The process can repeat several times along the same path in less than half a second, making lightning flicker.

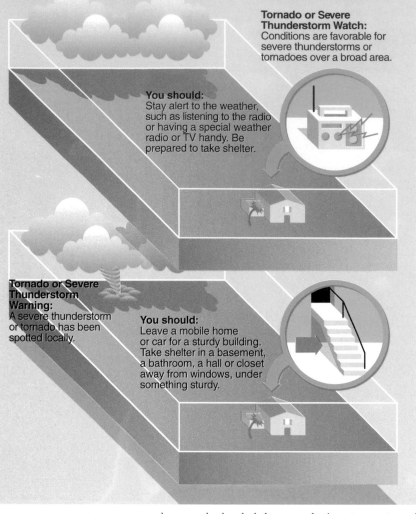

What 'watch' and 'warning' mean

The National Weather Service issues "watches" and "warnings" for life-threatening weather. If you know what the terms mean, you can take the right action.

Tornado or Severe Thunderstorm Watch: Conditions are favorable for severe thunderstorms or tornadoes over a broad area.

You should: Stay alert to the weather, such as listening to the radio or having a special weather radio or TV handy. Be prepared to take shelter.

Tornado or Severe Thunderstorm Warning: A severe thunderstorm or tornado has been spotted locally.

You should: Leave a mobile home or car for a sturdy building. Take shelter in a basement, a bathroom, a hall or closet away from windows, under something sturdy.

The largest number of tornadoes hit between the hours of 5 p.m. and 6 p.m. The smallest number hit between 5 a.m. and 6 a.m.

lent tornadoes in west Texas, including a monster one on June 2, 1995, near the small town of Dimmitt. It became the best observed tornado in history.

Using a high-tech computer program to track the storm, Rasmussen put his team in the tornado's path. Each vehicle was tracked by a satellite positioning system. Ahead of the storm, vans launched balloons to measure the low-, middle-, and upper-atmospheric pressure, temperature and humidity.

Cars, called "probes," spread out under different parts of the storm. The storm's structure and the tornado's location and appearance were observed from all sides. As many as three probes had video cameras on board to take movies of the tornado. Instruments sprouting from each car's roof measured temperature, pressure, humidity, wind speed and direction every six seconds.

Doppler radar mounted on a truck recorded the tornado's entire life, from tornadogenesis to dissipation. Images reveal the twister's hollow core surrounded by intense winds and a debris cloud. It looks like a miniature hurricane.

While the ground teams were making their records, two planes — a P-3 and an Electra — used Doppler radar to measure the storm's precipitation, and the tornado's winds and debris patterns. Using two planes to gather data at the same time gave scientists a complete picture of the tornado's winds.

Moments of exhilaration mixed with frustration, even fear, as some chasers lost the half-mile-wide Dimmitt tornado behind a curtain of blowing dust. Winds flowing into the storm from nearly 10 miles away measured up to 60 mph, strong enough to damage nearby trees and snap power lines. Inflow winds are rarely this strong. Rasmussen knew before seeing any damage that a violent tornado was on the ground.

When the tornado crossed state Route 86 between Dimmitt and Nazareth, it gouged a 30-foot-wide, 300-foot-long section of roadbed out of the concrete-hard Texas earth and threw it more than 600 feet into a farmer's field. Power poles snapped at the ground. Rasmussen says there were reports of two truck trailers missing entirely. The tornado was rated F-4 on the Fujita scale.

Despite the severe damage, the VORTEX chasers and their vehicles escaped unharmed and collected enough data to keep researchers busy for 10 years, maybe more. By intricately probing this twister, Rasmussen believes he and other scientists finally will understand the connec-

need extremely detailed data on what's going on in and around the thunderstorms that produce tornadoes.

They also need the same kind of data taken in and around storms that don't whip up tornadoes. By carefully analyzing their observations, researchers believe they'll unwrap the mystery behind one thunderstorm's success at spinning out a twister and another very similar storm's failure.

The Verification of the Origins of Rotation in Tornadoes Experiment (VORTEX) set out to try and solve this mystery in 1994 and 1995. Led across the flat plains of Texas, Oklahoma and Kansas by Erik Rasmussen, it was the biggest tornado chase in history. An armada that included 21 heavily instrumented vehicles and two research planes gathered mountains of data on 31 storms, some with large tornadoes and some that didn't produce any twisters.

During its final week, VORTEX intercepted four vio-

Persistence and good instincts pay off in chasing the tornado puzzle

On April 10, 1979, 22-year-old Erik Rasmussen found himself in a "bucket o' bolts" station wagon, chasing a rotating thunderstorm toward Wichita Falls in northern Texas. A funnel that looked like a narrow stovepipe exploded from the sky near Seymour, about 45 miles southeast of Wichita Falls. It licked the ground in a flurry of dust and debris before disappearing back into the roiling sky. Having snapped several pictures of the tornado, Rasmussen sat back elated, high-fiving his chase buddies on their prize catch of the day.

He heard sometime later that his storm had gone on to spawn a 1.5-mile-wide twister that flattened the southeast side of Wichita Falls killing 42 people, injuring 2,000 and destroying 3,000 homes.

"In 1979, we didn't know that a storm could produce one tornado after another in a cyclic fashion," Rasmussen says.

The Wichita Falls tornado was "a real eye-opener" for Rasmussen. Shortly after, he began a crusade to learn everything he could about tornadoes and how they spawn the most powerful winds on earth.

His research began with renowned tornado observer and professor Howard Bluestein at the University of Oklahoma. Rasmussen continued to chase tornadoes through the 1980s, while earning degrees in meteorology from Oklahoma and Texas Tech University.

"Early chasing provided us a wealth of information about what the cloud features look like, what they're doing," he says. But scientists still couldn't make sense of the seemingly jumbled atmospheric ingredients that mix together and result in deadly tornadoes.

In 1989, as part of his doctorate work at Colorado State University, Rasmussen traveled to Australia to observe another facet of severe weather, squall-line thunderstorms. He found field work satisfying, but he missed the familiar thrill of chasing those giant, roiling barrel-

Erik Rasmussen studys a miniature tornado generated by a simulator.

shaped thunderclouds called "supercells."

"After seeing well into the hundreds of tornadoes over the last 20 years, I still see supercells do things that are totally surprising. That was even true in VORTEX," he says, referring to a research project that became his baby.

VORTEX stands for the Verification of the Origins of Rotation in Tornadoes Experiment, a two-year intensive field study in 1994-95. The goal was to probe a tornado with sophisticated instruments and learn its secrets. More than 100 meteorologists and students were part of VORTEX, the largest and most successful tornado chase in history. Rasmussen designed and directed it.

After all, storm chasing was a longtime pursuit. By second grade, he was a professed

weather nut and in middle school he would listen to his weather radio and plot storm positions. Later, if he wasn't on his high school's roof watching storms roll in, he was on the road in his green VW "bug," driving long hours into the Kansas and Oklahoma prairies to see nature's awesome storm power.

"It seemed the natural thing to do," says Rasmussen, a native of tornado-prone Kansas.

Rasmussen has always prided himself on being a conservative chaser, never taking undue risks, such as "punching the core" of a storm — driving into the blinding rain and hail that can hide a twister. He didn't have to. He had a gift, the uncanny ability among chasers to pick the one storm out of several that would produce the day's only tornado.

He had especially good luck with "dryline" storms in western Texas and Oklahoma. The dryline can trigger tornadic storms for days at a time. An author writing about it in 1984 called Rasmussen the "Dryline Kid."

Though he doesn't enjoy chasing as much anymore — hordes of inexperienced chasers are taking over the serene plains — Rasmussen tries to keep his youthful spirit in the field. His work for the Severe Storms Lab forces him to spend whole days, sometimes more, on the road in search of nature's most violent wind.

As a pioneer in the small world of tornado research, his years of chasing have taken on new meaning. Not just tornado chasers. He and his colleagues are trying to unlock the secrets of the magnificently complicated tornado.

"It's something I really need to do," Rasmussen says. "It satisfies my soul."

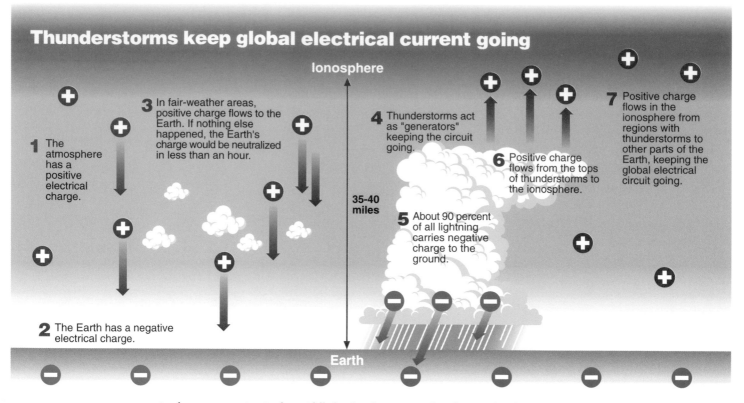

Thunderstorms keep global electrical current going

Ionosphere

1 The atmosphere has a positive electrical charge.

3 In fair-weather areas, positive charge flows to the Earth. If nothing else happened, the Earth's charge would be neutralized in less than an hour.

4 Thunderstorms act as "generators" keeping the circuit going.

6 Positive charge flows from the tops of thunderstorms to the ionosphere.

7 Positive charge flows in the ionosphere from regions with thunderstorms to other parts of the Earth, keeping the global electrical circuit going.

35-40 miles

5 About 90 percent of all lightning carries negative charge to the ground.

2 The Earth has a negative electrical charge.

Earth

Hurricanes often spawn tornadoes when they hit land, but compared to the other damage from the hurricane, their danger is relatively small. The tornadoes are usually concentrated in the right front quadrant of the hurricane. They're usually relatively small with winds less than 100 mph.

tion between rotation in the middle levels of a storm and the tornado. "VORTEX will give the first well-documented look at how a tornado forms and write the book on tornadogensis," he concludes. "There's no doubt about that."

A graphic on Page 129 shows how a storm's midlevel winds might tilt low-level spinning air upward, spawning a twister. To complete the picture, scientists also hope to find a different set of storm features that explain why some supercells don't spin up tornadoes.

The book on tornado formation might need to be updated if researchers such as Rasmussen have their way. Future scientific chases are likely to focus directly on the tornado's spin at the moment it's born. Two mobile radars would scan the air under the supercell for new clues. More cars with antler-like instrument racks are sure to be scanning storm-laden skies. Missiles and even tiny measurement-taking spheres, like those in the movie *Twister*, might be common in the future, Rasmussen says. "It sounds like science fiction, but the technology's there. The biggest hurdle will be how we get these things where we need them."

Downbursts: A new discovery

In addition to being the most destructive day of tornadoes on record, the Super Outbreak of April 3 and 4, 1974, led to the discovery of a source of thunderstorm

danger that had not been recognized. Theodore Fujita was flying near Beckley, W. Va., surveying storm damage, when he noticed what he calls a "starburst" pattern of tree damage, not the circular or semi-circular swaths left by tornadoes. It reminded him of something he had seen in Japan. (See story Pages 126-127.)

As he looked down at the damage, he later wrote, it was "natural to visualize a jet of downdraft which hard-landed at the center of the starburst." As Fujita continued studying damage from the outbreak, he found a number of other, similar damage patterns. The following year, while investigating the June 24, 1975, crash of an Eastern Airlines Boeing 727 at John F. Kennedy Airport in New York, Fujita hypothesized that the airliner had flown through the same kind of winds that had caused the starburst pattern he had seen in West Virginia.

Fellow scientists at first rejected Fujita's hypothesis, but various studies confirmed the theory and led to a decision to spend millions on special airport radars to spot such "downbursts" in time to warn pilots. The drawing on Page 127 shows what causes thunderstorms to produce such winds.

Derechos: Severe windstorms

Most people across eastern New York couldn't believe what hit them around dawn July 15, 1995. Severe nighttime thunderstorms are extremely rare in the East.

And thunderstorms with tops to 70,000 feet, spiking the ground with 3,000 lightning strikes an hour and racing at speeds of 85 mph, aren't supposed to happen anywhere, or so they thought.

The storms' 106 mph winds were nearly as strong as an F-2 tornado, but unlike a tornado's roughly quarter-mile-wide damage path, the winds tore apart small buildings and did more than $200 million in damage from Watertown to White Plains, about 300 miles south. Four hikers were killed as trees toppled across about 900,000 acres, including the Adirondack wilderness in northern New York.

The severe windstorms were what is known as "derechos," a name coined in the 1880s by Gustavus Hinrichs, director of the Iowa Weather Service. According to Robert H. Johns of the Storm Prediction Center, the name (pronounced day-RAY-cho) comes from a Spanish word meaning "straight ahead" or "direct." He says Hinrichs intended it to contrast with "tornado," which comes from a Spanish word for "turn."

Derechos are the result of squall-line thunderstorms that create one downburst after another as they move along. Occasionally they'll also spin out a small tornado or two. They occur most frequently at night from the central and northern Plains across the Midwest into the Ohio Valley in the late spring and summer.

Johns, who works on improving forecasts of derechos, says the windstorms are relatively common. He counted 70 during one four-year period in the 1980s. To qualify as a derecho, a storm must have winds stronger than 58 mph and spread damage across an area at least 280 miles long.

Very humid weather with strong winds in the middle atmosphere causes the storms. Dry air enters the storms a couple of miles above ground and is cooled when precipitation falls into it and evaporates. The cooled, and therefore heavier, air blasts down carrying the momentum of the fast upper-air winds with it, creating the strong gusts at the ground.

Hail, yet another thunderstorm danger

In most parts of the United States hailstones much bigger than peas are rare, which is something to be thankful for since hail can wipe out a field of corn, wheat or vegetables in minutes. The Great Plains from Texas north into Canada are an exception; large hailstones are relatively common.

Hailstones are balls of ice that grow as they're held up by thunderstorm updrafts while supercooled water drops hit and freeze onto them The faster the updraft, the bigger the stones can grow. Details of how hailstones grow are complicated, but the results, especially on the Plains, are irregular balls of ice that can be as large as baseballs, sometimes even bigger.

While crops are the major victims, hail is also a hazard to vehicles and the windows and roofs of homes and businesses. Blown by 75 mile an hour winds, hail, some the size of grapefruits, pulverized the Dallas/Fort Worth area on May 5, 1995, injuring 109 people, four critically. The storm also caused flash floods that killed 17 people in Dallas County. It was the costliest thunderstorm in U.S. history; damage was $1.2 billion.

Normally hail falls for a shorter time than rain from the same storm, since a smaller area of the storm has the right conditions to manufacture hail. While hail falls on any particular place for only a few minutes, in rare cases it has been known to fall for an hour or so, leaving several inches of icy balls on the ground. Since thunderstorms normally move, hail usually falls in swaths that can be 100 miles long (although five or 10 mile swaths are more common) and 10 or more miles wide.

Large hailstones can fall at a rate of 90 mph and in rare cases weigh more than a pound. Such a missile falling from the sky could be deadly, but the last known hail death in the United States was an infant killed in Fort Collins, Colo., in August 1979.

Lightning makes them thunderstorms

Thunderstorms are called thunderstorms for the obvious reason that they produce thunder. And thunder, in turn, is a product of lightning.

Benjamin Franklin showed with his famous kite experiment in 1752 that lightning is electricity. Since then scientists have learned much about lightning, but many questions remain. Efforts are still being made to discover what goes on at the molecular level and what happens in a miles-long flash.

Here's the key question: How do clouds build up the electrical charges that lead to lightning flashes? Scientists have developed two kinds of models to explain the electrical structure of clouds.

Precipitation theories hold that negative and positive charges are left on different size raindrops, graupel and hail as they collide in a cloud. Heavier particles carry a negative charge to the bottom of the cloud.

Convection theories hold that updrafts carry positive charges normally found near the ground upward while downdrafts carry negative charges from the upper air downward.

Both theories explain some observations, but neither explains everything that's known about lightning. Many researchers believe that some combination of the two theories will turn out to hold the answer.

They also believe that understanding the causes of lightning will clear up other questions about thunderstorms. More knowledge of how electrical charges develop is likely to shed light on how ice and water drops grow in clouds.

While researchers are seeking answers to explain the more common forms of lightning that appear as a bolt or flash, other types of lightning have been seen but remain unexplained.

Since the earliest days of written record, as far back as the ancient Greeks, people have been reporting seeing reddish glowing balls, up to about a yard in diameter. These glowing orbs slowly move around and then vanish, sometimes with a small explosion. Such "ball lightning" has been reported outdoors, in buildings and even in airplanes, usually after lightning hits nearby. Scientists are convinced ball lightning is real, but they don't know what it is or what causes it.

Recently, scientists using sensitive cameras discovered bursts of red and blue light and strange green flashes in the middle atmosphere, high atop the biggest thunderstorms. These blood-colored "red sprites" and "blue jets" shoot toward space while lightning arcs to the ground. Above them, "green elves" glow across the upper-atmosphere for tiny fractions of a second.

Scientists say red sprites, blue jets and green elves haven't been studied before because they are extremely difficult to see. Clouds often block the view above a storm's top, and lightning in or below the storm makes everything too bright to see them flicker dimly. Reports of these inner-space lights appeared in scientific journals more than a hundred years ago, but they were dismissed because very few people saw them.

In the 1990s, researchers in high-altitude planes took thousands of pictures and videos of jets and sprites. They've been observed from darkened fields and mountain tops often hundreds of miles away from the storms that produced them. On the ground, you need to be hundreds of miles away to see into the darkness above a storm's core.

Leading theories suggest red sprites, blue jets and green elves are caused by pulses of energy traveling up after being triggered by lightning. The energy hits gas molecules in the middle atmosphere forcing negatively-charged electrons to break free. Once one high-speed electron is loose, it triggers an upward-avalanche of freed electrons. This causes the molecules of gases that make up the air to glow red or blue and take on different shapes, depending on their altitude. When the energy pulse hits the electrically-charged layer of the atmosphere called the ionosphere, a ring of green light flashes outward like a beacon, expanding 250 miles wide in less than a second.

Unlike ball lightning that rarely, if ever, seems to do any harm, scientists at NASA are concerned these energetic lights, sometimes called "upward lightning," might interfere with rocket launches. Researchers are trying to find out more about the causes of these high-altitude flashes so that one day they can predict them.

Forecasting thunderstorms and tornadoes

Severe thunderstorm and tornado forecasting begins with the wide-ranging computer forecasts produced by the National Weather Service's National Centers for Environmental Prediction in Camp Springs, Md. Each night, while most of the country sleeps, forecasters at the Storm Prediction Center on the outskirts of Norman, Okla., comb the nationwide computer forecasts and other data to get a handle on where dangerous thunderstorms are likely in the next couple of days.

At 1 a.m. each day a forecaster sends out a severe weather outlook bulletin for the nation. This either announces the day will be quiet or gives a picture of conditions that could spawn severe storms. At 3 a.m., the following day's outlook is issued, giving forecasters more than a day to prepare if storms threaten. Both bulletins are updated throughout the day.

If conditions favor severe storms and tornadoes, the Storm Center's forecasters try to figure out where they might hit. They use an interactive computer system to combine many kinds of data on one computer screen, for example, radar images with satellite photos, computer forecasts of the day's weather features and the latest map showing every lightning bolt that has hit the ground in the last 15 minutes. The computer calculates information such as the air's stability and analyzes wind speeds and directions at various altitudes.

This interactive system "gives the forecaster a detailed look at the weather situation quickly," says the

Flying into dangerous weather to see what is really going on

"Stay away from thunder-storms" is one of the first weather lessons pilots learn. Even the most experienced pilots in the strongest airplanes try to avoid the severe turbulence, hail-stones, icing, downpours and lightning that may be hiding in any thunderstorm.

But no matter how many Doppler radars and other advanced instruments science invents, to understand thunderstorms, somebody has to fly into them and see what's going on.

Wayne Sand spent a good part of his career as a pilot and researcher deliberately flying into thunderstorms and other nasty weather. The thesis for his University of Wyoming doctorate was based on flights into thunderstorms in a single-engine military trainer.

Wayne Sand earned a living deliberately flying into thunderstorms.

The plane, a T-28, was armored with stronger-than-normal wings and windshield. It was one tough plane, but even with the extra protection, storm flying "never, never became routine." The armor worked; 20 years after Sand's 1970s flights, the T-28 was still probing storms.

Ora Lohse, who ran a crop-dusting business in Sand's hometown, Valier, Mont., taught him to fly. Sand credits Lohse with "instilling in me some fundamental ways of thinking and attacking problems" and flying that helped him both as a researcher and pilot.

As a pilot, Lohse believed "you needed to know how to fly in the wind," Sand says. "The more the wind blew, the more we flew." Lohse was also an inventor "with all kinds of unique ideas of how to do a lot of stuff. He motivated you to think about things before starting to do them. A little brain power would replace a

whole lot of back power. He gave me these ways of thinking early."

In addition to crop dusting, Lohse did some cloud-seeding experiments with Sand doing some of the flying. In 1963, Sand went to Colorado State University as a research pilot and graduate student. "Here I was about 23 years old and the expert on [cloud seeding]. I hadn't had much experience, but I knew more than anyone else."

When Sand left to become a Navy pilot, a professor, Richard Schleusener, told him, "Call me if you ever need a job."

In 1971, Sand was "struggling with a decision to say in the Navy or get out. I had orders to go to test pilot school. It was a very viable option to stay in the Navy." But Schleusener, who then headed the Institute of Atmospheric Science at the South Dakota School of Mines and Technology, offered him a

job developing and flying the T-28. Sand was attracted by the chance to return to the West and weather studies.

By 1980, Wayne Sand was at the University of Wyoming as a full-time research pilot and doctoral candidate.

A U.S. Bureau of Reclamation contract for cloud-seeding research meant flying into both thunderstorms and winter storms in the university's twin-engine King Air. Unlike the two-seat T-28, which Sand flew alone, the King Air carried two pilots plus a couple of scientists.

"The King Air wasn't armor-plated, but based on what we had done with the T-28, we could poke around in and under thunderstorms with some sense of what we're doing without getting into trouble."

From Wyoming, Sand went to the National Center for Atmospheric Research where he worked on a major microburst research program and then helped direct an aircraft icing project. After retiring from NCAR, he became a private aviation weather consultant.

Unlike many atmospheric scientists, Sand wasn't unusually fascinated by weather as a youngster. But, he explains, "I grew up in an agricultural community where whether it rained or not, whether it hailed or not, made the difference in whether people made a living."

Sand's interest in flying and weather "evolved together. I was never motivated to fly for an airline. I think those are great jobs, but I found the kind of flying I was doing more interesting than what the airline guys are doing. I felt it was a lot more fun."

Storm Prediction Center's Director Joe Schaefer. "It tells us if indeed we're getting conditions conducive to severe storms."

When a Storm Prediction Center forecaster decides that conditions will be ripe for either tornadoes or severe thunderstorms over a wide area, the forecaster issues a "watch." The computer helps here, too.

Years ago the forecaster would have to look on a map for the names of towns that would define the area covered by the watch. Now the forecaster outlines the threatened area on the computer screen and the computer adds the names of the affected towns.

A "watch" issued by the Storm Prediction Center means only that conditions are favorable for dangerous weather. The message goes out to National Weather Service offices in the threatened area. The local offices broadcast the information on NOAA weather radio and it's made available to news media.

In many areas, especially on the Plains, a tornado watch triggers alerts to local storm-spotter networks. These can be police officers or volunteers who are trained to recognize tornadoes and other dangerous weather.

They have quick communication, usually radio, tied directly to the local Weather Service office. Since conventional weather radar can't tell for sure if a thunderstorm is producing a tornado, volunteer weather watchers or spotters are vital to quick warnings.

Improvements in forecasting techniques should allow both watches and warnings to be issued further ahead of time over smaller, better-defined areas.

Living with dangerous thunderstorms

No matter how good the watches and warnings for severe thunderstorms and tornadoes, life-and-death decisions are up to individuals. Lessons learned during the April 3-4, 1974 Super Outbreak still apply.

Any time killer weather strikes the United States, the National Oceanic and Atmospheric Administration, which includes the National Weather Service, appoints a team to investigate the disaster. This survey team goes out quickly to try to discover what the Weather Service and others could do to save lives in the future. The team examines the forecasts, the warnings and the response. It then makes recommendations for improvement.

The team that investigated the April 3-4, 1974, tornado outbreak said in its report: "Everywhere the survey teams went — and especially in Guin, Ala., Brandenburg, Ky., Monticello, Ind., and Xenia, Ohio —

the question was the same: 'How could anyone have survived?' Education somewhere, at some time, through some means, had to be the answer. The people in those towns and elsewhere in 11 states heard weather watches and warnings over radio and television; were notified by their neighbors, relatives or friends; saw the tornadoes approaching; or heard the ominous roar of the tornado. In interviews with people in all the hard-hit areas, the team found very few who were unaware that they were under a severe weather threat.

"But most important, they knew what to do when the time came to take action. They all seemed to know that a basement, if they had one, was the best place to be. Others went into closets, under beds, in ravines, behind a sofa, under stairwells, in the center of the home, under dining room tables, and in the halls of large well-constructed brick buildings. They got out of gymnasiums and large open rooms, stayed away from windows, protected their heads from flying debris."

The team wrote about a Hanover, Ind., high school physics teacher who had talked about tornado safety in his class the day before. Among other things, he told his students of the danger of being hit by a tornado in a vehicle. He told them to leave their cars and take shelter in a ditch. The next day one of his students was on a school bus on a collision course with a tornado. He persuaded the driver to stop and everyone headed for a ditch. The tornado demolished the bus, but not a single student or the driver was hurt.

Warnings and education weren't the whole story, however. The team also cited the role of luck. For instance, one tornado destroyed downtown Monticello, Ind., killing two people. But lives were saved because businesses were closed, as they normally were on Wednesday afternoons.

In many ways, however, those in the path of a tornado or other severe thunderstorm do much to create their own luck. While it's almost never possible to do much about saving property in the path of a storm, the kinds of quick action that saved lives April 3-4, 1974, continues to allow people to survive the world's strongest winds.

Hurricanes are the only natural disasters with their own names. Andrew, Camille, Fran, Hazel, Hugo — each evokes its particular image of disaster. Hurricanes are the same in many ways; yet, like people, each has its own personality.

Names seem appropriate because we come to know hurricanes before they strike, unlike earthquakes, which hit without warning, or tornadoes, which quickly come and go with at best a few minutes warning.

Hurricanes are special. You can make a good argument that they are the Earth's most awesome storms.

Winds in the strongest tornadoes can top 300 mph, while hurricane winds above 150 mph are rare. But a tornado is much more concentrated than even the smallest hurricane: A mile-wide tornado is huge, a 100-mile-wide hurricane is small. Few tornadoes last even an hour, and a damage path of 100 miles goes into the record books.

Hurricanes easily can last more than a week and can devastate islands around the Caribbean days before slamming into the United States.

Extratropical cyclones, which we examined in Chapter 4, are larger than hurricanes and even a weak one will affect more people than most hurricanes. But extratropical cyclones rarely produce hurricane force winds — 74 mph or more.

A large hurricane stirs up more than a million cubic miles of the atmosphere every second. Hurricane winds can kick up 50-foot or higher waves in the open ocean. When a storm hits land, it brings a mound of water that can rise to a peak height of more than 20 feet near the eye and flood 100 miles of coast with 10 feet of water. A typical hurricane dumps six to 12 inches of rain when it comes ashore — some bring much more and have caused some of our worst floods.

Discussions of a storm's expected path normally focus on the eye. Maps show the eye's location. Forecast arrival times are often projections of when the eye will hit. Since even a small storm will have dangerous winds more than 100 miles out from the eye, damaging wind and waves may arrive six hours or more before the eye.

Sorting out the storms

Extratropical cyclones account for most of the United States' stormy weather. Large-scale storm systems generally come in two varieties: extratropical cyclones or tropical cyclones. One or two percent of the world's tropical storms are hybrids that have some characteristics of extratropical storms. In the last few years they have been called "subtropical" storms. Tropical storms also commonly merge with or develop into extratropical storms outside of the tropics.

How tropical and extratropical storms differ

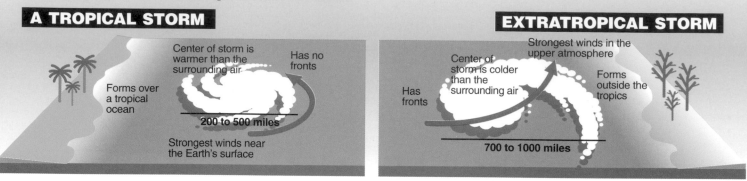

A TROPICAL STORM

Forms over a tropical ocean

Center of storm is warmer than the surrounding air

Has no fronts

200 to 500 miles

Strongest winds near the Earth's surface

EXTRATROPICAL STORM

Has fronts

Center of storm is colder than the surrounding air

Strongest winds in the upper atmosphere

Forms outside the tropics

700 to 1000 miles

Tropical cyclone: A low-pressure weather system in which the central core is warmer than the surrounding atmosphere. Storms called hurricanes or typhoons elsewhere are called tropical cyclones in the Indian Ocean and around the Coral Sea off northeastern Australia.

Tropical depression: A tropical cyclone with maximum sustained winds near the surface of less than 39 mph.

Tropical storm: Tropical cyclone with 39 to 74 mph winds.

Hurricane: A tropical cyclone with winds of 74 mph or more.

Typhoon: A hurricane in the north Pacific west of the International Date Line.

In July 1994, the remnants of Tropical Storm Alberto stalled, bringing 10 to 25 inches of rain to Georgia and Alabama, causing major floods that killed 32 people and did more than $1 billion in damage. When Hurricane Fran slammed into the North Carolina Coast in September 1996, it continued inland to cause major floods in the Appalachians.

Hurricane researcher Peter Black says Fran, like all tropical storms and hurricanes, "can be thought of as a moisture processor" that carries humid air over warm oceans for great distances. This humid air is "continually spiraling into the storm center at low levels, upward in the interior and outward at the storm top. During the upward moving part of the cycle the moisture condenses and falls as rain."

Hurricanes even affect the very depths of the ocean. In 1975 instruments dropped from research airplanes in the Gulf of Mexico showed that Hurricane Eloise disturbed the ocean hundreds of feet down and created underwater waves that persisted for weeks.

Figuring out the big picture

We're so used to seeing satellite photos of hurricanes, it's difficult to imagine a time when people didn't understand that such storms are huge masses of wind circling around a center. The center is the eye we see so clearly on most satellite photos of full-blown hurricanes.

Well into the 19th century a clear distinction wasn't

A giant whirlwind

A simple picture of a hurricane, somewhat like the one William Redfield presented in 1831, helps us understand what happens when a storm hits.

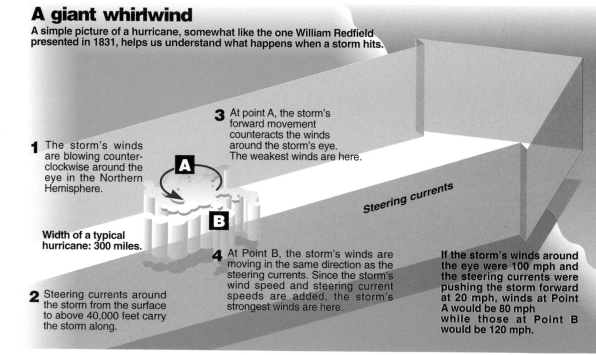

1 The storm's winds are blowing counter-clockwise around the eye in the Northern Hemisphere.

Width of a typical hurricane: 300 miles.

2 Steering currents around the storm from the surface to above 40,000 feet carry the storm along.

3 At point A, the storm's forward movement counteracts the winds around the storm's eye. The weakest winds are here.

4 At Point B, the storm's winds are moving in the same direction as the steering currents. Since the storm's wind speed and steering current speeds are added, the storm's strongest winds are here.

Steering currents

If the storm's winds around the eye were 100 mph and the steering currents were pushing the storm forward at 20 mph, winds at Point A would be 80 mph while those at Point B would be 120 mph.

always made between tropical cyclones and extratropical cyclones. Some summer and fall storms from the tropics, however, were known to be especially fierce. While a few sailors and scientists had suggested that storms were huge "whirlwinds," the general view was that storms were masses of air moving straight across the Earth. Winds that died and then blew from another direction were considered to be close, but separate storms. Today we know the calm is the eye of a hurricane.

William Redfield, a Connecticut saddle and harness maker — and a self-taught scientist — is credited with taking a key early step toward today's understanding of hurricanes. While traveling around Connecticut by foot — he couldn't afford a horse — after a strong hurricane that hit in September 1821, he noticed trees and corn in some areas were blown down pointing toward the northwest, while those only a few miles away were pointing to the southeast.

He gathered more information about the storm from other places along the East Coast and concluded in an article in the *American Journal of Science* in 1831 that: "This storm was exhibited in the form of a great whirlwind." Redfield's detailed observations and logical presentation convinced others in the United States and elsewhere that he was on the right track. Following his lead, they began the work of understanding hurricanes that continues today.

Where and how hurricanes begin

In terms of their potential for destruction, hurricanes and their cousins — typhoons and Indian Ocean cyclones — are the world's most violent storms. Yet, these most furious of storms are born in the most placid of climates — the tropics.

The tropics supply the key ingredients needed for tropical cyclones: wide expanses of warm ocean water; air that's both warm and humid; normally weak upper-air winds blowing from the same direction as winds near the surface.

In the big picture of how the Earth keeps its heat budget balanced, hurricanes are one of the ways the atmosphere moves excess heat from the tropics to the middle latitudes.

We can understand hurricanes as huge machines, heat engines, that convert the warmth of the tropical oceans and atmosphere into wind and waves. We can be thankful that hurricanes aren't very efficient machines. Even the strongest hurricane converts only a small

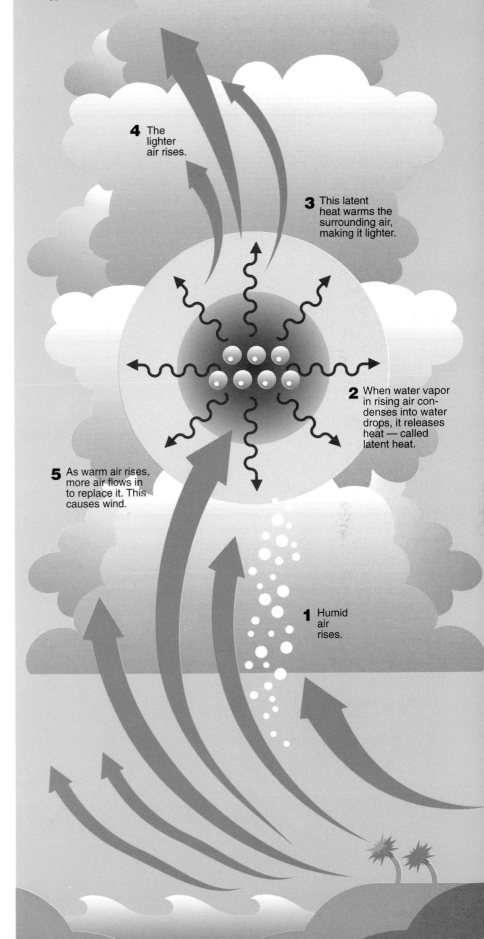

How water powers hurricanes

Heat released when water vapor condenses into water drops or turns directly into ice high in the air supplies most of the energy for hurricanes. The amounts of energy are huge: In one day a hurricane can release enough energy to supply all of the nation's electrical needs for about six months.

4 The lighter air rises.

3 This latent heat warms the surrounding air, making it lighter.

2 When water vapor in rising air condenses into water drops, it releases heat — called latent heat.

5 As warm air rises, more air flows in to replace it. This causes wind.

1 Humid air rises.

The ingredients a hurricane needs

Fewer than 10 percent of tropical weather disturbances grow into tropical storms because the right ingredients are relatively rare.

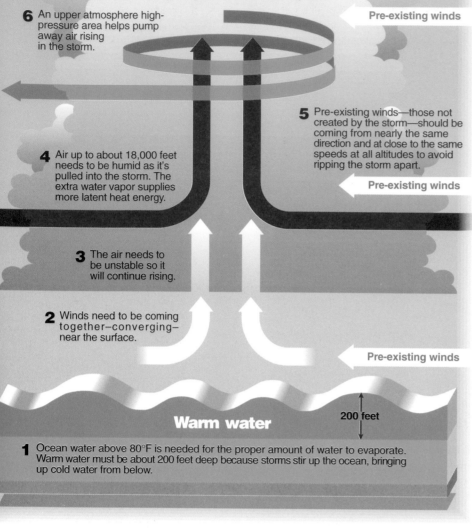

6 An upper atmosphere high-pressure area helps pump away air rising in the storm.

5 Pre-existing winds—those not created by the storm—should be coming from nearly the same direction and at close to the same speeds at all altitudes to avoid ripping the storm apart.

4 Air up to about 18,000 feet needs to be humid as it's pulled into the storm. The extra water vapor supplies more latent heat energy.

3 The air needs to be unstable so it will continue rising.

2 Winds need to be coming together—converging—near the surface.

Pre-existing winds

Pre-existing winds

Pre-existing winds

Warm water

200 feet

1 Ocean water above 80°F is needed for the proper amount of water to evaporate. Warm water must be about 200 feet deep because storms stir up the ocean, bringing up cold water from below.

amount — around three percent — of its available energy into wind and waves.

We can also be thankful that only a small number of

the disturbances with the potential to spawn hurricanes actually grow into storms. One study of satellite photos found that of the 608 weather systems that moved over the tropical Atlantic during a six-year period, only 50 developed into tropical storms with 39 mph or stronger winds.

In fact, winds high above the tropical Atlantic often blow from the west, opposite the direction of the low-level winds, or blow too fast to allow hurricanes to develop. This accounts for the relatively small number of disturbances that become hurricanes. Such unfavorable upper-air winds are probably why tropical storms don't form over the South Atlantic, between Africa and South America, and over the southeastern Pacific, off the South American Coast. These are the world's two tropical oceans that don't produce hurricanes.

Earliest tracks of hurricanes

Tropical storms that grow into U.S. hurricanes can begin far out in the Atlantic Ocean, in the Gulf of Mexico, in the Caribbean Sea or relatively close to shore off the Atlantic Coast. A large number of the strongest hurricanes form far out in the Atlantic. These are often called Cape Verde hurricanes after the islands near the coast of West Africa where they first begin. These can take more than a week to cross the Atlantic. If conditions are favorable, they grow into monster storms during that trip.

The average Northern Hemisphere tropical cyclone moves from east to west in the tropical trade winds. Many storms turn toward the north, sometimes far out in the Atlantic, sometimes farther west. After turning toward the northwest or west, storms tend to curve to the north-

Where the world's tropical cyclones form

Each year on the average 100 tropical storms form in the world. About two-thirds of them grow into 74 mph or stronger hurricanes, typhoons or cyclones.

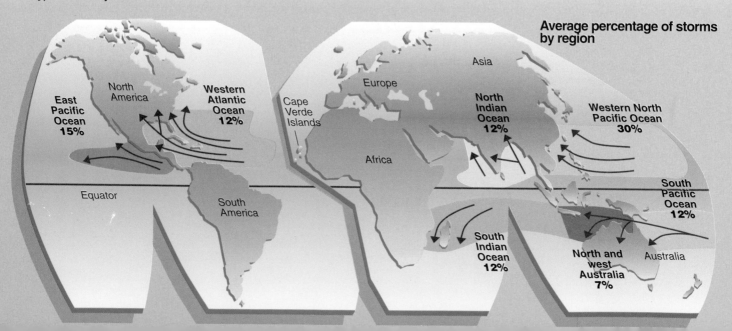

Average percentage of storms by region

East Pacific Ocean **15%**

North America

Western Atlantic Ocean **12%**

Cape Verde Islands

Europe

Asia

Africa

North Indian Ocean **12%**

Western North Pacific Ocean **30%**

South Pacific Ocean **12%**

Equator

South America

South Indian Ocean **12%**

North and west Australia **7%**

Australia

east. Where a storm begins turning toward the northwest or north determines whether it hits the United States or stays over the ocean until it dies.

As storms move north, the steering currents usually become faster because winds at all levels of the atmosphere tend to be faster in the mid-latitudes than in the tropics. Storms moving northeast along the East Coast usually speed up when they're about as far north as Cape Hatteras, N.C. This speedup makes getting warnings out in time difficult.

The fast-moving Hurricane of '38

Such a fast-moving storm killed at least 600 people in New England in 1938, making it America's fourth deadliest storm of the 1900s.

At 7 a.m. the morning of Sept. 21, 1938, the hurricane's eye was about 100 miles east of Cape Hatteras, N.C. By 2 p.m. it was crossing Long Island, N.Y., about 350 miles farther along, and by 8 that night the storm's center passed Burlington, Vt., another 275 miles. At times the storm was moving forward at a speed of 70 mph.

Quickly rising water — the storm surge — destroyed houses along the Long Island, Rhode Island and Connecticut shorelines and the hurricane's winds toppled trees across New England. Seventeen inches of rain fell on an area that had been soaked by rain for days, causing record-breaking floods. The hurricane set a record for damage that stood until Hurricane Carol in 1954.

How the Bermuda High helps guide hurricanes

The clockwise winds around the area of high pressure that dominates the Atlantic Ocean — the Bermuda High — establish the steering currents for many hurricanes.

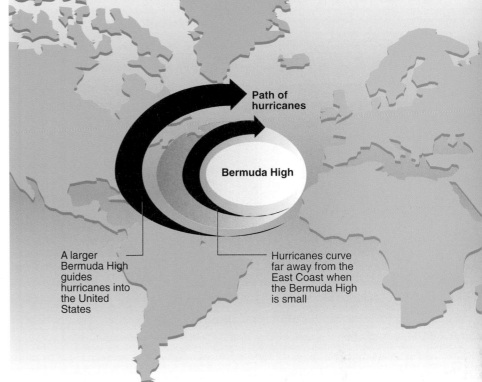

Path of hurricanes

Bermuda High

A larger Bermuda High guides hurricanes into the United States

Hurricanes curve far away from the East Coast when the Bermuda High is small

Such fast-moving storms lead forecasters to issue warnings when a hurricane is still hundreds of miles away from land. For instance, as Hurricane Bob moved up the Atlantic Coast in August 1991, the U.S. Weather Service

Deadliest 20th century hurricanes

1. Galveston, Texas, 1900, more than 7,200 deaths.
2. South Florida, 1928, 1,836 deaths.
3. Florida Keys and Corpus Christi, Texas, 1919, 600 to 900 deaths.
4. New England, 1938, 600 deaths.
5. Florida Keys, 1935, 408 deaths.
6. Louisiana and Texas, 1957 (Hurricane Audrey), 390 deaths.
7. East Coast from Virginia to Massachusetts, 1944, 390 deaths.
8. Grand Isle, La., 1909, 350 deaths
9. Galveston, Texas, 1915, 275 deaths.
10. New Orleans, La., 1915, 275 deaths.

While the U.S. has fewer hurricane deaths ...

... the cost of hurricane damage has increased

(1996 dollars in billions)

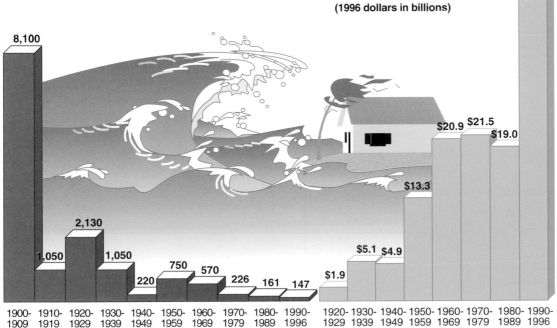

8,100
1,050
2,130
1,050
220
750
570
226
161
147

1900-1909 | 1910-1919 | 1920-1929 | 1930-1939 | 1940-1949 | 1950-1959 | 1960-1969 | 1970-1979 | 1980-1989 | 1990-1996

$40.9
$20.9 $21.5
$19.0
$13.3
$5.1 $4.9
$1.9

1920-1929 | 1930-1939 | 1940-1949 | 1950-1959 | 1960-1969 | 1970-1979 | 1980-1989 | 1990-1996

How hurricanes pile up water

A hurricane's winds push water, tending to pile it up in a mound. When the storm is at sea, the water easily flows away. But as the storm nears land, the water piles up to create a storm surge, which can drown places near the coast under more than 20 feet of water.

3 Unable to flow away, the mound of water piles up higher and higher to the right of the eye.

Mound of water

1 Winds continue pushing water inward toward the storm's center.

2 But the ocean floor blocks water from flowing away.

Worst surges hit places where the ocean floor slopes gradually, such as around the Gulf of Mexico.

What happens when the surge comes ashore ...

Ultimate height of the "storm tide" is a combination of the astronomical tide and the storm surge. The surge normally does not arrive as a "wall of water," but more like a quick rise in the tide to extremely high levels.

3 Surge's worst effect is bringing storm-whipped waves far inland; battering of the waves causes more damage than high water alone.

2 ...a 10-foot storm surge will push the water 12 feet above mean sea level.

12 feet — Normal high tide

1 A 2-foot, normal high tide plus ...

Mean sea level

What a super storm's surge would do to New Orleans...

The map below was created by National Weather Service's SLOSH model. It shows water depths of the surge from a Category 4 hurricane following the path shown. If the Category 5 Hurricane Camille had jogged just a little to the left in 1969, it could have followed a similar path. The storm shown would leave water more than 20 feet deep in downtown New Orleans. Mississippi River levees would block much of the surge from the west side of the river.

Water height above mean sea level

- 18-21 feet
- 15-18 feet
- 12-15 feet
- 9-12 feet
- 0-9 feet

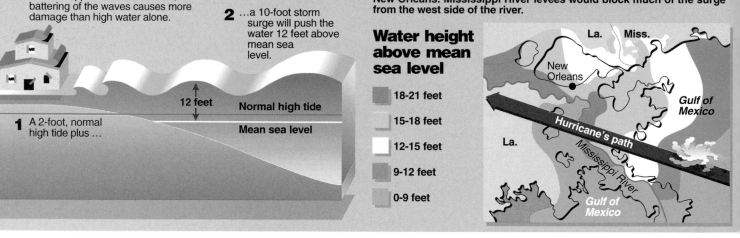

issued hurricane warnings for Long Island and parts of New England when the storm was still about 400 miles away off Cape Hatteras, but only around 22 hours from hitting the warned areas.

Slow-moving 'loop-the-loop' hurricanes

Slow-moving storms also cause forecasting problems. When the steering currents become weak, a storm seems to take on a mind of its own. If you look at maps of hurricane tracks, you'll see that storms sometimes follow "loop-the-loop" paths.

At the end of August 1985, Hurricane Elena moved into the Gulf of Mexico and headed toward the Florida Panhandle, where beachfront communities were evacuated. Then it turned east, heading toward the heavily populated Tampa Bay area of Florida, where thousands were evacuated as people in the Panhandle were heading back home. Then Elena did a loop and headed back toward the Panhandle.

People along the Panhandle beaches, as well as in Alabama, Mississippi and Louisiana were evacuated again. Elena eventually came ashore on the Mississippi-Louisiana border, causing about $1.4 billion in damage.

An engineer and a meteorologist team up to rate hurricanes

Hurricanes not only have names, they have numbers — based on the Saffir-Simpson Damage-Potential Scale.

Hurricane Hugo in 1989 was a Category 4. Hurricane Camille in 1969 was a Category 5. Any storm of Category 3 or stronger is considered "major."

Herbert Saffir, a consulting engineer in Coral Gables, Fla., and Robert Simpson, then director of the National Hurricane Center, developed the scale in the early 1970s.

Simpson says the center was having difficulty telling disaster agencies how much damage to expect from particular storms. So Simpson called on Saffir, who, he says, "was well known as the father of the Miami building code. He looked at it from the engineering point of view, I looked at it from the meteorological point of view."

Saffir already had worked out what kinds of damage to expect from different wind speeds. Simpson added meteorological details and figured the potential storm surge in different strength storms.

Both men had first-hand experiences with hurricanes.

Simpson's went back to the Category 4 storm that hit Corpus Christi, Texas, in 1919, when he was six years old. "I had just started school and I guess I'm here today because it happened on a Sunday, not a school day," he says. His school collapsed, killing many who had taken shelter there.

As the storm's surge washed in, Simpson says, "We floated my grandmother out the back door strapped in a cane wheel chair." He remembers his mother trying unsuccessfully to hold up a paper bag with the fried chicken and homemade donuts they were going to have for dinner.

After the family took refuge in the courthouse, Simpson watched the storm rip apart Corpus Christi. "It was pretty fearsome. Houses were coming off their foundations and floating intact until they ran into another obstacle."

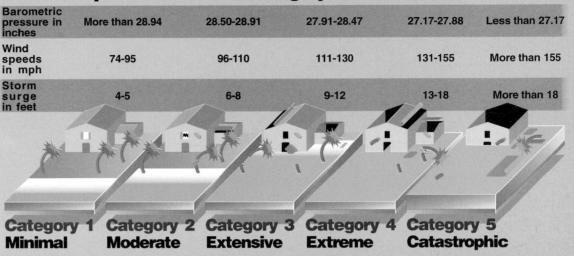

Saffir–Simpson hurricane damage potential scale

Barometric pressure in inches	More than 28.94	28.50-28.91	27.91-28.47	27.17-27.88	Less than 27.17
Wind speeds in mph	74-95	96-110	111-130	131-155	More than 155
Storm surge in feet	4-5	6-8	9-12	13-18	More than 18
	Category 1 **Minimal**	**Category 2** **Moderate**	**Category 3** **Extensive**	**Category 4** **Extreme**	**Category 5** **Catastrophic**

Saffir went through a hurricane soon after he moved to Florida in 1947 to become assistant county engineer for Dade County. "We moved here in September 1947 and had a major hurricane that month and another one in October," he says. "We were without power for about two weeks."

Those hurricanes, and others that hit South Florida in the 1940s, meant that Saffir's recommendations for a tough building code encountered little resistance. Over the years Saffir became widely recognized as an authority on designing buildings likely to withstand hurricanes. When Simpson became director of the National Hurricane Center in 1968, he was already one of the world's leading authorities on tropical cyclones and a veteran of scores of Air Force and Navy flights into storms. He says he stopped counting after about 200 hurricane flights.

In 1955 he was appointed first director of the National Hurricane Research Project after Congress increased Weather Bureau funding in response to the disasters of Hurricanes Carol, Edna and Hazel in 1954 and Hurricanes Connie and Diane in 1955.

The extra funds not only paid for hurricane research, but also accelerated work on computer forecasting and the purchase of a computer. "The Weather Bureau budget in 1953 was $27.5 million for everything," Simpson says. "In 1957 after we had gotten our computer and established the Hurricane Research Project, the Bureau had a budget of $57.5 million, more than double as a result of those hurricanes."

Strong building codes can reduce hurricane damage, says Herbert Saffir. But, he adds, "one of the key things we've found in many parts of Florida is if you don't have adequate building inspection, you're spinning your wheels. You're not going to get what you're planning for."

Storm surge: Hurricanes' big killer

Hurricane forecasters consider New Orleans the USA's most dangerous area for storm surge, since a storm could drive 20 feet of water into the city.

They consider southwest Florida — from Tampa Bay south to Everglades National Park — the next most dangerous because of the large number of people near the coast and the shallow slope of the ocean bottom, which makes storm surges higher.

On the Atlantic Coast, areas north and south of Savannah, Ga., are dangerous because the ocean floor slopes more gradually, which creates higher surges than steeply sloping floors.

We usually describe hurricanes in terms of their wind speeds, but over the years storm-surge flooding has claimed more victims than wind. Flooding also usually accounts for much of the damage within a few hundred yards of the shoreline. Boats ripped from their moorings, utility poles, parts of destroyed buildings and other debris crashing in the waves atop hurricane surge often destroy buildings that stood up to the wind. The lesson is being learned. A key reason for the drop in hurricane deaths since the 1960s is the emphasis on evacuating people from surge-threatened areas.

Since sea water weighs 64 pounds per cubic foot, ocean waves are like battering rams. But water does more than batter.

High water and pounding waves carry away the sand under sea walls, buildings and roads. As the water begins rising in advance of the storm — sometimes hours in advance — it erodes the beach and then the dunes or undercuts buildings behind the beach.

When Hurricane Eloise hit the Florida Panhandle in September 1975, it carried away an average of 45 cubic yards of sand for each yard of beachfront along about nine miles of the coast.

The worst hurricane storm surge recorded in the United States came ashore with Hurricane Camille on Aug. 17, 1969, at Pass Christian, Miss., where the water rose 24 feet above normal. While the highest storm surge was concentrated to the right of the eye, Camille pushed the Gulf of Mexico at least three feet higher than normal as far as 125 miles to the east of Pass Christian and 31 miles to the west.

Devastation was concentrated along 60 miles of coast in Alabama and Louisiana as well as Mississippi. A damage survey found that more than 5,500 homes had been destroyed and another 12,500 had major damage. Some 700 businesses were destroyed or left with major damage. After slamming into the Gulf Coast, Camille moved inland to Kentucky and then headed back to the east over West Virginia and Virginia where its rain brought devastating floods. About half of the 256 people killed by Camille died in the floods away from the coast.

Storm surge isn't a killer only along beaches facing the ocean; water is also pushed into bays and rivers. As the surge of water squeezes up a narrowing bay or river, it rises even higher. When Hurricane Carla hit the Texas Coast between Corpus Christi and Galveston in September 1961, the surge pushed water as high as 13 feet above normal on the coast, but more than 19 feet above normal 20 miles inland in Lavaca and Galveston Bays.

Today, when a hurricane threatens any place along the U.S. coast, officials use a "MEOW" from a computer to help decide who should evacuate. "MEOW" is the "Maximum Envelope of Water" likely to be pushed ashore by a hurricane. Local MEOWs are available for every point along the Gulf and Atlantic coasts from Texas to Maine.

The National Weather Service's National Centers for Environmental Prediction use a computer program called SLOSH for "Sea, Lake, and Overland Surge from Hurricanes" to produce maps showing what kind of flooding to expect from just about any hurricane. For each basin, the local geography, including the shape of the ocean bottom as far as 75 miles out to sea, is fed into the computer model. Then simulations are run for each basin showing the probable paths of hundreds (in some cases more than a thousand) possible hurricanes of different strengths. Each hurricane simulation requires from a half-billion to one billion calculations. The resulting calculations are used to create a MEOW for each area.

Maps are recomputed for major changes, such as levees being raised in New Orleans.

Local officials use MEOWs to make evacuation and shelter plans. In some cases, important escape-route roads have been raised onto embankments after models showed these roads would flood before evacuation was complete. Anyone who is thinking of moving to an area where hurricanes hit should contact the local emergency management office to find out whether the house they might buy is in a flood-danger zone or in a place where storm surge could block the evacuation route long before a storm arrives.

When hurricanes focus their violence

In 1975 researcher Peter Black was aboard a DC-6 airplane flying in Hurricane Gladys over the Atlantic. Even though it was daytime, the rain and clouds were so thick Black couldn't see the ocean 800 feet below. "We hit sudden wind shear," Black says. "In 10 to 15 seconds, the wind dropped from 100 knots to zero. The plane stalled and what really scared me was it was total vertigo. There was no visual reference and I didn't know whether we were right side up or upside down." The pilots recovered control, but they were reluctant to make any more trips into the storm's clouds. Data from the onboard weather instruments offered few details about what the

One muddy, angry woman started his hurricane crusade

During the 1970s and 1980s Neil Frank became known as "Mr. Hurricane."

National Weather Service meteorologists are rarely well known, but when hurricanes threatened, Neil Frank became a familiar face on television. With his 1950s flat-top haircut, he obviously wasn't just another slick television personality. His earnest manner combined with blunt honesty — "our forecast could be off by 100 miles" — made him believable.

Before he became director of the National Hurricane Center, Frank had appeared on local television programs in Miami. He says the experience was invaluable. For someone new to television, "all the mechanical things clutter up your mind, make it difficult to be free with what you want to say. You wonder, 'Where is the camera? Where do you want me to look?' You see this guy out there waving his arms, what does that mean?"

During the hurricane off season, Frank carried the gospel of hurricane preparedness to emergency management officials, civic groups and journalists along the Atlantic and Gulf coasts. To coastal developers he often sounded like an Old Testament prophet, warning of dangers they discounted. He said over and over that he wasn't against coastal development; just that anyone living near the coast needed to know the dangers.

Frank says his concern about getting the hurricane message to the public began in 1965, while investigating Hurricane Betsy's storm surge in the Miami area. The surge had flooded houses on the right side of one street while houses on higher left side stayed dry. "On my right here comes a lady out of a house; she was mud from head to toe. I walked up in my bubbly way and said, 'I'm a scientist and I'm collecting scientific data.' Well, she lowered the boom on me. She said, 'Why didn't someone tell me I would have been safe on the other side of the street?' That made a big impression on me. Somebody ought to be going out and telling people those kinds of things."

After becoming Hurricane Center director in 1973 his main responsibility was making sure the best possible hurricane forecasts were issued. But Frank knew the "best possible" forecasts weren't going to be good enough. Coastal population was growing faster than forecasting skill.

He put together a slide show of what the Mississip-

Neil Frank, now at a Houston television station

When Neil Frank first started talking to local officials in 1973, few appreciated the danger of hurricanes. "I found the blind leading the blind. Bad hurricanes are rare events and not many people living along the coast have ever experienced a bad hurricane." Now, he says, most are well informed.

pi Coast looked like before and after Hurricane Camille in 1969. He also "cleaned up my language, got rid of the meteorological jargon" and started speaking to any group that would listen. "I spoke to one group of five people; they got the whole show, the whole treatment."

Even though he was giving more talks, Frank knew he was reaching relatively few people along the coasts. He began offering background hurricane briefings to the press. When a hurricane became news, he welcomed reporters into the Hurricane Center work space. They could talk directly with him and other forecasters, see what was on their computer screens and listen to discussions as decisions were made.

Hurricanes didn't become a part of Frank's life until he was out of college.

Two typhoon seasons as an Air Force weather officer on Okinawa, Japan, led to a fascination with hurricanes that prompted Frank to begin graduate studies in tropical meteorology at Florida State University.

He retired from the government knowing that, thanks at least in part to his efforts, people along the coasts are more aware of hurricane dangers than they were 20 years ago. "There's been a true increase in understanding." He wants to see that understanding leads to better coastal management; to making sure that anyone allowed to live where hurricanes hit will have an escape route when the warnings go up.

HURRICANES: MONSTERS FROM THE TROPICS

In simple terms, a hurricane is a giant machine that converts the heat energy of warm, tropical ocean water into wind. When a hurricane hits, it seems like a solid mass of clouds and winds with alternating downpours and lighter rain. But when meteorologists first began looking at hurricanes with radar after World War II, they discovered that the storms are really made of bands of thunderstorms, spiraling in toward the center -- the eye. This structure usually doesn't show up well on satellite photos because solid clouds often cover the storms' tops.

A TYPICAL HURRICANE'S LIFE STORY

DAY ONE
Thunderstorms begin growing in a disturbed area over the tropical Atlantic.

INSIDE A HURRICANE

Air flowing out from the center of the storm begins curving clockwise.

Warm, humid air is spiraling inward, speeding up as it approaches the center.

Evacuation of barrier islands and other dangerous places should be done before the outlying rain bands hit. They bring winds above 39 mph and downpours that can make driving impossible. Storm surge can be flooding low-lying roads by this time.

As the storm approaches the coast, a mound of water — the storm surge — builds to the right of the eye.

Winds spiraling upward in the eye wall — thick clouds surrounding the clear eye — are the storm's fastest.

Flooding by the storm surge begins before the eye hits land.

DAY THREE
Thunderstorms are organized into a swirl, The Hurricane Center calls it a tropical depression.

DAY FIVE
Winds top 39 mph; the Hurricane Center names it as a tropical storm.

DAY SEVEN
Winds reach 74 mph, making the storm a hurricane.

DAY TWELVE
The hurricane begins to weaken after hitting land.

Air is sinking 20 to 40 feet a minute at the storm's center, warming it and suppressing clouds. Winds are mostly calm.

Wind reaching the top of the storm, above 40,000 feet, in the eyewall and in bands of thunderstorms flows out. Water vapor left in the air forms cirrostratus clouds, which cap the storm.

Air is also rising under the bands of thunderstorms that make up the storm.

A hurricane is made of bands of thunderstorms that spiral around the center. These "rain bands" or "spiral bands" are typically 3 to 30 miles wide and 50 to 300 miles long

Some scientists fly their 'laboratories' straight into the eye of the storm

Science laboratories are not always in buildings on solid ground; sometimes they're airplanes in storms. From NOAA's airplanes, Jack Parrish, a flight meteorologist, has studied weather over polar ice caps, tropical oceans and around Midwestern thunderstorms. But flying into hurricanes is his first love: "Hurricanes are incredibly spectacular. There's something about a well-formed eye that keeps people going back into them."

A hurricane eye is always extraordinary, he says. The plane punches through the eyewall, which has "the heaviest rain, the darkest clouds, the least visibility. It's turbulent. Then the whole airplane starts to brighten up. It starts

Jack Parrish with one of NOAA's two WP-3 airplanes

to look the most pleasant when you hit the big updraft right inside the eye." Then the plane begins circling in the eye's calm, smooth air.

"The eyewall goes up to perhaps 60,000 feet and you're in the bottom of the bowl. It's an incredible view of this stadium all around you. Of course, there's always the slight apprehension that the only way back out is the way you got in. But it is spectacular.

"I wouldn't have believed it until the first time I was in a big one how calm it really is in the middle of a 200 mph hurricane." Below, the ocean is boiling with the hurricane's huge waves. "Here you are up at 5,000 or 10,000 feet floating around drinking a cup of coffee."

On hurricane flights, Parrish works as flight director, the scientist who helps decide where the plane should go. "It's a joint decision between the flight director and the flight commander, the pilot. If I have doubts or he has doubts, we don't do it." Parrish says researchers on a flight "tell me the scientific

goals; my job is to make it work."

He says a 10-hour ride into a hurricane can be more comfortable than a long airline trip. "People imagine the whole time you're in the envelope of a hurricane, you've got to be strapped in and things are tumbling about. In reality, the turbulence is highly confined to the rainbands and even that turbulence is moderate at worse. Only in the eyewall itself can it be severe. We have very straightforward procedures to make this stuff safe. Certainly, well over 99.9 percent of the time those procedures work just fine. Once in a while nature throws a curve. Things happen. They bring mortality right back to home."

The only plane to be lost in a hurricane was a Navy P-2V, which disappeared in Hurricane Janet in the Caribbean Sea in September 1955. Three Air Force planes have been lost flying into Pacific typhoons in more than 50 years of such flying.

While the WC-130s and WP-3D turbo-

props are better than jets for flying around in hurricanes at relatively low altitudes, they can't do two important jobs for researchers and the Hurricane Center.

Researchers want to find out more about what goes on at the top of a hurricane, more than 40,000 feet above the ocean. Hugh Willoughby, head of the NOAA Hurricane Research Division in Miami, says measurements there should answer questions about why hurricanes strengthen and weaken unexpectedly. Also, new computer forecasting models could do a much better job of predicting where a hurricane is heading if data were available for winds high above the oceans for thousands of square miles around the storm.

To fill these gaps, in 1996 NOAA purchased a Gulfstream IV business jet and turned it into a research airplane. It can fly to 45,000 feet and go more than 4,600 miles without refueling. Willoughby says the jet won't fly directly into the most turbulent parts of hurricanes as the WC-130s and WP-3Ds will continue doing.

Pilots and researchers flying in the jet will have the satisfaction of knowing they're gathering information that will help produce better hurricane forecasts. But they'll miss the unique experience that Parrish and others have as their slower airplanes bore into a storm's eye.

Still, they will probably echo Parrish when he says, "As someone fascinated by weather, I couldn't ask for a better catbird seat. It's just not any old job."

plane had encountered.

On Sept. 15, 1989, Black was aboard the NOAA WP-3D airplane that made the first penetration of Hurricane Hugo, over the Atlantic about 300 miles northeast of Barbados.

As the airplane flew into the wall of clouds around Hugo's eye, the wind increased from 67 mph to 184 mph in less than two minutes. A series of strong updrafts and downdrafts shook the plane violently. As it flew into the edge of the clouds surrounding the eye, it encountered air moving upward at 45 mph and then, four seconds later, air moving downward at 18 mph. While all of this was going on, the flight engineer shut down one of the plane's four turboprops that suddenly started running too fast, out of control.

Black was strapped in, next to a bundled life raft that broke lose from the straps holding it to the floor. It flew up, denting a one-inch steel pipe that ran the length of the airplane's cabin as a handhold.

With only three engines running, the airplane slowly climbed for an hour to smoother air. As it made 14 tight circles in the storm's eye, it encountered the swirl of winds nine more times. Fortunately, the violence decreased with height. Through it all, radars, weather instruments and navigation computers on the crippled WP-3D were recording data that later allowed Black and other scientists to reconstruct in great detail exactly what had happened in Hugo.

"What could have been a disaster turned out to lead to some of the most fascinating data we've ever collected," Black says. His colleague Hugh Willoughby is reminded of what the English author Samuel Johnson said in the 18th century: "Depend upon it, Sir, when a man knows he is to be hanged in a fortnight, it concentrates his mind wonderfully." The Hugo encounter "focused our attention marvelously," Willoughby says. "It got us started thinking."

In 1990, Black and others told scientific conferences what they had discovered. They called the violent swirls of winds "mesoscale vortexes" or "mesovortexes." These areas of rising, swirling winds are about 10 miles in diameter. That is around 10 times smaller than the 100-mile diameter of the strongest winds in most hurricanes and 10 times larger than the mile-wide individual thunderstorms that make up hurricanes. Black says mesovortexes are more likely in storms that are intensifying.

Hurricane Andrew was intensifying when it slammed into Dade County, Fla., before dawn on Aug. 24, 1992. Radar images, satellite photos and surveys of Andrew's damage leave no doubt that mesovortexes were responsible for the hurricane's worst damage, which was in relatively narrow swaths.

In a 1996 paper on what happened in Andrew, Willoughby and Black note that photos of Typhoon Ida taken from a high-flying U2 aircraft in 1958 show eyewall clouds wrapping around a mesovortex. Ida, like Hugo when it tossed around the WP-3D and Andrew when it hit Florida, was an intense storm that was rapidly strengthening, they said.

Black says that during a flight into Hurricane Luis in 1995, he photographed mesovortexes. At the same time, the new GOES 8 weather satellite was capturing images of the storm at one-minute intervals. Black and NASA scientists blended his hand-held photos with images from the WP-3D's radar and the satellite photos to produce animated videos of mesovortexes. The hope is that understanding how mesovortexes work and exactly what causes them will eventually allow forecasters to give an early warning of a severe hurricane on the way.

Using photos with satellite and radar images to reconstruct mesovortexes is a long way from William Redfield noting which directions trees in Connecticut had been blown down by a hurricane in 1821. Both, however, represent human minds, using the best tools available, stretching to understand how hurricanes create their incredible violence.

Can technology outpace growth?

Hurricanes have become national spectacles in the United States. Through constantly updated television and radio reports, even people on the Plains, in the Rockies and on the West Coast, who'll never experience a hurricane at home, are caught up in the drama of what a storm will do.

When a hurricane threatens the United States, television takes us directly to forecasters at the Weather Service's National Hurricane Center in Miami. Still satellite photos — shown in rapid sequence — make the hurricane seem almost like a living thing as we follow its advance toward land.

Today's forecasts combined with instant communication of warnings, better evacuation techniques and publicity about storm dangers are saving lives. The dollar cost of hurricane damage continues to rise. But the death toll has steadily decreased in the United States in this century.

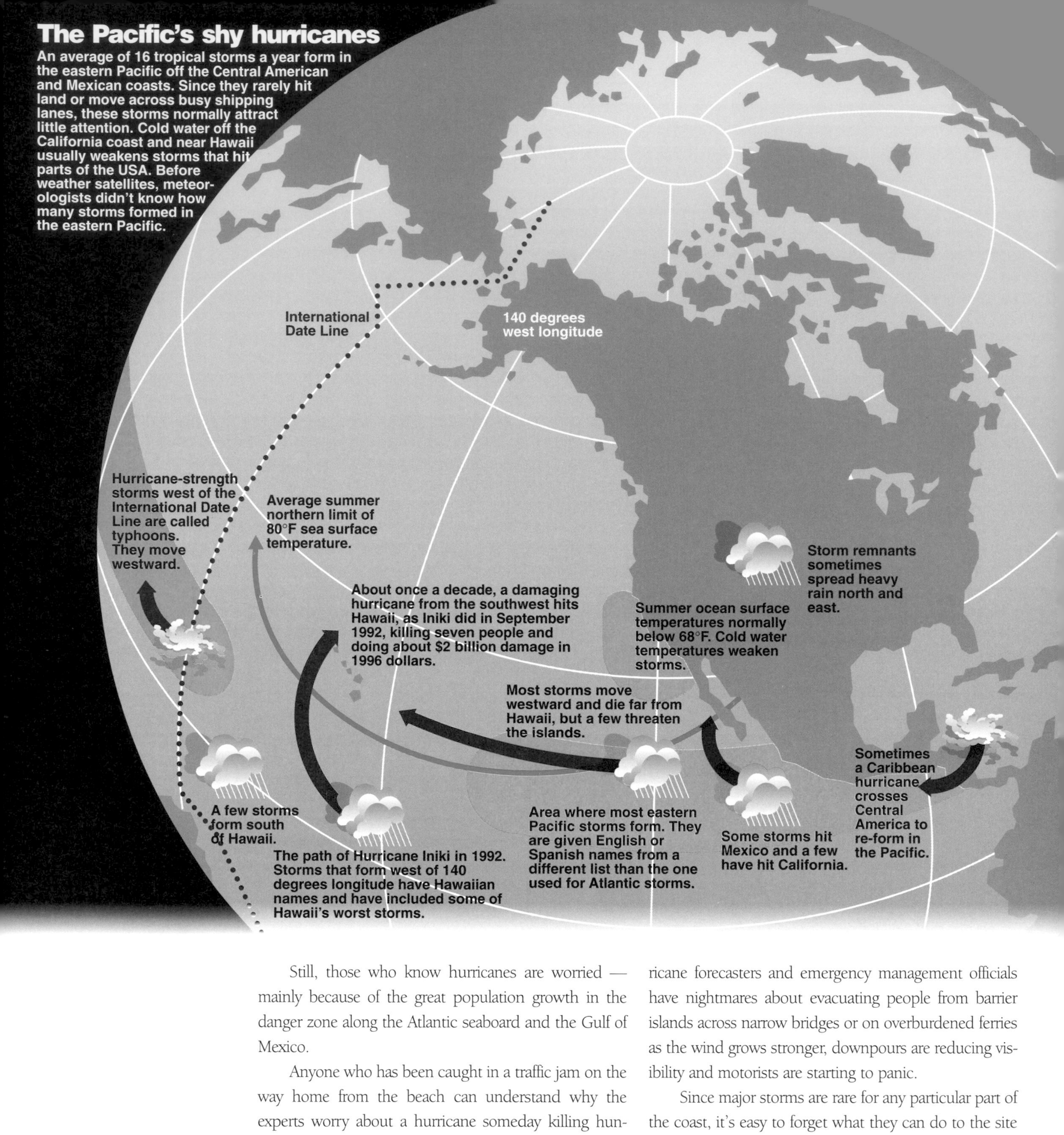

The Pacific's shy hurricanes

An average of 16 tropical storms a year form in the eastern Pacific off the Central American and Mexican coasts. Since they rarely hit land or move across busy shipping lanes, these storms normally attract little attention. Cold water off the California coast and near Hawaii usually weakens storms that hit parts of the USA. Before weather satellites, meteorologists didn't know how many storms formed in the eastern Pacific.

International Date Line

140 degrees west longitude

Hurricane-strength storms west of the International Date Line are called typhoons. They move westward.

Average summer northern limit of 80°F sea surface temperature.

Storm remnants sometimes spread heavy rain north and east.

About once a decade, a damaging hurricane from the southwest hits Hawaii, as Iniki did in September 1992, killing seven people and doing about $2 billion damage in 1996 dollars.

Summer ocean surface temperatures normally below 68°F. Cold water temperatures weaken storms.

Most storms move westward and die far from Hawaii, but a few threaten the islands.

A few storms form south of Hawaii.

Area where most eastern Pacific storms form. They are given English or Spanish names from a different list than the one used for Atlantic storms.

Some storms hit Mexico and a few have hit California.

Sometimes a Caribbean hurricane crosses Central America to re-form in the Pacific.

The path of Hurricane Iniki in 1992. Storms that form west of 140 degrees longitude have Hawaiian names and have included some of Hawaii's worst storms.

Still, those who know hurricanes are worried — mainly because of the great population growth in the danger zone along the Atlantic seaboard and the Gulf of Mexico.

Anyone who has been caught in a traffic jam on the way home from the beach can understand why the experts worry about a hurricane someday killing hundreds who are trying to flee. Many of the most populated beaches from Texas to Maine are on the 295 barrier islands that line the coast.

A barrier island is an especially dangerous place in a hurricane because it absorbs a storm's worst blows. Hurricane forecasters and emergency management officials have nightmares about evacuating people from barrier islands across narrow bridges or on overburdened ferries as the wind grows stronger, downpours are reducing visibility and motorists are starting to panic.

Since major storms are rare for any particular part of the coast, it's easy to forget what they can do to the site of your dream home. Not every place forgets. As shown in Chapter 1, the 1900 hurricane that killed more than 7,200 people in Galveston, Texas, is still much a part of local lore. In 1900 some people called Galveston "the New York of the South." History doesn't forget a storm that destroys such an important place.

But sometimes, new coastline residents don't known about their neighborhood's hurricane history.

In the 1970s and early 1980s, now-retired Hurricane Center Director Neil Frank traveled to hurricane-threatened areas with one message: storm preparedness. People in many places told him they didn't have a hurricane problem; a reef offshore or the curve of the coastline protected their beaches. South Carolina's Sea Islands, where golf courses and luxury houses and hotels were replacing the fields and tiny homes of farmers and fishermen, was such a place.

Frank says he asked an official on one of the islands "if they could evacuate and he said they don't see any reason to evacuate. And I said, 'If I told you 2,000 people died out there in 1893 when the islands went under water would you change your mind?' His face went white and he said, 'I never heard of that.' " The August 1893 hurricane, which seems to have been roughly the same strength as 1989's Hugo, hit between Charleston, S.C., and Savannah, Ga. At the time most of those living on the more remote islands were former slaves or their children and grandchildren. Estimates of the death toll range as high as 3,000.

Joel Chandler Harris, collector of the Uncle Remus stories, described the horror in two 1894 magazine articles: The islands, he wrote, "were exposed to the full fury of the tempest. And the winds fell about them as if trying to tear the earth asunder, and the rains beat upon them as if to wash them away, and the tide rose and swept over them twelve feet above high water mark. Pitiable as the story is, it may be condensed into a few words: near three thousand people drowned, between twenty and thirty thousand human beings without means of subsistence, their homes destroyed, their little crops ruined, and their boats blown away."

In 1893 the Weather Bureau had sent out alerts that a tropical storm was heading toward the Southeast Coast. But forecasters had no way to estimate where and when it would hit. In those days before radio, no one could warn people living on remote islands. And even if they had been warned, those in danger had no way to escape and nowhere to go.

Getting out the warnings today

While issuing warnings is easier today, the decisions are much more complicated. Hurricane warnings cause people to spend a great deal of money preparing for a storm.

Imagine that a strong hurricane is moving up the East Coast and computer models are hinting, but only hinting, that it could turn inland over Delaware Bay and southern New Jersey, pushing a storm surge as much as 11 feet deep into Atlantic City, N.J. If a warning is issued, the Atlantic City casinos would close and evacuations would start. If the forecast is correct, hundreds of lives could be saved. If it's wrong, Atlantic City would have a windy, rainy day that endangers no one. In either case, the casinos would lose around $8.8 million.

Forecasters want to avoid issuing hurricane warnings for too wide an area, not only because of the cost but because they fear warnings that turn out to be unnecessary will make residents less likely to heed the next warning. Like all other kinds of forecasts, hurricane predictions lose accuracy with time. A forecast of where a storm will be in the next 24 hours is much more accurate than one for where it will be in 48 hours.

By waiting as long as possible to issue a warning, a forecaster is less likely to warn areas that won't be hit. Yet, the warnings have to go out in enough time to allow those in danger to flee. Evacuating many threatened areas can take more than 12 hours.

Since hurricanes begin losing strength as they come ashore, evacuation orders or recommendations normally are limited to areas near the ocean, bays or rivers where storm surge threatens. A hurricane's strongest winds also are most likely to lash these areas. But deadly winds can reach far inland, as Hugo showed when it hit Charleston, S.C., in September 1989. Hugo's winds maintained 75 mph hurricane strength all the way to Charlotte, N.C., about 250 miles from the storm's landfall.

Watches and warnings issued for Hugo illustrate how forecasters consider both residents' reactions and the meteorology. "It's always a very narrow line you walk in trying to provide sufficient warning to protect life, but yet minimize the overwarning," says Robert Sheets, who was Hurricane Center director. "We try to use a process that slowly elevates the level of alert."

With Hugo still more than two days away from hitting the East Coast, "we started out by saying this hurricane will likely affect the Southeast Coast within the next two to three days somewhere between central Florida and North Carolina," Sheets says. "What we're trying to generate is that people who had planned to go on vacation to those areas have second thoughts."

At 6 p.m. on Sept. 20, when Hugo was less than a

Hurricane Iniki hit the Hawaiian island of Kauai on Sept. 11, 1992, killing seven people and doing an estimated $2 billion damage. It destroyed or damaged most of the island's tourist hotels and 10,000 homes. This made it the state's worst natural disaster.

The state's second and third worst disasters were also hurricanes that hit Kauai: Iwa in 1982 and Dot in 1959.

How Atlantic Basin hurricanes get their names

Long before forecasters made it official in 1950, people were attaching names to hurricanes.

In Spanish-speaking areas, storms were sometimes named for the saint's day on which they hit. This is why Puerto Rico had two San Felipe hurricanes — one, Sept. 13, 1876; the second, Sept. 13, 1928.

In 1941, George R. Stewart's novel, *Storm,* a popular book among meteorologists, featured a forecaster who used women's names for extratropical storms.

The idea took hold. During World War II military forecasters informally attached women's names to storms.

After the war forecasters continued describing hurricanes by their location. But twice in 1950 three hurricanes were being tracked at the same time; bulletins were sometimes confusing. The international phonetic alphabet — Able, Baker, Charlie, etc. — was used for storms from 1950-1952.

From Ana to Wilfred

Six-year cycle for names of Atlantic Ocean, Caribbean Sea or Gulf of Mexico hurricanes:

1997	1998	1999	2000	2001	2002
Ana	Alex	Arlene	Alberto	Allison	Arthur
Bill	Bonnie	Bret	Beryl	Barry	Bertha
Claudette	Charley	Cindy	Chris	Chantal	Cesar
Danny	Danielle	Dennis	Debby	Dean	Dolly
Erika	Earl	Emily	Ernesto	Erin	Edouard
Fabian	Frances	Floyd	Florence	Felix	(not chosen)
Grace	Georges	Gert	Gordon	Gabrielle	Gustav
Henri	Hermine	Harvey	Helene	Humberto	Hortense
Isabel	Ivan	Irene	Isaac	Iris	Isidore
Juan	Jeanne	Jose	Joyce	Jerry	Josephine
Kate	Karl	Katrina	Keith	Karen	Kyle
Larry	Lisa	Lenny	Leslie	Lorenzo	Lili
Mindy	Mitch	Maria	Michael	Michelle	Marco
Nicholas	Nicole	Nate	Nadine	Noel	Nana
Odette	Otto	Ophelia	Oscar	Olga	Omar
Peter	Paula	Philippe	Patty	Pablo	Paloma
Rose	Richard	Rita	Rafael	Rebekah	Rene
Sam	Shary	Stan	Sandy	Sebastien	Sally
Teresa	Tomas	Tammy	Tony	Tanya	Teddy
Victor	Virginie	Vince	Valerie	Van	Vicky
Wanda	Walter	Wilma	William	Wendy	Wilfred

From 1953 through 1978 women's names were used. By 1979 men's names made the list, too, and not just English ones. French and Spanish names were added.

A storm is named when its winds reach 39 mph, tropical storm strength.

Countries affected by hurricanes suggest names; they are approved by the World Meteorological Organization's Region 4 Hurricane Committee, which includes representatives of the affected nations.

day and a half away from land, the Hurricane Center issued a hurricane watch for the coast from St. Augustine, Fla., to North Carolina's Outer Banks. At that time, the likely landfall couldn't be pinned down any more precisely.

While a watch means people should be ready to evacuate on short notice, it also encourages many to leave without waiting for a warning. "We know these people who went out early are also the people that are least likely to complain after the fact," Sheets says. "They're the cautious people that we love."

When the warning was issued for a narrower area the morning of Sept. 21, Sheets says, people still on barrier islands "were not encumbered by those people who had already left or who were planning to come for vacation and didn't come."

At 3 p.m., the South Carolina governor's office reported that almost everyone was off the threatened islands. The fringes of Hugo began whipping the islands around 6 p.m. and the eye came ashore about midnight.

Evacuation also worked well in crowded South

Florida when Hurricane Andrew took aim on Miami in August 1992. Around three-quarters of a million people fled mobile homes, even those far inland, and areas threatened by storm surge.

Andrew wasn't as cooperative with forecasters as Hugo had been. On Friday, Aug. 21, Andrew seemed likely to turn northward over the Atlantic. "At worst it looked like it would be no farther west than the eastern Bahamas by Sunday," Sheets says. "We told emergency managers and people who called that if they've done what they should have done by June 1 to prepare for the hurricane season to enjoy the weekend and tune back in Sunday and Monday."

On Saturday, however, conditions changed and Andrew headed for a Monday morning date with Miami. "We had to bring the level of interest up quickly," Sheets said. "The response system worked quite well if you look at it in terms of loss of life compared with the damage." Andrew killed 58 people while doing more than $27 billion in damage — more damage than all of the hurricanes in the 1980s combined.

How hurricanes are forecast

In terms of saving lives, hurricane forecasts have been quite successful since the 1960s. While hurricane damage costs have skyrocketed, hurricane death tolls have remained low.

Forecasters have saved lives by announcing warnings early so people could evacuate.

The forecasts come from the National Weather Service's National Hurricane Center on the campus of Florida International University in Miami. Over the years, the Hurricane Center's director and many who work there have become public figures as television networks set up their cameras for live reports whenever a big storm threatens.

Forecasting hurricanes involves a lot more than looking at satellite photos and telling television viewers what's going on. Weather observations and projections from several computer models are flowing into the Hurricane Center to be analyzed. Conference calls are set up with emergency management officials in threatened areas. Statements and warnings are discussed, written and sent on their way.

Satellite photos have become an important part of broadcasts from the Hurricane Center because there's no better way to quickly show a storm's location and size. And showing satellite photos of a menacing hurricane can prod residents to evacuate.

Weather satellites provide invaluable information about where a storm is located, although it's not always possible to pinpoint a storm's center with only satellite views. Satellites also give clues about a storm strengthening or weakening. But satellite photos alone can't tell a forecaster where a hurricane is going to be in 12 or 24 hours, much less 48 or 72 hours. Satellite pictures also can't predict the strength of a storm at landfall. Yet that's just the information needed to issue warnings.

To predict hurricane movement, a forecaster needs to know what will happen to the speed and direction of the winds at different heights — from right above the ocean to more than 40,000 feet up for hundreds of miles around the storm. These winds are the storm's "steering currents."

Forecasts for upper-air winds are based primarily on readings from weather balloons. These go aloft twice a day all over the world.

Since the mid-1990s automated wind reports from large jet airliners have been supplementing balloon readings, greatly increasing the amount of data. But airliners cruise at about 33,000 feet and for most of the Atlantic, there is no data from lower altitudes.

Satellites also supply some information about upper-air winds as well as data to help calculate what the winds should do, but not always in the detail hurricane forecasters would like.

To gather information about winds for thousands of square miles around storms, NOAA has outfitted a Gulfstream IV business jet for weather work. Instead of a corporate jet's designer interior, the NOAA Gulfstream's cabin is a workplace for scientists and technicians who sit in front of utilitarian consoles filled with dials and cathode-ray tubes for monitoring the atmosphere. Besides taking weather measurements around the plane, the scientists drop instrument packages that radio back data as they fall to the ocean under small parachutes.

All the extra data the Gulfstream collects would be useless without the computer models that are the mainstay of today's weather forecasting.

From hand-drawn maps to super computers

When Miles Lawrence began at the Hurricane Center in 1966, forecasters were using data from weather stations, balloons and ships hundreds of miles around a storm to help them draw weather maps of what the atmosphere was doing. "They had techniques such as steering storms using [air pressure patterns] on weather maps. I have a great deal of respect for the people who were doing the forecast at that time. There were some super forecasters who did a really great job with what they had available."

The big change over the past decades has been "numerical forecasting," using computers to process data and produce forecasts. "I'm convinced numerical forecasting is the only way," Lawrence says. "There's just too much data for an individual to process. With data from all around the world at several levels of the atmosphere, a person can only do so much. A computer can make all of the calculations with great precision and come up with a forecast that people can't do by hand."

Even with today's computer models, forecasters "don't just look at a line on a computer map and say, 'That's where it's going, I'll issue warnings for 60 miles on each side of the line,'" says Mark DeMaria, chief of technical support at the Hurricane Center. "In certain cases none of the models are right."

He cites 1996's Hurricane Hortense. When it was south of Puerto Rico, none of the models had the center

Longest period on record without a hurricane hitting the United States: 3 years, 7 days from Hurricane Allen, Brownsville, Texas, Aug. 10, 1980 to Hurricane Alicia, Galveston, Texas, Aug. 17, 1983.

Second longest: Between an unnamed storm that hit the Florida Keys on Sept. 28, 1929 to one that hit near Galveston, Texas, on Aug. 14, 1932.

Amid the destruction of Hurricane Hugo, at right, Erica Martinez comforts Elaine Barto whose house was destroyed at Folly Beach, S.C., Sept. 21, 1989.

Hurricane Kate hits Key West, Fla., at left, on Nov. 20, 1985. The hurricane lasted from Nov. 15 to 23, causing seven deaths in the U.S. and $1 billion in damage.

Only two Category 5 hurricanes with winds over 155 mph hit the USA between 1900 and 1996: Camille in 1969 (256 deaths) and the unnamed 1935 Florida Keys storm (408 deaths).

crossing the island. But the Hurricane Center issued a warning anyway because forecasters knew the kinds of errors models had been making that year. Also, they knew the storm would bring downpours, flooding and maybe hurricane force winds even if the storm's center only brushed Puerto Rico.

As it turned out, Hortense blasted the island with wind gusts of 75 mph and more than a foot of rain. Towering waves ate away beaches and flooded the southern shoreline, closing roads and bridges. In San Juan, wind gusts hit 60 mph, while winds over 39 mph raked the island. Torrential rain unleashed devastating floods, killing eight people.

Sometimes warnings that are not issued can be as important as those that are.

Hurricane Emily is a good example. It was a Category 3 hurricane as it menaced North Carolina's Outer Banks at the end of August 1993.

Forecasters knew Emily was getting ready to turn. But when? If it turned sharply back out to sea, it would whip the Outer Banks, maybe areas as far north as Delaware Bay, with wind and rain, but not hurricane force winds. If the turn were slower it could rake crowded beaches along the New Jersey shore with hurricane

winds and storm surge on summer's last weekend. It would also threaten Long Island and New England.

Robert Sheets, then Hurricane Center director, knew that if he posted warnings for the New Jersey coast it would cost millions for the emergency services that would have to be set up. The Hurricane Center estimates that preparations cost more than $500,000 per mile of coast put under a hurricane warning.

If Sheets didn't issue warnings and Emily did hit, scores of people could die. Sheets, with confidence based on his airplane data and a new hurricane forecasting computer model, elected not to issue the warnings north of Delaware Bay. Emily did as forecast and turned away from the coast.

The airplane data came from flights on Aug. 30 and 31. NOAA WP-3D airplanes flew a few hundred miles around the storm collecting data. Such "synoptic flow experiments" in 10 other hurricanes since 1982 had shown the extra data could improve forecasts. These results were a major argument for buying the Gulfstream jet. The two flights for Emily cost about $80,000, much less than the cost of a hurricane warning for even a mile of coast.

The new computer model was developed at

NOAA's Geophysical Fluid Dynamics Laboratory in Princeton, N.J., by Yoshio Kurihara and collaborators Bob Tuleya and Morris Bender. The Hurricane Center had been gaining confidence in the new GFDL model during earlier storms. While conventional models were predicting Emily would hit Georgia, the new model forecast it would brush the Outer Banks and turn away, as it did.

Lawrence says, "Over the period I've been here, it's been getting easier to [make forecasts] because of greater confidence we're placing in models. Now I look at the GFDL output and ask myself 'Is there any reason I have to deviate the official forecast from what the model is showing?' "

Added trust in the new models isn't blind faith. Forecasters look at results of hurricane models run by not only the National Weather Service, but also the Navy and British Met office, as they plot the official track forecast. The size of an area put under a watch or warning depends on the forecasters' confidence in the models. Less confidence means putting a larger area under a watch or warning.

When they issue watches and warnings, forecasters "can't help but be aware that there's an impact," Lawrence says. "My personal attitude about that is to be as detached as possible. I feel that I'm a mechanic.

"The less emotionally involved you can be, the better the chance you will come up with the best decision. I try to stick to the meteorology and come up with the best I can based on the meteorology. Then, if the impacts are awesome, I take a moment to reevaluate."

Intensity predictions are the big challenge

While forecasters have been getting better at predicting where storms will go, "we have no skill at all in intensity forecasting," says Hugh Willoughby, head of NOAA's Hurricane Research Division.

According to Willoughby, a typical hurricane will grow slowly and stay as a Category 1 or 2, or maybe a weak Category 3, until it hits land or turns away to die over the ocean. But the most intense storms strengthen rapidly in a day or two. In 1992, Hurricane Andrew went from a Category 1 to a Category 4 in 36 hours. In 1969, Camille went from a Category 1 to a 5 in two days.

Most storms go through cycles of eyewall replacement as they strengthen and weaken. The eyewall is the ring of thunderstorms around the calm eye containing the fastest winds and heaviest rain. Often, a new ring of thunderstorms will form around and outside of the eyewall. As this ring grows, it begins stealing warm, humid air blowing in toward the eye, starving the original eye-

Homestead, Fla., was the scene of great devastation after Hurricane Andrew came ashore in August 1992.

Cycle could be swinging back to more big East Coast hurricanes

Number of Category 3, 4 and 5 hurricanes hitting the East Coast and Florida Peninsula seem to go in two-to three-decade cycles. The big map shows the number of storms in the 25 years from 1941 through 1965. The small maps show major hurricanes from 1966 through 1990 and from 1991 through 1996. Scientists think the U.S. could be already beginning a new cycle like that of the 1940s, '50s, and early '60s. Gulf of Mexico hurricanes do not seem to follow this pattern.

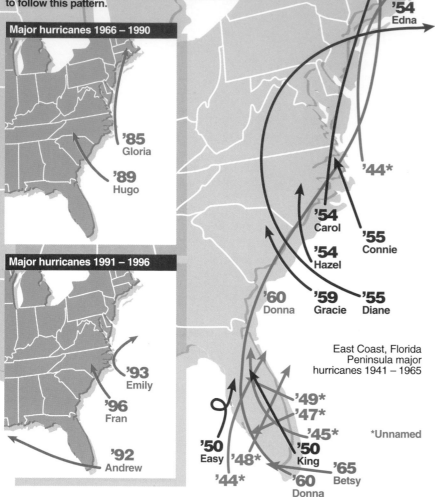

Major hurricanes 1966 – 1990

'85 Gloria

'89 Hugo

Major hurricanes 1991 – 1996

'93 Emily

'96 Fran

'92 Andrew

'54 Edna

'44*

'54 Carol

'54 Hazel

'55 Connie

'60 Donna

'59 Gracie

'55 Diane

East Coast, Florida Peninsula major hurricanes 1941 – 1965

'49*

'47*

'45*

*Unnamed

'50 Easy

'48*

'50 King

'65 Betsy

'44*

'60 Donna

Hurricane Joan in October 1988 was the first Category 4 storm on record to hit Nicaragua. Then it crossed Central America into the Pacific and changed names, becoming Tropical Storm Miriam. In 1971, Hurricane Irene crossed into the Pacific to become Tropical Storm Olivia.

wall of the humid air it needs to keep its thunderstorms going. Also, air is slowly descending in a storm's eye. As air begins sinking inside the new, larger eyewall, it suppresses the original eyewall thunderstorms.

Eventually the new outer ring of thunderstorms becomes the storm's eyewall. The eye itself is now larger and the highest winds slow down. The winds slow as the eye becomes larger for the same reason that an ice skater spins more slowly as she holds out her arms. Then the new eyewall contracts, the eye grows smaller, and the winds speed up again. The shrinking eye has the same effect as a skater pulling her arms in and spinning faster.

Sometimes, as with a Camille or Andrew, the strengthening part of this cycle goes wild. Some of the worst storms are ones like Andrew, which was growing

stronger as it slammed into Dade County, Fla. While Hurricane Opal in 1995 and Fran in 1996 did tremendous damage, they could have been much worse. Both were in the weakening part of the cycle when they hit land.

For a storm to grow into a monster like Camille or Andrew, it needs to be over very warm water. The theoretical strength a hurricane can reach depends on the contrast between the temperatures of the warm ocean at the bottom of the storm and the cold stratosphere at its top.

The greater the contrast, the stronger the storm can become.

Also, says Willoughby, the winds around the storm can't be trying to rip it apart. This wind shear, along with ocean and air temperatures, are easy to measure.

Willoughby says both theoretical work and data collected in hurricanes Iris and Luis in 1995 indicate that interaction between atmospheric waves above 40,000 feet and the hurricane itself can affect the intensity of the thunderstorms that make up the hurricane. This interaction can produce the energy that makes a storm strengthen.

The Gulfstream jet should help researchers learn more about these interactions. Then, if the theory turns out to be correct, measurements by the high-flying jet could give forecasters the data they need to predict when a storm is going to strengthen or weaken.

No one is eager to go charging into the top of a hurricane in the jet. "Our experienced storm pilots aren't jet pilots and experienced jet pilots aren't experienced storm pilots," Willoughby says. "There will come a day when there's a storm with a 40-mile-wide eye and blowing 100 knots. It's very close and everybody will look at each other and say, 'that's the one.' "

After such a tentative first step into a weak storm, more flights will be made into the tops of storms. Scientists will write research papers and develop new computer models based on what's learned. Eventually, Hurricane Center forecasters will begin making good predictions of how strong hurricanes will be when they hit land. "I hope that by the year 2005," Willoughby says, "we'll be able to say we can now forecast intensity changes because of research that began in 1995."

Keeping the death toll down

Sheets, who retired as Hurricane Center director in 1994, is now a hurricane consultant. He wishes local officials who allow development on coastal islands would spend more time worrying about evacuating new residents when a storm threatens. "They should think about

Marie Whitley and James Evan survey damage after Hurricane Andrew struck South Florida in August 1992.

extra bridges, raised bridges, four-lane roads," he says.

Hurricane Andrew added to officials' fears. Unlike many hurricanes, Andrew's winds, not the 16.9 feet of storm surge it pushed into parts of Biscayne Bay, did most of the damage. As a result, Sheets says, "everyone in Dade County is saying 'When the next one comes, I'm getting out of here.' If that happens we'll have a loss of life on the beaches because traffic will be clogged, we won't get everyone out."

Evacuating almost all of South Florida, not just mobile homes or areas threatened by surge, would take more than three days. To begin in time, explains Sheets, "we'd have to issue warnings when storms are out by Puerto Rico. We'd have at least 15 false alarms for each storm that hits." The only solution is to make local shelters able to stand up to winds like Andrew's — and then to make residents aware that the shelters will keep them safe.

Sheets says he finds it hard to imagine a hurricane killing thousands of people in the United States. Although he says that's certainly possible if you had people trapped on the barrier islands. "It's my belief that if we have a major loss of life in a hurricane it is going to be people in their vehicles trapped on the roads, not people in their houses."

Another era of intense hurricanes could be here

From 1941 through 1965, 16 major hurricanes hit Florida or the East Coast. Then, for some reason, the hurricanes nearly stopped. From 1966 through 1993, only four major hurricanes hit. And 1994 had no major hurricanes.

But then, the 1995 hurricane season arrived with a vengeance. From June through November, 19 tropical storms formed in the Atlantic, Caribbean or Gulf of Mexico. Eleven became hurricanes and five grew into major hurricanes, making 1995 the first year since 1964 with five major storms. It was the second most active season in 125 years.

Hurricanes continued brewing in larger-than-average numbers during 1996. Thirteen tropical storms produced nine hurricanes. Six were major.

The total of 32 tropical storms in 1995 and 1996 was the largest number in two consecutive years since the National Hurricane Center was organized in 1935. The two-year total of major storms, 11, was more than the five-year average.

What's going on? Are such wild swings in numbers of hurricanes mere statistical flukes, or is there a decades-long cycle in major Atlantic and Caribbean hurricanes?

William Gray of Colorado State University studies tropical cyclones all over the world. In the mid-1980s, he began asking why fewer major hurricanes were hitting the Florida Peninsula, the U.S. East Coast and the Caribbean.

As Gray compared the number of hurricanes with global weather patterns, he discovered statistical correlations between the number of hurricanes over the

Atlantic Basin and faraway events, including rainfall in West Africa and unusually warm ocean water temperatures in the tropical Pacific. The statistical links are strongest between these factors and Category 3 or stronger hurricanes that form far out in the Atlantic. These are called Cape Verde storms because they begin taking shape near the Cape Verde islands off Africa. They grow into monsters as they move west to the Caribbean Sea, Bahamas or the U.S.

While the statistical links are interesting, meteorologists look for physical links — "A causes B, which in turn causes C." It turns out that faraway factors such as Pacific Ocean temperatures, through complex atmospheric links, affect the ingredients that hurricanes need, especially winds blowing at about the same speeds and from the same direction at all levels of the atmosphere. All the needed ingredients are shown in the graphic on Page 142.

One of the most important of the faraway events influencing Atlantic hurricanes is an "El Niño" warming of the tropical Pacific. (Chapter 12 describes El Niño.) When unusually warm water spreads east along the equator in the Pacific during an El Niño, it helps create huge thunderstorms in the eastern Pacific. These pump large amounts of air upward, which end up blowing eastward over the tropical Atlantic as strong upper-atmospheric winds. These are strong enough to rip storms apart before they have time to grow into major hurricanes.

An unusually strong, long-lasting El Niño from 1990 well into 1994 created winds that destroyed most Atlantic hurricanes during those years. That El Niño finally disappeared before the start of the 1995 season.

Fortunately for those who live and vacation along the U.S. East Coast, the Gulf of Mexico and the Caribbean, the Atlantic Basin is not a particularly favorable environment for growing hurricanes. Over the long term, an average of only six hurricanes form in the Atlantic Basin during each year's six-month season, and only two of those become major. Many hurricanes never touch land.

The hurricane 'sluice gate'

Over the long term, Cape Verde storms account for about 85 percent of the Atlantic's major hurricanes. Stanley Goldenberg of NOAA's Hurricane Research Division in Miami notes that all of these storms pass through a rectangular area stretching from the Lesser Antilles, which separate the Caribbean from the Atlantic, east about 1,200 miles and from South America's northern

coast to the 20 degrees north latitude line. This is where storms usually grow into major hurricanes.

Goldenberg compares this area to a "sluice gate," like one that controls the flow of water in a channel. During years such as 1991 to 1994, global weather features, including upper-altitude winds linked to El Niño, close this sluice gate and no major hurricanes get through. Some major hurricanes form away from the sluice gate, but they are rare. In an average year, about three or four Cape Verde hurricanes make it through the gate and a couple of those become major. In some years, a half dozen or more make it through and most of them grow to 111 mph or stronger. That is what happened in 1995.

By the middle of the 1990s, Gray, Goldenberg and other researchers were beginning to think that water temperature differences could be the leading operator of the "sluice gate." From satellites and research ships, they knew that a great ribbon of water, a global "conveyor belt," moves warm water northward at the top of the Atlantic and cold water southward deep inside the ocean. Since the conveyor belt moves warm water north in the Atlantic, the faster it's running, the warmer the North Atlantic becomes. (How the conveyer belt works and its other effects on weather and climate are discussed in Chapter 12.)

In the late 1960s, the ocean conveyor belt slowed and the North Atlantic cooled slightly. But by the mid-1990s, the North Atlantic was warming, probably because the belt was speeding up. Scientists are just beginning to understand details of the conveyor belt; they can't say for sure it is why the North Atlantic warmed. But there's no doubt that the North Atlantic became warmer in the 1990s and that the North Atlantic also was warmer during the period of intense U.S. hurricanes from the 1940s into the 1960s.

"We're not 100 percent sure we've reached this new era," Gray says, "but the likelihood is high."

Small changes make big differences

By everyday standards these ocean temperature changes are tiny — one degree or less. But such a seemingly small difference spread over thousands of square miles of ocean represent a lot of heat.

He says ocean temperatures help determine the strength and direction of winds over the "sluice gate." When the North Atlantic is relatively cool, winds high above the tropical Atlantic tend to blow from the west and at high speeds. As the jet stream graphic on Pages 38 and 39 in Chapter 3 shows, the stronger the contrasts between air temperatures, the faster upper-air winds blow. In the "sluice gate," the faster moving winds tend to rip apart hurricanes before they can form.

When the North Atlantic is relatively warm, as it was by the middle of the 1990s, the upper-level winds and surface winds are weaker, allowing hurricanes to grow. Gray says maps of upper-air winds over the tropics in 1995 and 1996 reminded him of maps from the 1950s.

Is global warming affecting hurricanes?

By the early 1990s, some people were saying that global warming was responsible for an increasing number of hurricanes. But at the time these ideas were attracting attention, the number of Atlantic Basin hurricanes was in a 30-year decline. Most hurricane researchers and the UN's Intergovernmental Panel on Climate Change (IPCC) don't see any links between global warming and hurricanes.

The argument that global warming could increase the number and strength of hurricanes is based on the idea that the oceans would be warmer. It's true that warm water is an essential ingredient for hurricanes. Also, a hurricane's theoretical strength depends on the contrast in temperature between the warm ocean and the cold stratospheric air at the storm's top. The greater the contrast, the stronger a hurricane can grow. But global warming could change other ingredients, such as the speed and direction of upper-air winds, in ways that could work against hurricanes.

"Those of us who study these storms don't buy" warming as the cause of an increase, Gray says. "All we can say now is we can't say much" about how global warming would affect hurricanes. Adds Goldenberg: "The type of slow, gradual, small changes that would be expected from global warming are very different from the kinds of changes we're seeing."

Cycles aren't the whole story

Even "quiet" hurricane seasons can bring a few major storms. In 1983, Hurricane Alicia formed and grew into a major hurricane in the Gulf of Mexico only 200 miles south of the Louisiana Coast and then smashed into Galveston and Houston doing $3.1 billion damage in 1996 dollars.

In fact, the three strongest hurricanes to hit the USA from 1900 through 1996 were not Cape Verde hurricanes. Both the 1935 unnamed Labor Day hurricane that hit the Florida Keys (the strongest in 96 years) and Hurricane Andrew, which hit Dade County, Fla., in 1992, were what Gray calls "Bahamas busters." These are storms that grow into major hurricanes over the Atlantic north of 25 degrees north latitude, in the general area of the Bahamas. The second strongest hurricane ever recorded, Camille, which hit Mississippi's Gulf Coast in 1969, formed in the western Caribbean Sea south of Cuba and quickly grew into a Category 5 storm as it moved northward across the Gulf of Mexico.

A personal note about the terror of hurricanes

These storms, especially the 1935 monster, worry forecasters because their quick growth gives less time to evacuate vulnerable areas such as the Florida Keys and New Orleans.

No matter what hurricane cycles do, people who live along the USA's East and Gulf coasts are in danger of experiencing terror like that Stanley Goldenberg and his family survived in the pre-dawn hours of Aug. 24, 1992. Goldenberg, his three young sons, his wife's sister, her husband, and their three sons huddled under a blown-down kitchen wall as Hurricane Andrew ripped apart his house. His wife, Barbara, was in Miami's Doctors Hospital where she had just given birth to their daughter Pearl.

The storm's noise was so loud, Goldenberg says, "our mental noise meters were pegged." That is, their brains didn't register the added noise when the house's roof flew away and hit a neighbor's house. "When the roof came off, the wall between the kitchen, where we were, and the living room fell on us. The fridge braced the wall. We were under that little lean-to. It's incredible that nine people and one small kitten survived like that." No one was seriously hurt.

Combining theory and observation to probe hurricanes

For Hugh Willoughby, hurricanes are other-worldly places to be explored with mathematics and airplanes.

"The things happening near the center of a tropical cyclone are so exotic that conditions in the core differ from the Earth's day-to-day weather as much as the atmosphere of another planet does," says Willoughby, director of the NOAA Hurricane Research Division in Miami.

Intellectually, hurricanes represent complex problems in fluid dynamics, the science of the movement of fluids such as air and ocean water. Swirling vortexes — winds whirling into a hurricane's eye, whirling seas and winds around the hurricane — can be described mathematically.

These equations may distill the essence of a hurricane, but they aren't a hurricane.

That you find by getting into an airplane and flying into the core of a storm as Willoughby has been doing since 1970 when, as a Navy weather officer, he flew from Guam into Pacific typhoons.

"The whole process of getting into an airplane and flying out into a storm is compelling," he says. "How many people get to do that every summer? I'm interested in the physical problems of how a hurricane works. I find that what we see in the air is like looking in the back of the book for the answer. Casual visual observations give you hints about what's going on."

"In my own career, I've gone through times in which I'm lost in the equations." Flying into a storm restores a balance. But things can swing the other way. "In the airplane you have to worry about being too focused on things like how much gas is left" instead of the science of the storm.

While hurricanes are complex, Willoughby says "the thing that attracted me to hurricanes is that mathematically they are a lot simpler than day-to-day weather." The day-to-day weather with millions of interactions creating events as different as heat waves and blizzards

Hugh Willoughby, head of hurricane research, doesn't fly as often as he would like.

"just sort of overwhelms you."

Hurricane researchers try to discover things that will help forecasters do a better job of saying where a hurricane is going and how strong it will be when it hits. Thanks to such research, forecasters have been getting better at predicting hurricane paths. But they still have little skill at saying whether a storm will grow stronger right before it hits, as Andrew did in 1992, or weaker as Opal did in 1995.

Willoughby complains that as head of hurricane research he doesn't get to fly into as many storms as he would like. He did fly into Opal as it advanced across the Gulf of Mexico toward Florida Oct. 4, 1995. Shortly before takeoff, he was told that Opal had exploded overnight into a 150 mph monster. "As we were flying out, thoughts of Camille Two ran through my mind. I was wondering how I would explain this to the congressional inquiry we would have when a thousand people died." Hurricane Camille killed about 140 people when it hit Mississippi in 1969.

Instead of being another Camille, Opal weakened before landfall and did far less dam-

age than feared only hours before. "The forecasts were state of the art, but the state of the art doesn't handle these situations well."

Willoughby says he's "really kind of optimistic" that scientists will learn enough about how hurricanes work to give forecasters the tools they need to say when a storm is going to explode or weaken.

"This isn't like saying we're going to go to the moon in 10 years. One thing that's a constant source of hilarity is that we run into defense contractors looking for projects. They really think hurricane research will support procurement approaching a billion dollars. They're in for a big culture shock."

In 1996, NOAA and the Navy, which forecasts typhoons in the Pacific, were together spending around $50 million a year for hurricane research and forecasting. Willoughby's lab's budget was around $3 million.

"I'd like to see another million a year," Willoughby says. "With that I think we could do what we need to do" to begin making real improvements in forecasts for hurricane intensity. "Our wildest dreams are cheap."

Don't look out the window. If you're outside, don't look up. Now, describe the sky.

Is it clear or cloudy? If there are clouds, do they cover all of the sky or only part? Are the clouds mostly puffy, or flat, or a combination of the two?

If you failed this test, don't feel alone. Few people today pay much attention to the sky, which helps explain why many of its common objects, such as the planet Venus or lens-shaped clouds, sometimes are reported as "unidentified flying objects."

A few decades ago people looked to the sky often to pick up clues about what would happen with the weather. Today, we're more likely to look at a television screen for weather information. But if you get your forecasts only from radio or television, you're getting only part of the picture.

Combining weather information from television, radio or newspaper reports with what you see in the sky is the best way to understand the weather. Once you start looking at the sky regularly, you'll see a host of phenomena, not only clouds, but also various optical effects caused by the interplay of light with the atmosphere and clouds.

In this chapter we will first look at some of the sky's optical phenomena, beginning with why the sky is blue. Then we'll look at some of the many cloud forms and what they tell us about the weather. Finally, we'll introduce you to the atmosphere's layers.

Why is the sky blue?

If you could travel to Earth from somewhere else in the universe, the first thing you'd notice from far away is a blue dot surrounded by totally black space. As you drew closer you'd see that it's blue and white with some brown and green parts. Entering the atmosphere, you'd see the sky around you change from black to a

Primary and secondary rainbows over Tucson, Ariz.

You can't take a photo showing an entire rainbow with a standard camera lens.

This is as true for small rainbows made by a lawn sprinkler as for a rainbow that stretches across the sky.

All rainbows, large and small, make an arc of about 42 degrees. That would take a very wide-angle lens.

163

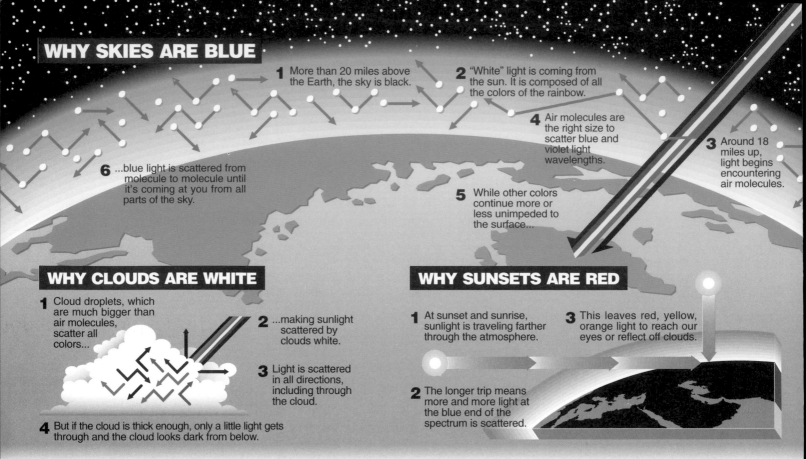

WHY SKIES ARE BLUE

1 More than 20 miles above the Earth, the sky is black.

2 "White" light is coming from the sun. It is composed of all the colors of the rainbow.

4 Air molecules are the right size to scatter blue and violet light wavelengths.

3 Around 18 miles up, light begins encountering air molecules.

6 ...blue light is scattered from molecule to molecule until it's coming at you from all parts of the sky.

5 While other colors continue more or less unimpeded to the surface...

WHY CLOUDS ARE WHITE

1 Cloud droplets, which are much bigger than air molecules, scatter all colors...

2 ...making sunlight scattered by clouds white.

3 Light is scattered in all directions, including through the cloud.

4 But if the cloud is thick enough, only a little light gets through and the cloud looks dark from below.

WHY SUNSETS ARE RED

1 At sunset and sunrise, sunlight is traveling farther through the atmosphere.

3 This leaves red, yellow, orange light to reach our eyes or reflect off clouds.

2 The longer trip means more and more light at the blue end of the spectrum is scattered.

Humid days are hazy because many particles in the air absorb water vapor and grow large enough to scatter all light waves as humidity increases.

When the humidity is less than 100 percent, the growth and added water absorption of a particle is called "deliquescence."

deep blue, and maybe to a hazy blue if you landed near the ocean or in a populated area. We take the blue sky for granted, but it's really quite unusual compared to the rest of the known universe.

Since the first experimental rocket planes began flying to the edge of space in the 1950s, we've been seeing photos of the black space high above Earth. The photos of Earth with the most impact, however, were taken from 240,000 miles away on Dec. 24, 1968 as the Apollo 8 spacecraft carrying astronauts Frank Borman, James Lovell and William Anders swung around the moon. Their photos showed Earth floating in the total blackness of space. "The Earth from here is a grand oasis to the vastness of space," Lovell radioed back.

The blue and white we see from space are caused by the atmosphere, clouds and the oceans. Yet, there's nothing with a blue pigment in the air or water. And if you looked at the individual water droplets or the simple ice crystals that make up clouds they'd be clear, not white.

The blue sky, white clouds and most of the other colors we see in the sky result from light being sent off in different directions as the light collides with air molecules, other substances or the water and ice of clouds. More than 12 miles above the Earth, the sky begins turning dark because there are fewer molecules

to send light off in different directions. From space, if you look toward the sun you see its light; if you look elsewhere you see only black sky, stars and maybe the moon.

To understand why the sky is blue and clouds white, you have to realize that the "white" light from the sun consists of all the colors of the rainbow. Light travels as waves of different lengths; each color has its unique wavelength.

You also have to know that light will travel in a straight line unless something sends it off in a different direction. As it turns out, air molecules are just the right size to send the shorter wavelengths of light, mostly blue, off in different directions. Longer waves, such as red, are not scattered by air molecules.

As "white" sunlight enters the atmosphere the blue light is scattered by air molecules. The blue light waves spread all over the sky and down in all directions. No matter which way you look, blue light (the scattered blue light waves) is coming at you from that direction.

On hazy days large numbers of relatively large particles in the air make the sky white or gray. Larger particles scatter more wavelengths of light. Hazy air has plenty of moisture and condensation nuclei. Moisture condensing onto nuclei makes drops large enough to scatter all wavelengths. Sometimes pollution can make the sky

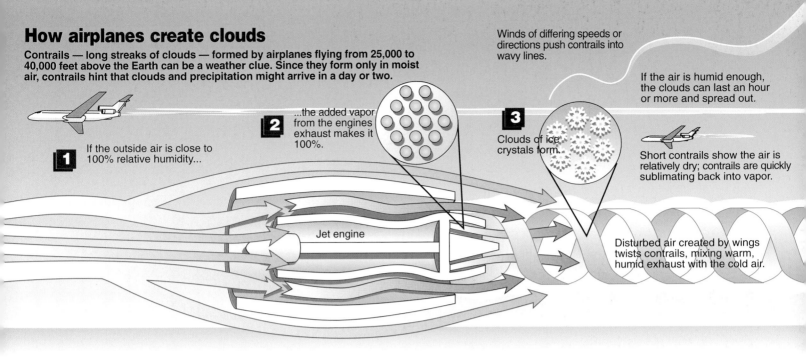

How airplanes create clouds

Contrails — long streaks of clouds — formed by airplanes flying from 25,000 to 40,000 feet above the Earth can be a weather clue. Since they form only in moist air, contrails hint that clouds and precipitation might arrive in a day or two.

Winds of differing speeds or directions push contrails into wavy lines.

1 If the outside air is close to 100% relative humidity...

2 ...the added vapor from the engines exhaust makes it 100%.

3 Clouds of ice crystals form.

If the air is humid enough, the clouds can last an hour or more and spread out.

Short contrails show the air is relatively dry; contrails are quickly sublimating back into vapor.

Jet engine

Disturbed air created by wings twists contrails, mixing warm, humid exhaust with the cold air.

---

appear brown or yellow because the particles are the right size to scatter those colors. When cold, dry air moves in, the sky turns a deep blue because there are few large particles to scatter other colors.

Clouds are white because their water droplets or ice crystals are big enough to scatter light of all wavelengths, which combine to produce white light. Clouds appear dark when they're in the shadow of other clouds, or when the top of a cloud is casting a shadow on its own base. How dark a cloud appears also depends on the background; the same cloud will look dark when it's surrounded by bright sky, lighter when it's in front of darker clouds.

Dark clouds aren't necessarily rain clouds. Rainy or snowy days are often dark because clouds block sunlight.

The sky at twilight

The sky turns red, orange or yellow at sunrise and sunset because light is traveling through more air when the sun is low in the sky. The long path through the air means most of the blue and colors with shorter wavelengths have been scattered in different directions. Few reach your eye. You see the yellow, orange and red colors, which pass more freely through the air.

Clean air brings orange sunsets because both yel-–low and red light travel through the air. Sunsets are especially vivid when red light reflects off clouds overhead. Volcanoes, and sometimes forest fires, bring the most vivid sunsets, however. Both produce particles that are just the right size to scatter light at the red end of the spectrum. The drawing at the bottom of the next page shows how this works.

Twilight is the period when the sun is below the horizon but we still see light in the sky. "Civil twilight" is the time from sunset — when the top of the sun drops below the horizon — until the sun is 6 degrees below the horizon. In general, the end of civil twilight is when lights are needed for outdoor activities. Street lights come on. "Astronomical twilight" lasts until the sky is dark enough to see faint stars; no sunlight is seen on the western horizon. The end of astronomical twilight occurs when the sun is 18 degrees below the horizon.

You continue seeing light in the sky long after it's dark on the ground. As anyone who has ever been in an airplane at sunset has seen, night doesn't "fall," it rises from the ground as the sun dips farther below the western horizon.

In the tropics, twilight is relatively brief because the sun is following a path nearly perpendicular to the horizon. It takes little time to reach 6 degrees and then 18 degrees below the horizon.

In the summer, twilight becomes longer the farther from the equator you go because the sun is setting at an angle to the horizon. It takes longer to reach 6 degrees and then 18 degrees below the horizon. In the far north or south during the summer, when the nights are very short, evening twilight lasts long enough to merge into morning twilight, creating the "white nights" of Scandinavia and Russia.

Vivid, normal sunset and sunrise colors result from sunlight traveling long distances through the atmo-

A glory around an airplane's shadow on clouds

Crepuscular rays streak across the sky from the sun, which is hidden behind cumulonimbus clouds.

this chapter we'll discuss clouds. For now, let's consider some optical phenomena you see from the air or ground.

If you are sitting on the opposite side of the plane from the sun and you can see the plane's shadows on clouds below, you might see a pale ring of colored light around the shadow. This is the glory. The only other likely place to see a glory is from high on a sunlit mountain when you're looking at your shadow on fog below. The glory is caused by sunlight hitting water drops, traveling around the drops and then coming back toward the sun. Light waves of different colors are bent at slightly different angles, giving the glory its colors. If the plane is high enough above the clouds, you might see only the glory, not the plane's shadow.

Sometimes you also see the shadow of your plane's contrails — the long cloud that sometimes trails behind high-flying airplanes.

Recognizing what you see in the sky takes some practice. Even a Ph.D degree in meteorology won't guarantee that you'll understand everything. Margaret LeMone, a National Center for Atmospheric Research scientist, offers an example: "I was flying in a jet and the glory was just a blob of light without the rainbow color. And you could see the shadow of the contrail. It looked like this ball of fire with black smoke coming out of it. It was wonderful. I said, 'Wow, is this a UFO?' "

Sometimes you can see rainbows as well as glories from airplanes. While a rainbow from the ground is an arc, from an airplane you might see the rainbow's entire circle.

When the sun is on your side of the airplane, or when you're looking in the sun's direction from the

sphere with various colors being scattered by different kinds of particles in the air and reflected from clouds.

The sun's long beams

Often you'll see bright rays, or "sun beams" crossing the sky. These are "crepuscular rays," created by particles in the air scattering light back to you. Sunlight scattered by dust in a darkened room creates the same effect. Sometimes you also see dark bands crossing the sky. These are shadows from clouds or mountains, sometimes ones below the horizon when the sun is low. Crepuscular rays seem to spread out, but they are really parallel. It is a matter of perspective. The rays seem to come together in the distance for the same reason railroad tracks converge at the horizon.

The glories of sky watching

One of the best places to watch the sky is from the sky. An airplane is the perfect sky observatory. Later in

How volcanoes create colorful sunrises, sunsets

A few months after the Mount Pinatubo volcano erupted in the Philippines in June 1991, sunrises and sunsets turned more colorful. The volcano shot 16 million to 18 million tons of sulfur dioxide high into the stratosphere. It turned into tiny droplets of sulfuric acid. These drops are about one micrometer across, just the right size to scatter red light.

5 The sulfuric acid haze scatters red and other colors at that end of the spectrum back to point A, making the sky look red, yellow and orange.

1 Sulfuric acid haze is from around 68,000 feet to 85,000 feet above the ground.

2 The haze is thin enough to allow most sunlight through.

3 The farther light travels through the atmosphere, the more red it becomes.

4 For someone at Point A, the sun is below the horizon; for them, the ground is growing dark.

A

Without the high-level haze, light would not be scattered downward, the sky would look dark to someone at Point A.

What a cloud's appearance tells us

When you know what to look for, a glance at a cumulus cloud can tell you which parts are growing and where the updrafts have died down.

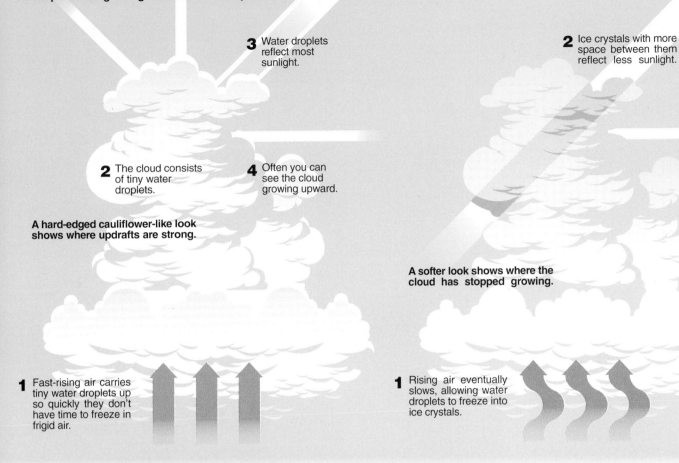

3 Water droplets reflect most sunlight.

2 Ice crystals with more space between them reflect less sunlight.

2 The cloud consists of tiny water droplets.

4 Often you can see the cloud growing upward.

A hard-edged cauliflower-like look shows where updrafts are strong.

A softer look shows where the cloud has stopped growing.

1 Fast-rising air carries tiny water droplets up so quickly they don't have time to freeze in frigid air.

1 Rising air eventually slows, allowing water droplets to freeze into ice crystals.

How clouds grow pouches

The underside of a thunderstorm's anvil often grow pouches, called mamma, which look threatening and are often described as a sign that hail or even a tornado could be likely soon. But Margaret LeMone of the National Center for Atmospheric Research says they aren't necessarily a sign of trouble, despite their ominous look. In fact, mamma often form when a thunderstorm is beginning to weaken.

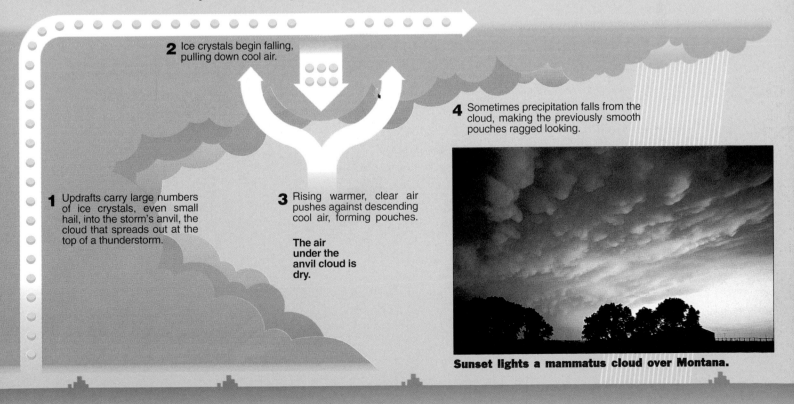

2 Ice crystals begin falling, pulling down cool air.

4 Sometimes precipitation falls from the cloud, making the previously smooth pouches ragged looking.

1 Updrafts carry large numbers of ice crystals, even small hail, into the storm's anvil, the cloud that spreads out at the top of a thunderstorm.

3 Rising warmer, clear air pushes against descending cool air, forming pouches.

The air under the anvil cloud is dry.

Sunset lights a mammatus cloud over Montana.

Why sunny skies turn showery

A day often will start out sunny. Then small cumulus clouds will begin popping up around mid-day. Sometimes those clouds will cover much of the sky, maybe with a few causing showers by late in the afternoon. As the sun sets, the clouds will fade away and the night will be clear. This is most common in the spring because air aloft is cold while the sun is warming the ground. This makes the air unstable as the day goes on.

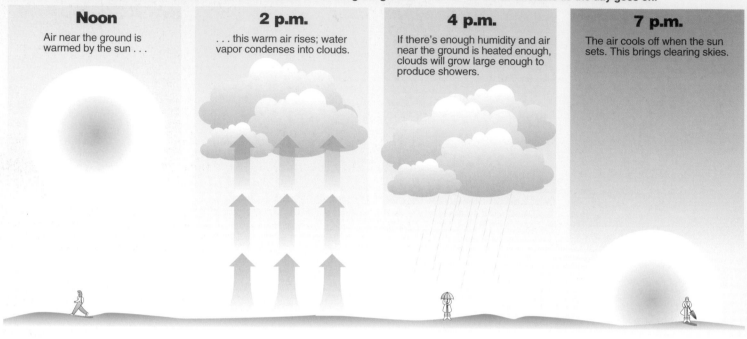

Noon
Air near the ground is warmed by the sun . . .

2 p.m.
. . . this warm air rises; water vapor condenses into clouds.

4 p.m.
If there's enough humidity and air near the ground is heated enough, clouds will grow large enough to produce showers.

7 p.m.
The air cools off when the sun sets. This brings clearing skies.

The World Meteorological Organization defines various "meteors" as:

<u>Hydrometeors:</u> Liquid or solid water from the air, such as rain and snow.

<u>Lithometeors:</u> Solid, dry particles, such as dust.

<u>Photometeors:</u> A luminous phenomenon produced by light, such as rainbows.

<u>Electrometeors:</u> A visible or audible manifestation of atmospheric electricity such as lightning or thunder.

ground, you might see coronas, halos, or "sun dogs."

The play of light on water, ice

Sun shining through clouds with water droplets creates circular splotches of light around the sun and moon known as coronas. Clouds of ice crystals create rings of light known as halos. You often see them when a layer of high, thin clouds covers the sky. Ice crystals can also create other kinds of rings and spots of light in the sky.

The most common kinds of spots of light, which are seen on one or both sides of the sun when it's low in the sky, are known as mock suns, sun dogs or parhelia. Halos and mock suns are relatively common. But if you point them out to most people, they'll tell you they've never noticed them before. How often you see halos or sun dogs depends on the cloudiness of your sky. But across the eastern United States you can expect to see them at least once a week on the average especially in winter and early spring. Other kinds of ice crystal displays are relatively rare. Some of these are shown in the drawings on the next page.

LeMone tells about once seeing a rare circumhorizontal arc — a rainbow-like ring running around the sky just above the horizon. "I screamed, 'Oh, look at that.'

People around me wondered what was going on." Most of those people probably thought that circumhorizontal arc was a low rainbow. But you have to be looking away from the sun to see a rainbow. To appreciate the sky's rarest displays, you have to be familiar with its more common sights.

Meteorology, more than weather

Today the word "meteorology" has come to mean the study of weather, but the word goes back to the Greeks, who used it to mean the study of natural phenomena above the Earth's surface and below the sphere of the moon's orbit. The philosopher Plato is often credited as the first to use the word.

Big research projects to learn more about clouds are justified by the benefits they'll bring in improved weather forecasts or a clearer understanding of how the climate might be changing. No one launches expensive research projects to understand more about glories, halos or rainbows.

While "meteorology" today means study of weather and climate, the word "meteor" commonly refers to the streaks of light created when meteoroids, tiny particles from space, enter the atmosphere and burn up. But such "shooting stars" are really just one kind of meteor.

Strictly speaking, a meteor is any natural phenomena in the atmosphere.

The Greeks recognized three types of "meteors": "Aerial" referred to the winds; "aqueous" referred to rain, hail, sleet and snow, and "luminous" referred to rainbows, halos, northern lights and meteoroid trails. The Greeks also included comets, which we know are far out in space, because they thought comets were within their meteorology boundaries between the Earth and moon.

Rainbows are always special

Rainbows are the most noticeable "meteor" next to clouds, rain and snow. Many cultures have attached a religious significance to rainbows. In Genesis the rainbow is described as the sign of a covenant between God and Noah after the Flood that "the waters shall no more become a flood to destroy all flesh."

Greeks associated rainbows with the goddess Iris. In the *Iliad,* Homer tells how Aphrodite, wounded by Diomedes, fled to Olympus along the rainbow, carried by Iris. The Greek goddess Iris gives us not only the word "iris" for the colored part of the eye and a flower, but also the word "iridescence." While in Genesis, the rainbow is a sign of God's mercy, the Greeks saw a sinister side. Iris is sometimes a hostile force that brings war and turbulence.

Most of us have heard the legend that if you can get to the end of the rainbow, you'll find a pot of gold; this is an old European tradition. But in some African mythology, the rainbow is a giant snake that brings bad luck to the house it touches. Many peoples have seen rainbows as bridges to heaven. According to a medieval German tradition, presumably based on the account in Genesis, no rainbows would be seen 40 years before the end of the world. Every time you saw a rainbow, you could be sure the world would last at least another 40 years.

The beginning of what we could consider scientific explanations of rainbows goes back to the Greek philosopher Anaximenes, who lived around 575 B.C. He thought rainbows resulted from a mixture of sunlight and the blackness of clouds. Various explanations were tried over the years with little success. French philosopher and mathematician Rene Descartes (1596-1650) worked out the the geometry of how water drops bend light waves, one of the important ideas that led to our understanding of rainbows. Isaac Newton (1642-1727) discovered that different colors are bent different

Coronas, halos and streaks of light

Light from the sun and moon passing through clouds can create many effects including splotches, lines or circles of light. These are some of the most common kinds.

The corona

A corona — it means crown — is a disc of light seen when the sun or moon is shining through clouds. Since the sun can blind you, it's easier to see a corona around the moon. If you look closely, often you'll see a corona is slightly blue nearest the moon and a reddish brown around the outside. Coronas are caused by the slight bending – called diffraction – of light passing through clouds.

Halo and arcs

Unlike a corona, halos don't touch the sun or moon; they're circles of light, caused by the bending — refraction — of light rays as they pass through ice crystals. Different shapes of crystals pointing in different directions account for the wide range of shapes you see including halos, arcs and various kinds of splotches of light.

How ice crystals create 'sun dogs'

One of the most common ice crystal displays is what appears to be copies of the sun on either side, often both sides, of the sun when it's low in the sky. These are known as sun dogs, mock suns or parhelia. Here's how ice crystals create them:

1 If clouds are made of six-sided ice crystals with the large, flat sides down ...

Light rays from the sun

2 ... light entering a crystal is bent at a 22-degree angle ...

3 ... and the redirected light means we see mock suns 22 degrees on either side of the real sun.

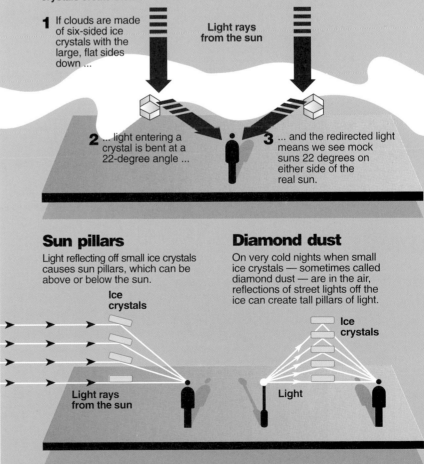

Sun pillars

Light reflecting off small ice crystals causes sun pillars, which can be above or below the sun.

Ice crystals

Light rays from the sun

Diamond dust

On very cold nights when small ice crystals — sometimes called diamond dust — are in the air, reflections of street lights off the ice can create tall pillars of light.

Ice crystals

Light

THE ENTIRE ATMOSPHERE

The storms and clouds we think of as weather are concentrated in the lower seven or so miles of the atmosphere known as the troposphere. To understand the sky, however, we need to look higher, into the stratosphere where jet airliners fly and where the ozone layer protects us against the sun's ultraviolet radiation and into the mesosphere and thermosphere where some of the sky's most beautiful sights — the auroras — display themselves. The atmosphere fades into space about 180 miles above the Earth.

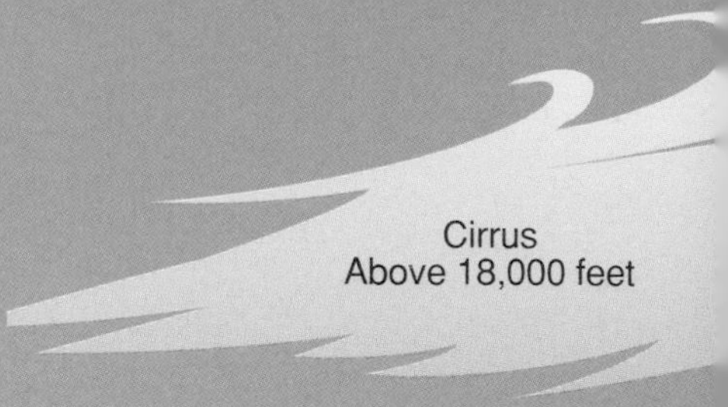

Cirrus
Above 18,000 feet

THE ATMOSPHERE'S LAYERS

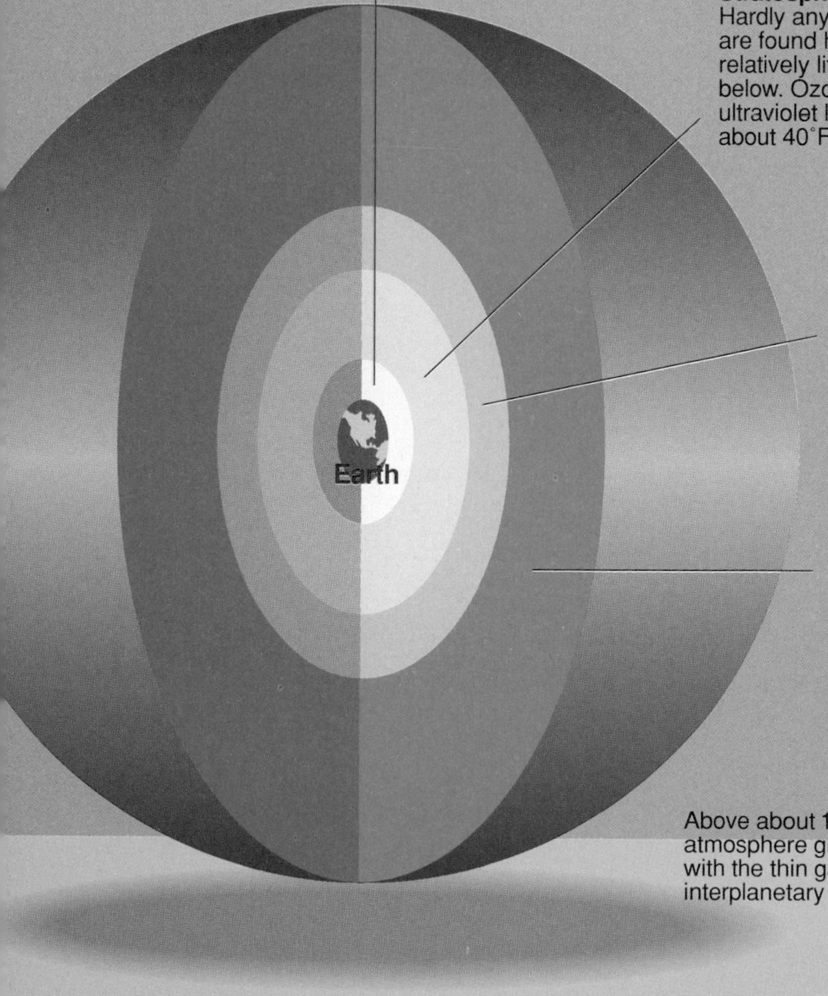

Troposphere (0 to about 7 miles up): Where most clouds and weather are located.

Stratosphere (7-30 miles): Hardly any water vapor or dust are found here since there's relatively little mixing with the air below. Ozone's absorption of ultraviolet light warms the air to about 40°F.

Mesosphere (30-50 miles): Temperatures begin falling again to around -90°F, making this the coldest part of the atmosphere. No aircraft and only the largest helium balloons reach this high.

Earth

Altostratus
6,000 to 20,000 feet

Thermosphere (About 50 miles up): Also called the ionosphere. The temperature can range from 930°F to more than 3,000°F depending on solar activity as various atoms and molecules absorb solar radiation.

Above about 180 miles the atmosphere gradually merges with the thin gases of interplanetary space.

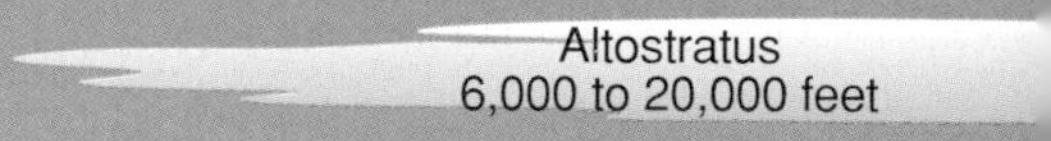

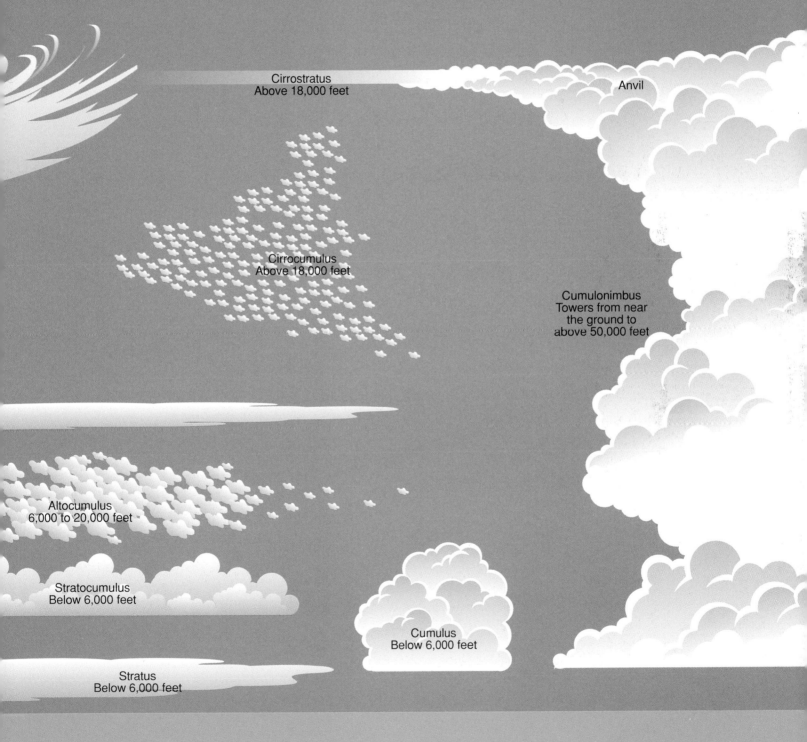

WHERE WEATHER IS BORN

These are the most common types of clouds in the troposphere. Elsewhere in this chapter we look at why various clouds have their unique shapes. Altitudes are averages for the middle latitudes.

Cirrostratus
Above 18,000 feet

Anvil

Cirrocumulus
Above 18,000 feet

Cumulonimbus
Towers from near
the ground to
above 50,000 feet

Altocumulus
6,000 to 20,000 feet

Stratocumulus
Below 6,000 feet

Cumulus
Below 6,000 feet

Stratus
Below 6,000 feet

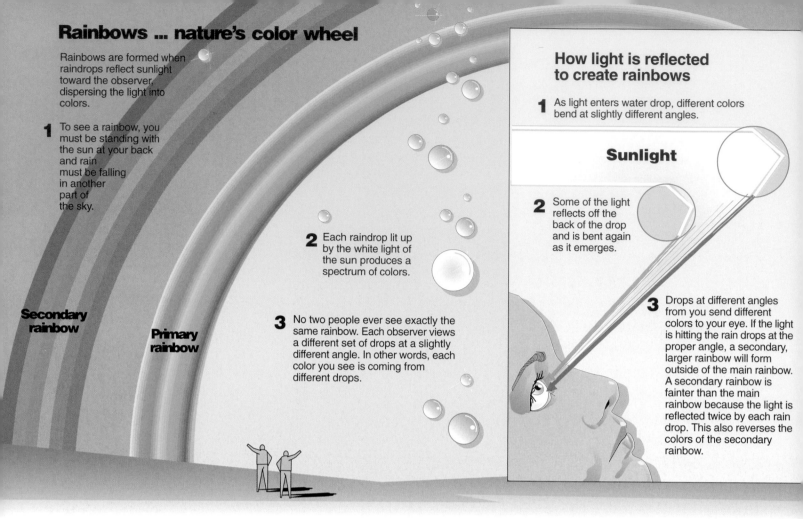

Rainbows ... nature's color wheel

Rainbows are formed when raindrops reflect sunlight toward the observer, dispersing the light into colors.

1 To see a rainbow, you must be standing with the sun at your back and rain must be falling in another part of the sky.

Secondary rainbow

Primary rainbow

2 Each raindrop lit up by the white light of the sun produces a spectrum of colors.

3 No two people ever see exactly the same rainbow. Each observer views a different set of drops at a slightly different angle. In other words, each color you see is coming from different drops.

How light is reflected to create rainbows

1 As light enters water drop, different colors bend at slightly different angles.

Sunlight

2 Some of the light reflects off the back of the drop and is bent again as it emerges.

3 Drops at different angles from you send different colors to your eye. If the light is hitting the rain drops at the proper angle, a secondary, larger rainbow will form outside of the main rainbow. A secondary rainbow is fainter than the main rainbow because the light is reflected twice by each rain drop. This also reverses the colors of the secondary rainbow.

In the middle latitudes where storms travel from west to east, rainbows are usually seen when a storm is over.

In the tropics where storms travel from east to west, rainbows are often seen before a storm arrives.

amounts by a prism, such as a drop of water. It is this prism effect of water that explains how raindrops break "white" sunlight into the colors of the rainbow.

Sorting out clouds

While conditions have to be exactly right to see a glory, a halo or a rainbow from a plane, you are always sure to see clouds. If you know even a little about weather, clouds help tell you what the atmosphere is doing along your plane's route.

Flat clouds indicate stable air. Piled up clouds, unstable air. If you're flying toward a warm front, you'll see the clouds grow thicker as shown in the drawing on Page 49.

Since airliners avoid thunderstorms, a flight is often the best way to see spectacular clouds such as a supercell or a row of towering thunderstorms in a squall line. The photo on the bottom left of Page 176 shows what a supercell looks like from an airliner. Haze, other clouds or precipitation usually hide the full magnificence of such clouds from the ground.

In 1803, English scientist Luke Howard devised

the basic system of cloud classification we still use. While some people today might find the Latin names difficult, the system has stood for nearly 200 years as a logical way to sort out clouds. It distinguishes clouds in two ways: by their general appearance and by their height above the ground.

There are two parts to understanding how clouds get their names. Start with the basic shapes:

• Cirrus are curly or fibrous;
• Stratus are layered or stratified;
• Cumulus are piled up or lumpy.

The second part of understanding cloud names is based on the height of the clouds:

• Cirro is the prefix given to high clouds, those with bases above around 20,000 feet in middle latitudes.

• Alto is the prefix for middle-level clouds with bases from around 6,000 feet to 20,000 feet.

• Nimbo added to the beginning or nimbus added to the end of a cloud's name means that cloud is producing precipitation.

As you can see, the system isn't uniform; for example, there's no prefix for low clouds. And there are some odd mixtures; the name "stratocumulus" combines words for two different shapes. But, that is a good description of these clouds.

Using the system, we see that a "cirrocumulus" cloud is a high cloud that's lumpy while a "cirrostratus" cloud is a high cloud that's flat.

A "cumulonimbus" cloud is producing rain. These clouds make thunderstorms.

"Nimbostratus" are low, layered clouds that are producing precipitation.

Clouds are important to understanding weather because they don't just happen. A particular cloud has a certain shape and location because of the movements of

Her career began with a roar of thunder and a cloud of smoke

By the time she was in the seventh grade Margaret LeMone was on her way toward becoming an atmospheric scientist.

"I had started keeping weather records" she recalls. "I was making forecasts, I had a barometer named Earl, I could guess the winds from looking at trees, but I mostly watched clouds. I spent hours drawing cloud formations."

Her interest in the sky and the things going on in the sky began one night at the supper table when she was in the third grade. "It sounds like a religious thing, 'Lightning struck and my fate was made.' I guess it was like that in a way," she says. "It was the loudest noise I ever heard; the whole house filled with smoke. Later we found out the lightning traveled down the chimney, cleaned it all out. It exploded the chimney and a three-by-ten foot section of roof was turned into splinters about two inches long and a half inch in diameter." Pouring rain put out the fire the lightning started.

She also credits living in Columbia, Mo., for her career choice. "Missouri has such great weather. Well, I don't know if you'd call it 'great,' but it's severe, and interesting."

Today LeMone is a leading researcher in what is known as boundary layer meteorology, the study of the first several hundred feet of atmosphere where many clouds have their beginnings.

Among other things, she has worked on developing better, more detailed pictures of the formation of clouds and storm systems, such as squall lines.

Margaret LeMone, sky watcher

Her studies have taken her to the tropics and on research flights into hurricanes, but she's also spent time closely analyzing the weather on the Great Plains.

In early 1993 she expects to go to Guadalcanal in the Solomon Islands as a part of a study of ocean-atmosphere interactions in the South Pacific.

The reason for collecting data around the world is to ensure that the theories LeMone and other scientists work out will explain equally well what's going on in the South Pacific and in Kansas.

LeMone says she finds the variety of atmospheric research one of the big attractions. "A lot of travel is involved, which I like at times, don't like at times. I've found it's a very flexible job; it's not something you have to be in the office every day for. If you're tired you can come in at noon, if you have a sick child you can stay home and work."

Unlike the post-World War II experi-

ences of Joanne Simpson described in Chapter 1, LeMone says women have a smoother time today in meteorology. Atmospheric science "is small field," she says. "It's kind of like a small town. I once did a survey on women in meteorology and found they didn't have problems like those of women in chemistry or medical research where you're part of a large community. In a small field like meteorology, people start thinking of you as yourself, for better or worse."

She says problems for women in meteorology now "are not institutional, they're the sorts of things like being excluded from social events or bull sessions."

One concern among those who worry about the United States having enough scientists in the future is ensuring that junior high school and high school girls study math and science despite social pressures.

LeMone says that was never a problem for her. "I was kind of an oddball in high school and junior high school. I hung around other girls who were as nerdy as I was. There were a few girls who kind of supported each other. We ignored it."

The problem isn't faced by girls alone. LeMone says "my son is in the seventh grade, he's going through all of this, telling me about the popular kids."

One thing she's told him is: "By the time I was in college I found all the super-social high school kids were already reminiscing about the good days in high school. This was at the time my life was beginning to get really interesting."

SOME OF THE THINGS CLOUDS TELL US

The sizes, shapes and locations of clouds are the results of the movements of air and changes in water from vapor to liquid or ice in the air. Even a basic understanding of these processes will help you visualize the movements of air when you look at clouds. The drawings and photos on these four pages are a few examples of the many shapes of clouds and the atmospheric movements that cause them.

CIRRUS CLOUDS

Cirrus clouds are made of ice crystals. They are found in cold air above 20,000 feet. Such clouds often are being carried along by winds blowing faster than 50 mph. But the movement looks slow from the ground since the clouds are so high, just as a 500-mph airplane at a high altitude seems to be moving slowly.

High speed winds

Slower winds

The "mares' tails" part of cirrus clouds are falling ice crystals. They trail behind when they fall into winds that are slower than those pushing the main part of the cloud.

LENTICULAR CLOUDS

These lens-shaped clouds are found at the crests of wind waves, such as the mountain waves shown in the drawing on Page 44. While the wind and the water vapor that create the cloud are moving, the cloud stays in the same location.

Rising air cools. When it cools to the dew point or frost point, water vapor in the air turns to cloud droplets or ice crystals.

Water droplets or ice crystals move along with the wind.

As the air descends, it warms, eventually turning the droplets or crystals back into invisible vapor.

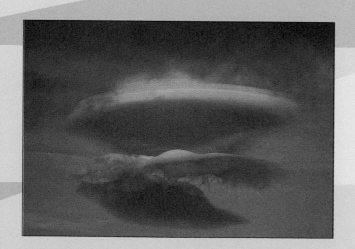

Lenticular clouds show a typical "flying saucer" shape. Such clouds are most common downwind from the West's high mountains, but are also often seen elsewhere.

Cirrus clouds with mares' tails.

BILLOWS CLOUDS

Clouds that look like breaking waves are created by winds blowing at different speeds in different layers of air, usually at a boundary between air masses with different densities, such as a layer of warm air atop cold air.

Differences in wind speeds cause a turbulent, rotary motion of the air between, which gives clouds their unique shape.

HOW TO WEIGH A CLOUD

Clouds are heavier than you might think. Here's how scientist Margaret LeMone went about "weighing" a fairly small cumulus cloud drifting over the Plains east of Boulder, Colo.

4 Our cloud turns out to weigh more than the heaviest version of the Boeing 747-200 jetliner with a full load of passengers and fuel, 416 tons.

3 A cloud like this has 0.5 gram of liquid water in each cubic meter. If we multiply 1,000 x 1,000 x 1,000 x 0.5, we find the cloud's liquid water weighs about 500 million grams. This is about 550 tons.

1 At midday, when the sun is the nearest to directly overhead, note the cloud's shadow on the ground. From the Appalachians westward, the United States is divided into square sections, one mile on a side, usually marked by roads.

2 Our cloud is about six-tenths of a mile wide and long. It's also roughly as high as it is wide. We'll use metric measurements as scientists do. Since 0.6 mile is about 1,000 meters, our cloud is 1,000 meters wide, 1,000 meters long and 1,000 meters high.

Altocumulus clouds are lined up in rows. In such cases, the air is rising where the clouds are located and sinking where the air is clear between the clouds.

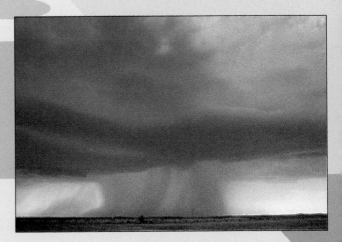

A wet microburst blasts down from a cloud. The foot-like appearance, with the 'toes' to the left, is typical of microbursts. Drawing on Page 127 shows what causes them.

CUMULONIMBUS: KINGS OF THE SKY

Cumulonimbus clouds, also called thunderheads, are by far the Earth's most awesome clouds. Such a cloud's base may be only a few thousand feet above the ground while its top pokes into the lower stratosphere, maybe 60,000 feet up. The most dangerous cumulonimbus clouds are supercell thunderstorms or squall line thunderstorms that produce heavy rain, hail and deadly winds, including tornadoes, as well as lightning. The photos and drawings on these two pages illustrate just a few aspects of these fascinating clouds.

PILEUS CLOUDS

Wind

- The upward movement of air — updrafts — that cause a cumulus cloud to grow into a cumulonimbus cloud push up the air above the growing cloud. The upward-moving air becomes an obstacle to upper-level winds, much like the top of a mountain. If the high-altitude winds are humid enough, new clouds form over the cumulus clouds in a way similar to the formation of lenticular clouds as shown on Page 174.

Updrafts

The word "pileus" is Latin for "skullcap." Similar clouds that form over mountains are called "cap" clouds.

A supercell photographed at 45,000 feet over Oklahoma.

A pileus cloud atop a growing cumulus cloud.

Upper-atmosphere winds push the air at the top of the cloud downwind, forming the anvil characteristic of large thunderstorms. As shown in the drawings on the bottom of Page 167, pouches called mamma, may form on the anvil's bottom. As the anvil spreads out, it may become cirrus clouds that drift for scores of miles. As these clouds evaporate, they create a layer of humid air that can supply moisture needed for a pileus cloud atop another growing cumulus cloud miles away.

As updrafts reach the stratosphere, where the air is stable, momentum carries the upward-moving air and the cloud upwards for a while before the cloud begins spreading out to the sides.

Air spiraling up into the bottom of a cumulonimbus cloud can form a wall cloud, which hangs down from the bottom of the cloud. The largest tornadoes usually come from such wall clouds.

A wall cloud looms over the High Plains near Akron, Colo., shortly before producing a tornado.

Telling the interwoven tales of weather and history

Forecasting the clouds became vital during World War II when air power came into its own as a weapon.

"Every hour on the hour we had to be ready to give the forecast. Wind direction, rain, of course, but mostly clouds," recalls David Ludlum who ran a mobile weather station in Europe during the war. "They needed to know cloud level and visibility for flying and for bombing. There were minimums that they needed, minimum levels of visibility, and they couldn't operate unless they had them."

During the campaign in Europe, Ludlum's weather station, which was on trucks, "moved right along with the front. I set up in 48 different places in Europe and forecast the weather there."

The war gave Ludlum personal experience in the field that he dominates: weather in history. He's a professional historian with a Ph.D. in history from Princeton University who was trained as a weather forecaster after joining the Army a year before the Pearl Harbor attack because "I knew a war was coming on and I wanted to get the training.

"People don't really realize the way that weather has influenced historical events, especially the famous battles, the invasion of Normandy, the Battle of Waterloo, and Revolutionary War battles. I did a lot of research on the Revolutionary War and the weather attending each battle." His book *The Weather Factor* deals with the American Revolution. He's also written several other books on American weather history, one on Vermont's weather and the *Audubon Society Field Guide to North American Weather.*

After leaving the Army Air Forces as a lieutenant colonel, "I had to support myself, so I started a weather instrument business in 1946. And then we needed some place to advertise the business, so I started a magazine in 1948." He sold the magazine, *Weatherwise*, in 1978 but continues writing for it as well as for other magazines.

A saying of Benjamin Franklin, "some are weatherwise, some are otherwise," supplied the magazine's name. "Historically, I'd have to say Ben Franklin influenced me most," says Ludlum, who headed the meteorology section at the Franklin Institute in Philadelphia for five years. "I wanted to carry on his tradition. He was quite a student of meteorology."

Ludlum was also one of the first television weather-

David Ludlum, *Weatherwise* founder

men, appearing on WCAU in Philadelphia in 1946-47. "On-air weather broadcasting was just starting at the time. But I didn't stay with it, well, actually I got fired. They wanted somebody to croon a weather ditty to open the program. And I said 'No, that's not for me.' So they hired this guy who was a crooner. He didn't know a thing about weather, but he stayed on for five or six years. And he sang a weather ditty every night." While it lasted, his television career paid well for the time, $50 a week.

Researching America's weather history is aided by the amount of written material dealing with weather in archives, Ludlum says. "The weather played an intrinsic role in the survival of the colonies, so they wrote about it frequently in their diaries and notes. They were so much more aware of weather then. It was an outdoor life that they lived. Even if you were a doctor or lawyer in town you usually had a backyard barn with a horse and cow. And, too, people were engaged in shipping, sailboats, and all of that depended completely, entirely on the weather. People live an indoors life today. Sort of an artificial, enclosed life. We're not nearly so dependent on the weather as we once were. Of course, it's still important in our lives, in bringing droughts, floods and destructive storms. To lead well-ordered lives, we should keep one eye on an up-to-date TV weather map indoors and the other eye on the changing sky outside."

David Ludlum describes how George Washington used his weather knowledge to escape when the British apparently had his army trapped on Jan. 2, 1777:

"He was hemmed in. There was mud behind him and an icy river on the other side and the British on the third side.

"But he knew there was a northwest wind blowing and it would bring a freeze.

"He started the retreat at midnight after the (mud) had frozen. They were able to make an end run, around the British and back up to Princeton. They had a small skirmish there and then they were safe."

the air and the amount of water vapor and condensation nuclei in that air. As we saw in Chapter 5, stable air will tend to produce stratus clouds and unstable air will tend to produce cumulus clouds.

Most sky guides give pictures of clouds with their Latin names. But when you try to compare what you see in the sky with the photos in the books, you may find that little of what you're seeing seems to match the pictures.

The trick is to look for general characteristics — not precise matches. Are the clouds mostly lumpy or flat? Do they seem to be high in the sky or almost on the ground? Learning how to tell one kind of cloud from another is like learning how to tell one kind of bird from another, or one kind of tree from another. Before you start learning, you notice only a few broad differences between different birds, or trees, or clouds. But then as you look more closely, you will discern tiny differences you never saw before.

What goes on high overhead

While most of the clouds and other "meteors" we see in the sky are in the "troposphere," — the layer of air from the ground up to around 40,000 feet in the middle latitudes — the higher reaches of the atmosphere produce their own important and fascinating phenomena.

The boundary between the troposphere and the stratosphere is known as the "tropopause." This boundary, which varies from around six to 10 miles above the surface, is the altitude at which the air's temperature stops falling and begins rising. Temperatures here range from about -50°F down to more than -100°F with the coldest temperatures over the equator, not over the poles as you might expect. In the tropics strong convection pumps warm air high into the sky. The air cools as it rises and pushes the tropopause to its highest altitude.

Since the stratosphere's temperatures are warmer than those below, the air is stable. In Chapter 5 we saw how an inversion, a layer of warm air over cold air, is stable, which means rising air stops rising unless it's being forced upward. While some thunderstorms push their tops into the stratosphere, little air from lower altitudes mixes with air at that level. Thunderstorm tops push through the tropopause because of the momentum of air racing upwards in their updrafts. It's like a rocket that keeps on rising for a while after its fuel burns out.

In Chapter 2 we saw how the sun's energy warms the ground and ocean, which in turn warm the air next to the ground. The greater the distance from the ground, the colder the air usually becomes. The exceptions are when the air near the ground cools more quickly than air above, for example on clear nights, or when extratropical cyclones or thunderstorms push up warm air and bring down cool air. So, if the air cools as it goes up, what warms the stratosphere?

The interaction between some of the sun's ultraviolet rays and ozone warms the stratosphere. In Chapter 12 we'll take a closer look at what's going on in this critical ozone layer. In brief, ozone absorbs ultraviolet light. Anything that absorbs energy is heated, which accounts for the temperatures in the +40°F range at the top of the stratosphere, around 30 miles up.

While the stratosphere and the upper troposphere would be deadly places for an unprotected human, we travel through them on almost all jet airplane trips. The thin air that would kill us in a few minutes if we didn't have a supply of oxygen, is friendly to jet engines; it's the region where they perform most efficiently.

Above the stratosphere, in the mesosphere, which extends from about 30 miles to around 50 miles, the temperature falls again, cooling to below -150°F at its top. Another warm layer, the "thermosphere," comes next. In this layer various atoms are absorbing ultraviolet and other kinds of solar energy.

Temperatures can range above 3,600°F in the thermosphere if the sun is in one of its periods with many sun spots and solar flares. This activity creates charged particles, which affect radio transmissions. That's why this layer, along with the higher level of the mesosphere, also is called the "ionosphere" — ions are electrically charged particles. The lowest region of the ionosphere, called the D Region, absorbs many radio waves while other regions reflect radio waves. The D Region weakens and disappears at night, allowing radio waves to bounce off the ionosphere. This is why you often can pick up distant radio stations at night.

We, like the ancient Greeks, can see the auroras and shooting stars as meteoroids burn up as they enter the atmosphere. Most of the things going on at the top of the atmosphere, however, are known to us only because of their effects on radio waves or through measurements made by satellites and research balloons.

However, most of the phenomena that make the sky a day-to-day entertainment — the wide variety of clouds, rainbows, halos, coronas and glories — take place in the troposphere, the bottom layer of the atmosphere where we live.

Meteorology Professor Stanley Gedzelman of the City College of New York offers this advice for distinguishing stratocumulus, altocumulus and cirrorcumulus clouds — all of which are lumpy:

Hold your hand at arm's length and sight over your hand.

Low, stratocumulus cloud lumps will be about the size of your fist or larger.

Middle-level altocumulus cloud lumps will be the size of your thumb.

High-level cirrocumulus cloud lumps will be the size of your little finger nail.

To make a quick, uncomplicated weather forecast, look out the window.

"If the visibility is 15 miles and there's not a cloud in the sky, you can bet it's not going to rain in the next 30 minutes," says William Bonner, an authority on forecasting with the University Corporation for Atmospheric Research in Boulder, Colo. "But if just over the horizon, 45 miles away, there's a squall line moving along at 40 mph, one hour from now you can get a lot of rain."

Many people, especially in past centuries, studied clouds and winds and were sometimes able to say what the next day's weather would be like.

But for good forecasts you need to look farther than over the horizon, and you need to know more about the atmosphere than clouds and winds alone can tell. Bonner says that if you think in terms of how far you have to "see" to forecast the weather, a view of most of North America is needed to say what's likely tomorrow on the East Coast.

To have any hope of a good forecast for three days from now, you need a view of the weather all over the Northern Hemisphere. A good five-day forecast requires a view of the weather across the whole Earth, Bonner says.

Technology, such as weather radar or satellite pictures, expands the weather horizon. A radar set shows a squall line as much as 200 miles away and helps measure how fast and in what direction it's moving. Satellites can show storms all over the world.

More needed than a weather view

By the mid-1990s Doppler weather radars were giving forecasters a more detailed view of weather than ever before and new advanced satellites were sending them sharper images from space. But even the best view of local and worldwide weather doesn't give enough information

A shower moves over West Palm Beach, Fla. New technology is improving forecasts of such events.

What cloud
forecasts mean:

Partly sunny/Partly
cloudy: No difference,
both mean clouds should
cover 30% to 70% of the
sky.

Cloudy: Clouds should
cover 90% or more of
the sky.

Fair: Clouds should cover
40% or less of the sky,
no fog likely.

to make a sure prediction of tomorrow's weather, much less the weather for a few days ahead.

There are too many things going on. Imagine that a storm heading your way is like a bowling ball heading down the alley for the pins. In a regular bowling lane, you could set up sensors to measure the exact path and speed of the ball as it leaves the bowler's hand and feed the data into a computer. Program in a few basic laws of physics written as mathematical equations and before the ball is halfway down the lane, the computer could say which pins are likely to be knocked down.

But this is the Weather Bowling Lane where the ball's motion can change the lane itself, point it in a totally different direction. The ball can knock down pins three lanes over. And the ball can grow or shrink as it rolls along. How do you predict all that? And yet that is simple compared to what is needed for a weather forecast.

Earlier in this book we looked at the building blocks of weather forecasting:

• How the sun heats the Earth unequally, forming warm and cold air masses

• How air-pressure differences cause winds

• How the Earth's rotation and friction change the winds

• How water's phase changes add or take heat away from the air

All of these actions, plus the possible interactions among them, go into making the weather. All are condensed into mathematical equations and programmed into weather forecasting computers.

Producing 'skilled' forecasts

No one can predict day-by-day detailed weather weeks in advance. In fact, accuracy generally drops off quickly four or five days ahead. Anyone who claims to be able to predict a year in advance, say that June 21 will be rainy while the 22nd will be sunny, is pulling your leg.

Meteorologists describe forecasts as "skilled" or "unskilled." A skilled forecast is one that does better than methods such as flipping a coin, forecasting that the weather will be the same as the 30-year average for a particular day, or predicting that a day's weather will be the same as yesterday's weather. (That's called a persistence forecast.)

For example, you could run up a respectable forecasting score in the summer in Los Angeles by using persistence. A forecast for no rain and a high in the 80s would work day after day. Enough unskilled forecasts that take normal conditions into account — no July snowstorms for Los Angeles — can add up to what seems like a respectable score of 50 or 60 percent correct.

Scientific forecasters don't merely add up the number of times they're correct. They see how their forecasts compare to various kinds of unskilled forecasts. Success

In forecasts, a "slight chance of precipitation" means there is a 10% to 20% probability of precipitation occurring.

A "chance of precipitation" means a 30% to 50% probability.

"Occasional precipitation" means there is a greater than 50% probability of precipitation — but for less than half of the forecast time period.

How far ahead can meteorologists accurately predict the weather?

By the mid-1990s, meteorologists generally agreed on the following assessment of forecasts:

• **Up to 12 hours.** Forecasts of general conditions and trends are good, but small, short-lived, severe local storms are predictable for only minutes to an hour ahead. New radars and satellites are improving these forecasts and thus saving lives, lowering costs for businesses such as airlines, and making some jobs, such as fighting forest fires, easier. Still, small-scale events, such as tornadoes or flash floods, can hit with little warning.

• **12 to 48 hours.** Predictions of the development and movement of large extra-tropical storms and the general, day-to-day weather changes they bring are forecast well. While forecasters can't say exactly when and where severe storms will hit, they do a good job of forecasting which areas will be threatened with snow and other hazards as much as 48 hours ahead.

• **3 to 5 days.** Big events, such as large storms or major cold waves, usually are forecast up to five days ahead. But where hurricanes will make landfall cannot be predicted with any accuracy this far ahead. Daily temperature forecasts for the third to fifth day of a forecast are more accurate than those for precipitation. Generally, five-day forecasts in the mid-1990s are as accurate as three-day forecasts in the early 1980s.

• **6 to 10 days.** Average temperatures and precipitation for the entire period can be forecast fairly well; temperature can be forecast better than precipitation. Day-to-day weather can't be forecast well.

• **Monthly and seasonal.** These offer odds for temperatures and precipitation averages being higher or lower than normal or near normal. Often forecasters can't give any odds for large areas of the USA, especially for precipitation. No attempt is made to predict day-to-day or even week-to-week weather changes or to say when and where storms will form. Winter forecasts are for precipitation, not for rain or snow.

THE MANY ROLES OF WEATHER SATELLITES

Cloud photos seen on television and in newspapers are the best-known weather satellite product. While the photos are important, making pictures is only part of the work done by satellites. Satellites also provide cloud-top, ocean and land temperatures, upper-air temperatures and humidities, wind speeds derived from cloud movements, and even locations of invisible water vapor. In addition, they monitor the sun, pick up emergency signals from downed aircraft and foundering ships, and relay data from weather instruments on land and at sea.

Geostationary satellites give the big picture

A west-to-east orbit 22,238 miles above the equator means that a satellite's speed matches the Earth's rotation, keeping it over the same place, or "geostationary." The USA's two satellites in such orbits are "Geostationary Operational Environmental Satellites" or GOES.

Sounder: Measures temperature, humidity at 19 levels of atmosphere

Imager: Takes regular and infrared photos

Length: 88.3 feet

Weight: 4,600 pounds

TWO SATELLITES COVER THE USA

GOES 9
Over equator at 135° west longitude

GOES 8
Over equator at 7.5° west longitude

THE VAPOR CHANNEL

Forecasters are making more use of the "vapor channel," which shows otherwise invisible water vapor in the air. For instance, plumes of vapor from the tropics have been linked ...

Thunderstorm

Water vapor

... to thunderstorms that cause downpours and flash floods in the northern United States.

POLAR ORBITERS STAY LOWER

North

Equator

South

1 Two NOAA TIROS satellites orbit over the North and South Poles, about 530 miles above the Earth. Two Defense Department weather satellites follow similar orbits.

2 The Earth is always giving off infrared and microwave radiation.

3 Clouds, air temperatures and water vapor affect how much radiation at various frequencies reaches satellites.

4 Satellites detect tiny differences in Earth's radiation, and radio the data to forecasting centers.

5 Among other things, the data supply forecasting computers with upper-air temperatures in areas without weather balloons.

requires doing better than the unskilled forecasts.

This complexity makes it easy to see why Edward Lorenz wrote in his book *The Essence of Chaos:* "To the often-heard question, 'Why can't we make better weather forecasts?' I have been tempted to reply, 'Well, why should we be able to make any forecasts at all?' " Lorenz is the Massachusetts Institute of Technology atmospheric science researcher whose work led to the development of the idea of chaos in physical systems. Chaos gives scientists good reasons to think that no matter what improvements are made, detailed weather forecasts for more than about two weeks ahead will never be possible. We look at chaos and how it applies to weather forecasting at the end of this chapter.

Computers move into forecasting

When Ronald McPherson joined the National Weather Service in 1959 — it was then the Weather Bureau — "the operations manual prohibited us from issuing forecasts beyond 36 hours. You could issue an 'outlook' for another day. Now we issue forecasts for five days and an outlook for six to 10 days. All of that came about largely because of numerical weather prediction."

"Numerical forecasting" means using mathematical equations to predict the weather — a job that can be handled only by the world's most powerful computers. In the United States, this is done at the National Centers for Environmental Prediction (NCEP) in Camp Springs, Md. This is the heart of weather forecasting in the United States. Ronald McPherson is the director.

Since coming to NCEP in 1968, McPherson has seen computers grow more powerful, allowing for improvements in the forecasting models. This, along with improved observation systems, has led to better predictions. Supercomputers, which can handle more than two billion operations a second, are necessary for numerical forecasting. The graphics on Pages 186 and 187 show how the NCEP's computers work 24 hours a day, seven days a week digesting weather information from around the world and producing forecasts.

The output includes more than 20,000 kinds of text and approximately 6,000 graphics, mostly maps, each day. The forecasts and other data go to local National Weather Service offices, other government agencies, private meteorologists (such as those for television stations and airlines) and other countries. The material includes discussions by NCEP meteorologists comparing a day's run from the computer models with previous runs and how the models compare with each other. These interpretations help other forecasters make their predictions. James E. Hoke is the director of the NCEP's Hydrometeorological Prediction Center, which provides basic analyses and forecasts. Computers are needed, he says, because "meteorology is based on very hard science."

Hoke says the equations for weather forecasting models are more complicated than those used in aerodynamics. "I have friends who are aerospace engineers and I like to tell them that we use many of the same equations they do, but we start where they stop." Designers of airplanes and rockets use equations for movement of air, just like meteorologists. But, Hoke says, "they don't deal with things such as heat radiating away at night and the effects of cumulus clouds." In many ways, meteorology is harder than rocket science.

Improving computer forecasts

While the basic science of weather goes back to laws of motion that Isaac Newton formulated around 1700, today's scientists are still learning the details of how these and other basic scientific laws apply to weather. The more researchers learn, the better job they can do of writing the equations computers need. There is a special skill to translating atmospheric knowledge into the kind of mathematics computers need to produce forecasts.

Computer models calculate what the atmosphere will do at particular points over a large area, maybe the entire Earth, and from the surface to the top of the atmosphere. The more of these "grid points," the better the model will do. But adding grid points balloons the number of calculations needed. Luckily, computer speed is expected to continue growing, making it possible to handle even more detailed models.

Having more grid points allows models to represent the surface of the Earth better. Eugenia Kalnay, former director of NCEP's Environmental Modeling Center, says a model with more realistic land forms had "a profound impact on forecasts" of the January and March 1995 floods in California. That year's precipitation forecasts were much improved over previous California predictions. Mountains help determine where heavy rain falls; computer models with realistic mountains and valleys will do a better job of predicting it.

Computer models need a lot of accurate information about what the weather is doing now to predict what it will do in the future. Kalnay says "a staggering amount

The world's first weather satellite photo was transmitted from the U.S. TIROS 1 on April 1, 1960.

People on airplanes aren't the only ones endangered by microbursts.

A microburst can fan a forest fire quickly into an inferno.

Microbursts also overturn boats. On July 7, 1984, one overturned a 90-foot excursion boat on the Tennessee River near Huntsville, Ala., killing 11 of the 18 people aboard.

and diversity of atmospheric observations are gathered every day" for numerical prediction. In the past, most information came only from weather balloons and surface observations. Twice a day, 1,100 weather balloons are launched around the globe to record temperature, humidity, air pressure, and winds aloft. Each day, researchers at weather stations make 28,000 observations of surface air pressure, temperature, wind and humidity. Since the mid-1990s, NCEP also has received daily about 6,000 temperature profiles and 11,000 wind measurements from satellites, and about 16,000 temperature and wind observations from aircraft, mostly large jets.

McPherson says the additional data from satellites, radar and aircraft was one of the "breakthroughs" in weather forecasting. "This kind of development doesn't get much attention — it's sort of buried in the weather forecasting infrastructure — but its influence is felt throughout the forecasting system," he says.

Ensembles help forecasters rate predictions

Ever since meteorologists started predicting the weather, they've known that sometimes the weather "behaves well." Most of the time, storms move along as expected, offering few surprises. But at other times, the weather seems to have a mind of its own, breaking many of the rules it's expected to follow. Predictions would be more accurate if forecasters had some way of knowing if the weather is likely to act according to the rules.

Since 1992 both NCEP and the European Center for Medium Range Weather Forecasting in England have been producing "ensemble forecasts."

Instead of running just one forecast for 10 days using one model, the center runs several forecasts, called an ensemble.

If each run looks pretty much the same, the weather is likely to do as predicted. If they don't look the same at certain places, that shows that something is going on in the atmosphere to make the weather act up.

Miles Lawrence, a National Hurricane Center forecaster, calls ensemble forecasting "the premier contribution that NCEP meteorologists are making to this whole business of saving lives and property." In the mid-1990s, ensemble hurricane forecasts were moving from experimental products to standard ones. "If we can say that we have a great deal of confidence in a forecast, this is important information for people who have to make decisions based on our forecasts," Lawrence says. "I'm really pleased to be a part of this advance."

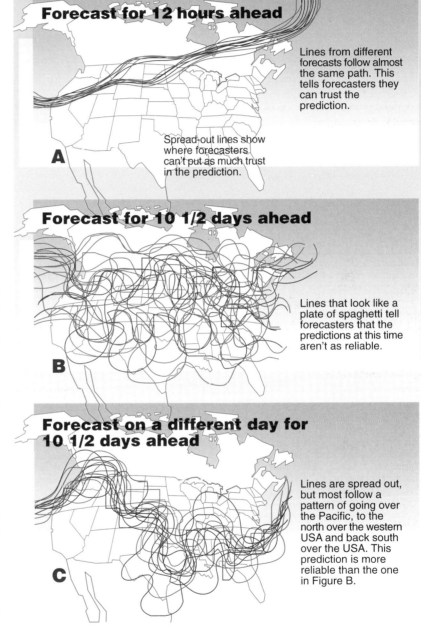

How "spaghetti plots" help forecasters use computer predictions

"Ensemble forecasts" help meteorologists decide when computer predictions are likely to be reliable. All computer forecasts start with the weather at a certain time and calculate what it will be in the future. For ensemble forecasts, meteorologists run the same forecast several times, but with slightly different data for each starting point.

In the figures here, you can think of the lines as being forecasts of where the jet stream will be.

Forecast for 12 hours ahead

Lines from different forecasts follow almost the same path. This tells forecasters they can trust the prediction.

Spread-out lines show where forecasters can't put as much trust in the prediction.

A

Forecast for 10 1/2 days ahead

Lines that look like a plate of spaghetti tell forecasters that the predictions at this time aren't as reliable.

B

Forecast on a different day for 10 1/2 days ahead

Lines are spread out, but most follow a pattern of going over the Pacific, to the north over the western USA and back south over the USA. This prediction is more reliable than the one in Figure B.

C

While a supercomputer is needed to make ensemble forecasts, anyone can use the same idea to help figure out how reliable a particular forecast for a few days ahead might be. Say you have something big planned for the coming Saturday that depends on good weather. On

IMAGINARY CENTER SHOWS HOW COMPUTERS MAKE FORECASTS

To see how computers forecast the weather, let's imagine computers had never been invented, but instead people did what computers do now. Our imaginary, humans-only system would need more than 123,000 people working in a seven-story building to do what the National Centers for Environmental Prediction's supercomputer, which would fit into a 15-by-20-foot room, does today. These drawings represent a simplified version of models like those the NCEP supercomputers run each day.

WEATHER OBSERVATIONS FLOW IN FROM ALL OVER THE WORLD

Weather balloons
Aircraft
Weather stations
Ships
Weather buoys
Satellites

QUALITY CONTROL

Junior quality controllers correct obvious errors such as wind measured at 212 mph instead of 21 mph.

Senior quality controllers ensure reports make sense when compared to earlier reports and those from nearby stations.

ANALYSIS

Analysts use data to draw separate weather maps for the seven levels of the atmosphere used in forecasting.

INPUT

Using the maps, analysts figure out the weather in each box of the seven-layer grid.

GRID SYSTEM

The forecasting building mimics forecasting computer models, which divide the world into a grid with several layers from the ground to around 12 miles up. This drawing shows a grid like that of the simplest model, with seven layers, the NCEP uses.

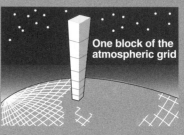

One block of the atmospheric grid

Our human-computer office has seven stories, each corresponding to one of the model's seven levels. Each of the 3,000 grid points on each level has its own computation desk with six people.

3 Keeps track of air entering or leaving the box. If more is coming in than going out, decides how much air rises or sinks.

2 Calculates north-south component of the wind.

1 Calculates east-west component of the wind.

HOW FORECASTS ARE DONE

 Schedule is keyed to the twice-a-day weather balloon launches. About three hours after balloon-launch time, all data has been sent to desks.

Each desk calculates how the weather would change at its location from noon, the time that weather balloons were launched, until 12:10 p.m. Each desk has only seconds to do this.

When time is up, each desk passes data on weather conditions to the four desks on each side plus the ones above and below.

Using the new data, each desk repeats this routine, figuring the weather for 12:20, then 12:30 and for each 10-minute step for 12 hours. The desks then send data to the output department before continuing to calculate forecasts for 24, 36 and 48 hours after the initial time. This is done twice each day. Once each day forecasts are carried out to 10 days.

Forecasting building is seven stories high. Each story contains 3,000 desks and 18,000 employees.

THE DESKS

Each person at each desk handles the equations describing various aspects of the weather. Here's what they do:

4 Calculates the effects of adding or taking away heat.

5 Keeps track of water in all forms and how much is changing to or from vapor, liquid or ice.

6 Calculates the air temperature, pressure and density.

OUTPUT DEPARTMENT

Clerks plot data from the desks on blank maps. Expert analysts use the data to draw various kinds of weather maps. Other analysts compose messages that add details not on the maps. The maps and information go to weather forecasters all over the Western Hemisphere.

The creative touch sparks better models for forecasting

Eugenia Kalnay says growing up in Argentina gave her a head start on becoming a leader in computer weather forecasting.

At the University of Buenos Aires in the early 1960s "40 percent of the students in the School of Sciences were women," she says. Even though "we were quite poor — I had holes in my shoes," no one questioned the idea of her wanting to become a scientist.

Her father died when she was 14, but her mother continued to encourage Kalnay's dream. "She had never finished school," but she encouraged her daughter to earn a doctorate.

The only hitch in Kalnay's plans came when "my mother discovered a scholarship in meteorology and she changed my major. I had wanted to study physics." Now Kalnay is happy about the switch. "Meteorology is wonderful because it allows you to do science and something useful."

She had a chance for graduate study at the Massachusetts Institute of Technology. "I had heard women in the U.S. were well treated by their husbands. I assumed women would be 50 percent of my graduate school class at MIT." Instead, the only women she found at MIT were wives of her fellow graduate students.

Her thesis was a computer model of the atmosphere of Venus. Her advisor was Jule Charney, one of the pioneers of computer forecasting.

In 1995, when the American Meteorological Society honored her with its Jule G. Charney Award, Kalnay said, "I am overwhelmed by this award because Professor Charney was for me much more than my thesis advisor, he was my hero. He was an incredibly good scientist. There is no doubt that if there were a Nobel Prize awarded in atmospheric science, he would have been everybody's first choice."

Kalnay won the AMS award for her use of computer models to explain how the atmos-

Eugenia Kalnay's biggest accomplishment was in gathering "extremely good people."

phere works and for improving the models used in forecasting.

After earning her Ph.D., Kalnay taught at MIT and the National University in Uruguay, then headed the Global Modeling Branch at NASA before the National Weather Service named her director of its Environmental Modeling Center in Camp Springs, Md., in 1986.

The Modeling Center is one of the Weather Service's National Centers for Environmental Prediction, headed by Ronald McPherson.

He credits Kalnay with "returning the [Modeling Center] to world eminence in numerical modeling. This in spite of shrinking budgets and smaller staffs, inadequate facilities, and procurement rules that condemn us always to lag other centers [in the United Kingdom, Canada and France] in computing capability."

In 1996, she decided to retire because "one shouldn't be director more than 10 years." Pointing to a photo of about 60 people, the Modeling Center's staff, she says,

"This is what I'm most proud of. They are people from all over the world, both sexes, all ages, all overworked. My biggest accomplishment was being able to attract extremely good people."

She talks with pride of how her group has led the world in important advances in computer forecasting, such as ensemble forecasting. "Meteorology is a story of success. A story of the government doing the right thing."

Kalnay plans to continue research and teaching. One day she would like to teach in Argentina, a way of repaying her native country for helping make her a scientist.

McPherson says, "Eugenia is an excellent scientist in her own right, but she also has a special touch for leadership of other scientists. It has been said that leading researchers is much akin to herding cats — they all want to go their own way, independently. One can only lead by example and by gently ... encouraging their creative energy into a broad channel. Eugenia has the touch to do this."

Monday, you can start checking the forecast for the weekend. If the forecast for Saturday keeps changing (for example, Monday predicts rain; Tuesday, dry weather; and Wednesday, rain), put less trust in the forecasts.

Computer forecasts show their stuff

Two major East Coast snowstorms (one in 1993, the other in 1996) showed how accurate weather forecasts are becoming. Paul Kocin, a research meteorologist at NCEP, says the March 1993 storm "affected so many people that they couldn't help but notice how well we did." It was immediately named "The Storm of the Century" because it dumped snow as far south as Mobile, Ala., set records for low barometric pressures up and down the East Coast, paralyzed the big cities from Washington, D.C., to New England, and closed every major airport in the East from Atlanta northward.

Computer models, both at NCEP and in Europe, began pointing to a major weekend storm five days before the storm began. Computers started forecasting the storm when they "saw" two upper-atmospheric disturbances, one north of Hawaii, the other over Alaska's Aleutian Islands. The computer predicted the merger of these two disturbances, which eventually triggered the massive storm.

Three days before the storm hit, on March 10, the computer models continued to agree on the storm's general picture. The Weather Service began issuing warnings. Kocin says ensemble forecasts give meteorologists the confidence to issue warnings two days before the storm had even begun forming over the Gulf of Mexico. When the first warnings went out, the storm existed only in the computers.

The 1993 storm "changed public expectations and brought an extraordinary change in forecasters' confidence," Kocin says. "Ten years ago, a five-day forecast was fiction. Now we're getting five days of lead time on a storm."

Much the same thing happened in January 1996 when one model saw the conditions for a blizzard five days ahead of time. "Within two or three days beforehand, the models zoomed in on it, warning of a 20-inch event. We made forecasts with confidence," Kocin says. "A lot of people were saying, 'I can't believe I'm saying this'" as they issued the kind of forecasts they would never have dreamed of a few years before.

"For many of them, it was what they got into the field for," Kocin says. "I never had more fun. At times I was the first person to see the forecast. I had my 15 minutes of fame on *The Today Show* and CNN. I ate the whole thing up. I'm normally flustered on TV. But I was in my element."

Computers aren't replacing humans

Successful computer forecasts aren't chasing humans out of weather offices. People are a long way from having programmed all of the equations needed for weather forecasting into computers. People have a way of working out new equations and finding better ways to use the old ones. Then too, even though weather forecasting is built on science, there's still enough art to it to require a human touch.

"The field is a fine-tuning between scientists and weather enthusiasts; between people with a feel of the math and physics and people with a feel for weather," Kocin says. "They often clash. It's rare to have both in one person. There are forecasters who couldn't do the math for beans and scientists who don't understand how the weather works. The models are king. But, if left up to the models, the forecasts would be no good."

Sondra Young, who is a "lead forecaster" (shift supervisor) in the NCEP division that produces national precipitation forecasts, says human precipitation forecasters on average "have never lost to a computer model over the course of a year. We do lose on particular days. At times you feel the model is on target, and you go with it. But sometimes you don't trust the models and go with what you think is going to happen." She says forecasts for the amounts of precipitation improved markedly in 1995. But the "big breakthrough was in the display of information." Forecasters went from looking at black and white paper maps to using computer workstations that let them use different colors to make a clearer picture. This graphic improvement helps forecasters do a better job of evaluating what the computer models are saying.

Satellites see the big and smaller picture

Since the TIROS I satellite began sending back the first regular photos of the Earth's clouds in 1960, satellites have been giving forecasters the big picture of what the weather is doing globally. Those first photos, which showed cloud patterns, were a revelation. For instance, meteorologists learned that more hurricanes formed and died in the eastern Pacific, far from land and shipping routes, than anyone had realized. Seeing the patterns of clouds in storms helped scientists develop the more com-

The Oklahoma's Mesonet is a network of 114 stations that measure air temperature and humidity, wind speed and direction, air pressure, rainfall, solar radiation and soil temperature. This data goes to the Oklahoma Climate Survey at the University of Oklahoma in Norman, which verifies data quality and sends it to users around the state.

Forecasters, farmers, news media, air traffic controllers and teachers use the Mesonet. Enid sixth-grade teacher Lori Painter has her students go on-line to use the data to post weather conditions every two hours. In the spring, they track the dryline, a moist air-dry air boundary that often triggers severe thunderstorms as it moves across Oklahoma.

How radar gives a better view of weather

Since World War II radar has been helping track weather. All weather radars detect the location of storms, track them and give data on strength. The latest radars, using the Doppler principle, also detect wind speeds and directions and can often "see" winds outside of storms. The Weather Service is making Doppler pictures available to television stations.

How radar works

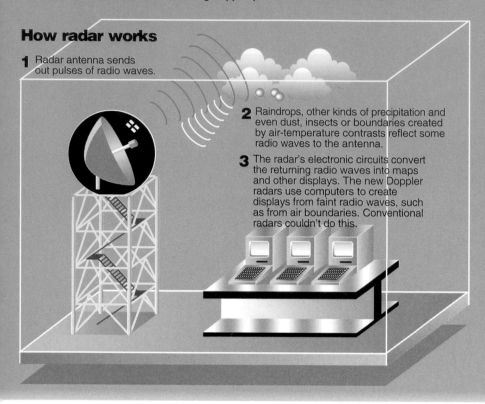

1 Radar antenna sends out pulses of radio waves.

2 Raindrops, other kinds of precipitation and even dust, insects or boundaries created by air-temperature contrasts reflect some radio waves to the antenna.

3 The radar's electronic circuits convert the returning radio waves into maps and other displays. The new Doppler radars use computers to create displays from faint radio waves, such as from air boundaries. Conventional radars couldn't do this.

Radar just shows precipitation

A map-like display seen on the National Weather Service Radar at Norman, Okla., at 6:33 p.m. CDT, April 26, 1991. Colors show rain and hail and their intensity. Blue areas are the lightest rain, green and then yellow heavier rain. Red areas show hail. This display, like those on older weather radars, gives no indication that the thunderstorms shown are spinning out tornadoes.

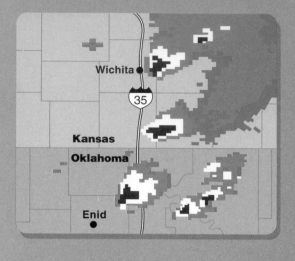

The word "radar" is an acronym for "radio detection and ranging."

The German physicist Heinrich Hertz demonstrated in 1888 that metal objects will reflect radio waves. In 1904, Christian Hulsmeyer, a German engineer, obtained patents for a device based on Hertz's findings.

Little development occurred until the late 1930s, however. All the major nations involved in World War II rushed to perfect radar and the resulting breakthroughs led to weather radar.

plete picture of storms described at the end of Chapter 4.

Satellites soon went beyond sending back only ordinary "visual light" photos. Researchers developed imagers that detected and measured the strength of various frequencies of infrared energy. At the most basic level, infrared images continue "seeing" storms and other clouds after sunset. Thanks to infrared images, hurricane forecasters no longer must await the first satellite photos after sunrise to see what a storm had been up to overnight.

Infrared images do more. Since the amount of infrared energy a cloud emits depends on its temperature, infrared images show cloud temperatures. Infrared images also show the temperatures of the land and the ocean surface. They can measure the average temperature of layers of the clear atmosphere and detect invisible water vapor in the air. Infrared "vapor channel" images not only show where the air is dry or humid, they also show otherwise invisible jet streams and areas of high and low pressure. Time-lapse movies of vapor images help forecasters see changing patterns. They show the otherwise invisible plumes of moisture from the tropics that can supply the water for downpours over the USA.

The most familiar satellite images are from geostationary satellites orbiting over the equator at just the right speed to keep up with the Earth's rotation. These stay over the same spot as the Earth rotates. Orbital speed depends on a satellite's altitude; to keep up with Earth's rotation, a satellite has to be 22,238 miles up. Two U.S. geostationary satellites plus European and Japanese satellites give nearly complete global coverage.

By 1996, the U.S. had two "next generation" geostationary satellites keeping watch over most of the USA (parts of Alaska aren't seen) and far out into the Atlantic and the Pacific. These satellites look for storms that can threaten the USA. They capture more detail and send more images than older satellites. With these new satellite images, forecasters can keep a closer eye on developing severe thunderstorms and see details of hurricanes that were only blurs in older images. Improved infrared imagers also show some things, such as fog forming at night, that older satellites couldn't see at all.

Complementing the geostationary satellites are two U.S. polar-orbiting ones. These circle the Earth from north to south, crossing over the South Pole and the North Pole. They are in much lower orbits than the geo-

How the Doppler principle adds vital information

Christian Doppler, an Austrian scientist, explained in 1842 why the whistle of an approaching train has a higher pitch than the same whistle when the train is going away. The frequency of sound waves from an approaching source is shifted to a higher frequency or pitch while those from a receding source shift to a lower frequency. The principle applies to the frequency of radio waves returning to a radar antenna.

How Doppler radar reads weather

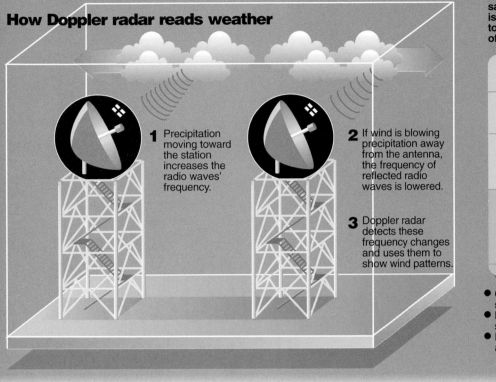

1 Precipitation moving toward the station increases the radio waves' frequency.

2 If wind is blowing precipitation away from the antenna, the frequency of reflected radio waves is lowered.

3 Doppler radar detects these frequency changes and uses them to show wind patterns.

Doppler shows possible tornadoes

This is a Doppler radar wind velocity picture of the April 26, 1991, Oklahoma storms. The radar automatically sounds an alarm when it detects a mesocyclone that could cause a tornado. On April 26, 1991, the Doppler saw a mesocyclone. Nine minutes later, forecasters issued a warning. Another 19 minutes later, the tornado touched down. The tornado near Enid left a 63-mile path of damage, killing one person.

- Green colors show winds blowing toward radar; lightest shade shows 50-knot winds.
- Blue colors show winds blowing away from radar; lightest shade shows 50-knot winds.
- Dark red circles show mesocyclones, where strong winds are blowing in opposite directions close together.

---

stationary satellites, only 517 miles up. As the satellites are flying in their north-south orbits, the Earth is rotating under them, which means each satellite captures images from every part of the Earth each 12 hours. With two satellites, data from any place on Earth is never more than six hours old. Infrared readings from the polar-orbiting satellites are an important source of upper-air data from areas with few or no weather stations. Satellites also measure ozone in the atmosphere, solar energy being reflected from the Earth, and infrared energy leaving the Earth. They relay radio signals from remote ground-based instruments, such as buoys far out in the ocean, and emergency signals from aircraft, ships and hikers.

Doppler radars blanket the country

On Dec. 2, 1996, the Weather Service turned off the 1950s radar at Charleston, S.C., the last of the 66 old radars (designed in 1957) that had been helping forecasters guard the country against severe storms for 35 years. The old models were replaced by a network of 161 Doppler radars. The old models didn't cover all of the U.S., and many mountain areas in the West had no radar coverage.

The new network covers all of the USA, including Alaska and Hawaii. Graphics above and on the previous page show the basics of ordinary weather radar and Doppler radar. The new Weather Service model is called 88D — 1988 is the design year and "D" is for Doppler.

These Dopplers detect wind motions inside storms. They are so sensitive they can "see" insects and other small objects in the air, which is how they can detect wind patterns in air with no raindrops or snowflakes to reflect radar waves.

During winter storms, the new radars see bands of heavy snow with lighter snow between. The old radars showed winter storms as mostly blobs, leaving forecasters to guess the locations of bands of heavy snow. Computers enhance the radars' power. For example, computers can estimate rain intensity and keep track of how much rain has fallen on particular areas. This helps flood forecasting.

The 88D radars can be programmed to sound an alarm when weather patterns are beginning to appear dangerous. With the old radars, someone had to watch the screen constantly when storms were possible to make sure nothing important was missed. Once a storm is

A key to weather forecasting is the launch of weather balloons from approximately 500 stations around the globe. They are launched at the same time all over the world, at noon and midnight Greenwich Mean Time, which is 7 a.m. and 7 p.m. Eastern Standard Time.

The balloons rise more than 15 miles with packages called radiosondes, which measure air pressure, temperature and humidity. The data are radioed back. The balloons are tracked to determine wind speeds and directions.

Reports of his death led to safer flying for everyone

John McCarthy has every right to feel proud of the nation's Terminal Doppler Weather Radar network. The $340 million network installed at busy airports by the Federal Aviation Administration will help pilots avoid microbursts and other dangerous weather. McCarthy was head of the team that developed the system. Installation was completed in 1997.

John McCarthy, microburst researcher

tually, the job's over. I feel very proud that I had something to do with it. Most scientists don't get a chance to start a process and finish in a few years."

Since concluding work on wind shear, McCarthy has devoted his attention to the need for improved weather information to reduce other hazards. Thunderstorm formation, conditions that form ice on airplane wings, heavy snow at major airports, and in-flight turbulence are all part of his focus.

"The most exciting thing for me as a scientist [was] to conceive of the problem, address it in detail, and come up with a solution all well inside of a career," says McCarthy, the special assistant for program development for the director of the National Center for Atmospheric Research (NCAR) in Boulder, Colo.

The first airplane accident to be definitely blamed on a microburst was the June 24, 1975, crash of an Eastern Air Lines Boeing 727 at Kennedy Airport in New York City. Of the 124 passengers on board, 113 were killed.

"I became immediately attracted to the accident," McCarthy says. "There was a John McCarthy on that airplane, and that plane was nonstop from New Orleans [where McCarthy grew up] to New York, so my name appeared in the New Orleans paper indicating that I was dead. My parents got numerous phone calls." The victim wasn't related.

McCarthy helped organize the large-scale radar experiments that proved microbursts exist and documented their potential danger to aircraft. He was also a leader in experiments that honed the warning technology. McCarthy was part of the team that developed the pilot microburst training now required of all U.S. commercial pilots. The new technology and the training have helped reduce microburst wind shear accidents dramatically since 1985.

"We treated the microburst as a wartime thing; it was a fast-track, let's solve this problem," McCarthy says. "We know it's saving lives, saving airplanes and intellec-

"What drives me," McCarthy says, "is I'm a pilot and I'm a meteorologist. I want to bring [better] weather products to the people, to the pilots, to the controllers, to the general public."

McCarthy recalls being fascinated as a young child by the movie tornado in *The Wizard of Oz*. When he was in the fifth grade, school was dismissed one day because of a tornado warning. "I went home and started studying in the encyclopedia about tornadoes and just got more and more interested in the weather.

"I used to hang around the Weather Bureau on Camp Street in New Orleans. There was a wonderful old guy named Steve Lichblau, who was the meteorologist in charge. He took me under this wing." Nash Roberts, a TV meteorologist, also influenced him.

After earning a degree in physics from Grinnell College in Iowa, McCarthy earned a master's degree in meteorology at the University of Oklahoma and a doctorate at the University of Chicago. Along the way, he also earned a pilot's license. He taught at the University of Oklahoma and then joined NCAR in 1979.

"By the year 2005, we're going to have twice as many airplanes in the sky," McCarthy says. "We've got a new, modernized air traffic control system, but we need to have a new, modernized aviation weather system to go hand in hand with that or we'll be paralyzed due to weather."

John McCarthy on the need for better aviation forecasts:

"By the year 2005, we're going to have twice as many airplanes in the sky. We've got a new, modernized air traffic control system, but we need to have a new, modernized aviation weather system to go hand in hand with that. Or we'll be paralyzed due to weather."

spotted, computers linked to the radars can draw maps showing where it's headed. That cuts the time it takes to get warnings out.

Another separate system of Doppler radars was also being installed at most of the nation's biggest airports in the 1990s. These are Terminal Doppler Weather Radars. While the Weather Service's 88D radars are designed to keep a weather eye on the entire nation, the Terminal Doppler radars are designed to give a close-up view of the area within a few miles of busy airports. These are watching, in particular, for signs of potentially deadly microbursts. Terminal Doppler radars are operated by the Federal Aviation Administration. If the computers linked to this radar find early signs of microbursts, they are programmed to alert air traffic controllers who then warn pilots.

Weather for those who fly

Microbursts are just one of the weather dangers that can threaten aircraft. In Chapter 7, we saw how researchers and forecasters are working to improve warnings of conditions that can cause dangerous ice to form on airplanes. Turbulence, thunderstorms, poor visibility, and low clouds add to the danger, delays and costs of flying. Good forecasts of these hazards are only the beginning of reducing hazards. Pilots, airline dispatchers and air traffic controllers need the information in an easy-to-understand form.

Getting improved aviation forecasts to those who need them is a different challenge. In February 1977, the Weather Service's Aviation Weather Center in Kansas City started the Aviation Gridded Forecast System. Computer forecasting models produce predictions for "grid points." (The graphic on Pages 186-187 shows how a grid is used for computer forecasts.)

Say, for example, a point 6,000 feet above Dayton, Ohio, has a relative humidity of 59 percent at 6 p.m. "For a pilot, the fact that the humidity will be 59 percent doesn't mean much," says David Sankey, head of weather research for the FAA. "As a pilot, I want to know whether there will be a cloud there or not and whether there will be icing in it." The Aviation Gridded Forecast System is so detailed, it can predict locations of clouds, icing danger, turbulence, poor visibility, and head winds. The grid point forecasts aren't only for particular airports as the old ones were, but also for various altitudes, about every 2,000 feet to above the highest altitude at which planes fly.

The Weather Service makes the information avail-

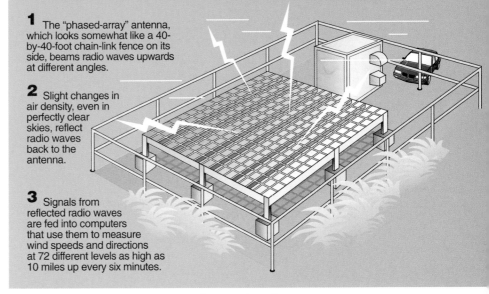

Radar profilers read the winds

The strange-looking devices — called wind profilers — shown here are being installed in fields in the central United States and eventually will be scattered around the country. They measure the speed and direction of high-altitude winds, a job that has been done mainly by weather balloons in the past.

1 The "phased-array" antenna, which looks somewhat like a 40-by-40-foot chain-link fence on its side, beams radio waves upwards at different angles.

2 Slight changes in air density, even in perfectly clear skies, reflect radio waves back to the antenna.

3 Signals from reflected radio waves are fed into computers that use them to measure wind speeds and directions at 72 different levels as high as 10 miles up every six minutes.

able. Knowledgeable pilots can use the raw data, but Sankey expects companies or individuals to produce software that will put the information into useful forms, such as cross-section drawings of the weather along a flight path.

The limits to forecasting

In 1904, Vilhelm Bjerknes stated the basic idea behind using computers to forecast weather: If we know enough about what the atmosphere is doing now, and if we know the physical laws that govern its movements, we can predict what will happen in the future. During World War I, Lewis Fry Richardson developed the key mathematical concepts needed to forecast weather. But until the development of electronic computers after World War II, scientists had no practical way to solve the equations.

By the early 1950s, researchers working under John von Neumann and Jule Charney at the Institute for Advanced Studies in Princeton, N.J., were making simple forecasts with computers. These forecasts were showing promise and scientists had every reason to expect their forecasts to become better as they improved their computers and learned more about the atmosphere's laws.

But even though scientists were learning more and the forecasts were getting better, they weren't improving as fast as expected. They weren't keeping up with the growth in knowledge and computer power.

One of those looking into the problems of comput-

The National Weather Service — then the Weather Bureau — first became involved in computer forecasting in 1954, when it set up a research office headed by Joseph Smagorinsky.

This office was the forerunner of both the National Meteorological Center and NOAA's Geophysical Fluid Dynamics Laboratory in Princeton, N.J., which conducts basic research on atmosphere and ocean models.

How TV weather forecasts get from the studio to the living room

What you see in studio ...

TV weather forecasts use a process called color-separation overlay or chroma key to change the maps behind the forecaster. In chroma key a single color, here it's green, forms a hole in the background, so that maps and photos put together on computer can be put in that hole.

... on TV

Since there's no map behind him, Doug Hill watches monitors on both sides of the green screen to see where to point.

While the weather forecast is an important part of local TV news, it's the only news on The Weather Channel, a 24-hour cable service based in Atlanta. The Weather Channel started with 2.5 million subscribers in 1982. In less than 10 years it had grown to 50 million. The Weather Channel produces more than 1,000 different computer-generated graphics every day.

The five-minute television weather forecast looks simple enough. One forecaster standing in front of a changing map, pointing to Highs and Lows. A quick look ahead at tomorrow's weather. Maybe some satellite pictures. A five-day forecast. A little chatter with the news anchors and it's over. It looks so easy.

But behind the camera, putting together a TV weather forecast is a marathon event. Today's TV weather reporters have to be part forecaster and part graphic artist.

In Washington, D.C., Doug Hill does the weather on WUSA-TV. This is how he puts together his report for the 5 p.m. news.

It begins just after noon. "I very quickly review the data from the U.S. Weather Service. We rely on a variety of computer models. We get quite a few. By working every day," Hill says, "you develop your own consistency. You become quite familiar with weather patterns."

Next comes a study of radar information (the station has its own Doppler radar), satellite pictures (still photos come in via computer and Hill picks the ones he wants) and the written discussions among Weather Services offices. "It's like a great meteorological stew." This takes about 45 minutes to an hour.

"Now I take off my meteorologist's hat and put on my graphic artist's hat." The station buys most of its weather data from a private company, but Hill and the other weather reporters must draw all the maps and build satellite and radar animations.

On a computer, Hill and his WUSA weather team puts together:

- Local and national maps for the next two days
- A six-day outlook. "We can forecast (ahead) 48 hours," Hill says. "After that it's trends."
- Several satellite photos run together so it appears that the weather systems are moving — in animation. "If you transpose pictures, you can have the exciting appearance of clouds moving backwards."
- Composite Doppler radar images.
- "Finally, we have 78 weather observers near and far who religiously call in observations on certain days at certain times." Names, temperatures and weather conditions are put on yet more maps.

"By midafternoon I start 'stacking the show.' I write a budget of everything [that will appear].

"Then I record our phone call-in weather line, do a couple more graphics of satellite photos and radar, race downstairs and feed the parking meter."

He checks the latest reports from Washington's National Airport — temperature, humidity, barometric pressure, wind speed and direction plus air quality.

"By 5:05 the tempo is picking up. The News is on the air. This is a very intense time. I assemble everything in sequence. If my timing is right, I cycle through the pictures to make sure they're in the right order.

"I run down the hall, bang the elevator, go down five stories, walk through the doors into Studio 12, [check in with] the floor director and go on the air.

"I haven't missed my page in nearly eight years. I've come close. It all comes together in the last second."

er forecasts in the early 1960s was Edward Lorenz, a theoretical meteorologist at the Massachusetts Institute of Technology (MIT). He had intended to be a mathematician when he graduated from Dartmouth College in 1938, but World War II made him an Army Air Corps weather forecaster. At MIT after the war, he turned his attention to mathematical and theoretical problems involving weather systems.

Lorenz developed a set of equations to mimic some aspects of weather. He programmed them into a Royal McBee computer and found they created changing patterns much like the weather's.

One day in 1961, Lorenz decided to rerun some results from his computer model, but instead of starting at the beginning, with original figures, he started about halfway through by entering the numbers he had on a printout. Soon the computer's "weather" began deviating from the first run. Eventually the two patterns hardly resembled each other. Yet both had started from the same point.

At first Lorenz thought something was wrong with the computer. Lorenz eventually found that a tiny difference in the numbers fed to the computer made a big difference in the results. The computer took calculations out to six decimal places, but printed them out only to three places. When Lorenz started the new run at the halfway point, he programmed in numbers only to three decimal places. This tiny change, for example the difference between 3.461453 and 3.461, was enough to change the output drastically.

The unexpected result showed that Lorenz's weather-like system had a "sensitive dependence on initial conditions." Tiny changes in the data eventually result in big changes in the output.

Lorenz says his results were like those obtained by the University of Chicago "dishpan experiments" that were attracting scientific attention at the time.

In these, a pan is filled with water and put on a turntable. The center of the pan is cooled while the rim is heated. Water in the cold center sinks and flows toward the rim while warm water at the rim rises and flows toward the center. When the pan is rotated, the moving water eventually develops wavy patterns resembling jet-stream winds.

As the speed is increased, the water's movements begin flipping from one kind of wavy pattern to another in no apparent order — just like real jet streams, the weather, and the results Lorenz found in his computer's output.

Lorenz's work was the beginning of the mathematical theory of chaos, which has become important not only in meteorology but in most other sciences.

The theory tells weather forecasters they can never expect to know all of the conditions that will affect a prediction. But there always will be conditions on too small a scale to be observed. If you try to forecast for a long enough period, these conditions eventually will make the forecast diverge further from reality.

Lorenz says that in the 1960s "we knew we weren't making good forecasts. The greatest practical use of these results was that they told us where to put our efforts. By working on things that will give us improvements in shorter range forecasts, there's a better likelihood we will break through." Today, meteorologists believe the ultimate limit to detailed, day-by-day forecasts is about two weeks.

While chaos means we shouldn't hope for day-to-day forecasts for more than two weeks, it doesn't mean we can't forecast anything about weather or climate weeks, months or even years from now.

If instead of day-to-day details, you're interested in average conditions, researchers have every hope of making useful forecasts for coming seasons.

These hopes are reasonable because large-scale processes, not small details, determine averages. Here's a simple example of how this works. If in the fall the ocean near a city is 5°F warmer than normal, and if the wind usually blows from the ocean, the winter should be warmer than normal. Oceans cool slowly and each winter day with an ocean breeze will be warmer than normal. But this knowledge won't help forecast a particular day's temperature.

Astronomers, who don't have to worry about chaos in eclipse predictions, are sure there will be a long, total eclipse of the sun across the Caribbean on June 13, 2132. If today's thinking about the limits chaos puts on forecasting are correct, the National Weather Service of the 22nd century might be able to tell people a year ahead that the summer of 2132 will be hot. But the San Juan, Puerto Rico, forecast office won't be able to tell tourists where to find clear skies for the eclipse any earlier than two weeks ahead.

No matter how much forecasts improve, meteorologists know that chaos will always keep their predictions from being perfect. Some intense but small events such as tornadoes will continue to hit with no or little warning because no system of observing and forecasting the weather can catch every event.

Meteorologists use the term "butterfly effect" to describe the effects of chaos on forecasting.

They might say: "If a butterfly flaps its wings in Bejing, it sets people to shoveling snow in New York." This is a way of saying small atmospheric events can have large consequences.

"We don't really know if anything as small as a butterfly would do the trick," says Edward Lorenz. But under certain conditions, tiny atmospheric changes can cascade, eventually affecting the size and paths of storms.

Olympics weather forecasting wins a gold medal

Lans Rothfusz made an extra effort to get a good night's sleep before the 1996 Olympic Games began. It didn't work.

As the meteorologist in charge of the Olympic Weather Support Office in Atlanta, the most advanced facility of its kind, he knew all eyes would be on him the following night if things went wrong.

Adding to the pressure, his team was making one of the costliest decisions of their careers — a weather forecast for the $30 million opening ceremony of the Centennial Olympic Games, an event organizers had been working six years to pull off. "That was a faith-tester right there," he says.

At 5 a.m., his forecasting team had made this prediction: zero percent chance of rain for the 8 p.m. lighting of the Olympic Torch.

From lunch time on, high above Atlanta's Olympic Ring, cumulus clouds grew in the tropical afternoon heat and humidity. The threat of rain grew with the clouds. Hot temperatures, high humidity. "You've got to start thinking thunderstorms," Rothfusz says. Yet he stuck by the forecast.

At 3 p.m., the pressure was too great. Rothfusz had to go outside and take a walk. He knew in his weather-educated mind that his forecasting team was "excellent." Their confidence was high and the technology was in place. "We had done everything we possibly could," he says. Still, he was nervous.

But by 4 p.m., clouds weren't so high. Others appeared wispy and not as solid as before, a sure sign of weakness. By show time, the sky was clearing, temperatures in the low 90s.

A perfect forecast.

After that, his team hit forecast after forecast, day after day. Their skills and success attracted attention to their state-of-the-art forecasting setup. The Olympic Weather Support Offices in Peachtree City, southwest of Atlanta, and on Wilmington Island near Savannah brimmed with cutting-edge technology. It is the same technology that will be in every forecast office in the U.S. by the 21st century.

Lans Rothfusz

Forecasting for the Olympic Games began with two goals: (1) give forecasters mountains of data to paint a complete weather picture and (2) combine all of this information into one package that can be viewed easily and quickly.

On any Olympic day, a network of 56 weather instruments scattered across Georgia and neighboring states brought the Rothfusz team the latest ground-level weather data every 15 minutes. Meanwhile, balloons carrying instrument boxes the size of milk cartons were launched every six hours. Temperature, humidity and pressure were radioed back from different levels of the atmosphere. Aircraft automatically radioed back temperature, pressure and wind information at the highest levels.

Doppler radar near each office scanned the lower and middle atmosphere for thunderstorms and the cool air downdrafts that thunderstorms bring just above the ground. These "outflow" boundaries often bump into each other, Rothfusz says, as they scamper away from their parent storm. Computer enhancements of the Doppler images helped forecasters decide which collisions would erupt into storms.

Eight times every hour satellites gave a high-altitude view of Georgia and the offshore waters sending back images of growing storm clouds. For the yachting competition, forecasters at the Wilmington Island office, headed by Steve Kinard, accessed buoy data close to shore every 10 minutes and used Doppler's sensitive software to detect sea breezes.

The National Centers for Environmental Prediction provided as many as five computer weather models every three hours. Some included ground features — hills, mountains, forests, sandy coastlines and shallow ocean water. This made the predictions more complicated but more accurate.

Taking this avalanche of output and observations and turning it into a quick, clear picture of what the weather was doing was the second challenge. To begin processing the continuous stream of data, the team relied on the Local Analysis and Prediction System (LAPS) developed by NOAA's Forecast Systems Laboratory in Boulder, Colo. Every hour LAPS provided detailed graphics of current weather conditions and a detailed 24-hour forecast for an area the size of Georgia. This smaller scale model showed weather features national computer models often miss.

Forecasters had access to 10 different computer programs to decipher, manipulate and display data in ways the forecasters easily understood. Using a supercomputer, the team had satellite and radar images, and graphics of temperatures, humidity, winds and cloud cover.

A computer system that ran time-lapse color movies of the day's weather became one of the most sought-after items at the site. "It would literally show you clouds forming, rain falling through the clouds and puddles of water on the ground, and show you this in three-dimensional space as a 48-hour prediction," Rothfusz says.

To get warnings out quickly the forecasters used the Warning Decision Support System (WDSS), which Rothfusz calls the next generation of the national Doppler radar network. High-tech computer software helped the team look inside and track storms.

Finally, an interactive computer actually wrote the forecast. The forecasters edited the words. But the Interactive Computer Worded Forecast put together data, wrote the forecast, ran a final check and sent it out.

Rothfusz recalls, "We were in hog heaven." He remembers seeing more than one forecaster walk out the door, turn and look back, not wanting to leave the office at the end of the day. "That was a testimony to me of what the system was and what it will do."

It was a taste of the forecasting power that is coming to the whole country.

Air pollution hangs over Chicago

Everybody expects the weather to be changeable. We know some winters will be more frigid than average and some summers will be scorchers. Droughts will come and go and so will floods.

Until the 1970s, however, most of us expected that over the long haul, the average temperature or precipitation would change little. We knew that the Earth had been warmer at some times and colder at others, but these changes occurred over thousands of years. We didn't expect to see changes in climate. A region's climate is its average weather — including the normal variations — measured over decades, usually 30 years.

Climate was like the height of the Rocky Mountains; we didn't expect it to change during our lifetimes or even the lifetimes of our grandchildren.

Today, however, we regularly see news stories about the possibility of climate change in the near future. Congressional committees call upon scientists to testify about the climate. International conferences work out treaties designed to address potential climate changes. Talk radio programs and letters to the editor include loud, often uninformed, debates about climate change. Books and television programs present scenarios of a "greenhouse" world of rising seas and food shortages by the middle of the 21st century. A cartoon of a man shoveling deep snow has the caption: "Where is global warming when we really need it?"

Two of the largest issues involving the Earth's atmosphere have been in the news since the 1970s: destruction of ozone in the stratosphere, and the possibility that human activities will cause the Earth's average temperature to increase, usually called the "greenhouse effect."

While both issues are complex, scientists have gone much further in understanding ozone depletion than climate change.

The Swedish chemist Svante Arrhenius reportedly was the first to use the term "greenhouse effect." He used it in an 1896 article describing how carbon dioxide in the air keeps the Earth warmer than it would otherwise be. His article helped solve the scientific mystery of why the Earth is as warm as it is.

U.S. national annual average temperature

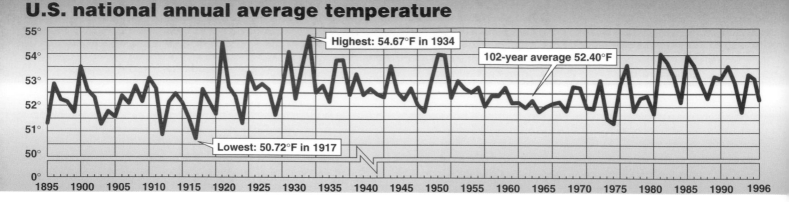

Annual temperature averages from 1895 through 1995 across the contiguous United States do not show a strong long-term upward or downward trend for the country as a whole. As the chart above shows, the beginning of the century was cool. The 1930s through the 1950s were warm while the 1960s and 1970s tended to be cool. Another generally warm period began in 1986, with seven of the 10 years through 1996 above the long-term average.

By the early 1990s, scientists showed conclusively that human-made substances called chlorofluorocarbons (CFCs) and some other chemicals destroy ozone in the stratosphere. The evidence convinced policy makers around the globe that the only practical way to save the ozone layer was to stop releasing these chemicals into the atmosphere. By the mid-1990s, treaties had led to reductions in releases of ozone-destroying substances, and scientists were saying that as long as the treaties are followed, the ozone layer should begin recovering early in the 21st century.

In contrast, while a greenhouse warming seems to be a likely result of carbon dioxide and other "greenhouse" gases being added to the atmosphere, much remains to be learned about the Earth's climate system and its response to potential warming. Although understanding continues to grow, scientists are not able to give the firm answers about the greenhouse effect that policy makers want.

Understanding how Earth's climate works — and being aware of the literally millions of interactions that produce it — is the start of understanding how humans can affect our atmosphere.

Looking for climate clues

To know whether the climate is changing, records are needed. The National Climatic Data Center in Asheville, N.C., keeps and studies climate records, not only for the U.S., but also for other parts of the world. Comprehensive records for the U.S. go back to around 1895. These include data not only from official weather stations, but also from an extensive network of volunteers, which by the mid-1990s numbered more than 11,500.

But the CDC has only spotty day-to-day temperature and precipitation data for America before 1895. There are some exceptions. For example, it has Thomas Jefferson's records of Philadelphia temperatures the week of July 4, 1776.

In the 1800s, the Army began keeping weather records. But the Civil War ended a lot of record keeping and even destroyed some data. "When Atlanta burned, a lot of old weather records burned with it," says Richard Heim, a scientist at the Asheville center. Authorities were slow to resume extensive weather observations after the Civil War.

"I'd love to see what the last fifty years of the 1800s were like," Heim says. "We know, based on anecdotal information, that the middle of the country was very wet in the 1870s and 1880s as it was being settled. The humid, wet climate [in states such as Kansas] was being promoted. In the 1890s, a major, major drought hit. "

If the data were available, Heim would be able to compare the USA's generally wet weather since the mid-1970s with the weather of the 1870s and 1880s to see just how unusual the late-20th century wetness is. He could also find just how unusual the USA's major 1930s and 1950s droughts were.

As for global weather records, Tom Karl, the CDC's senior scientist, says that while the world's nations share data needed to make daily weather forecasts, information that climatologists need is often difficult to obtain. Some nations "have a lot of concern about releasing their data because they fear it may go to others who could sell it, taking away potential revenue." As a result, Karl and other scientists have to make individual arrangements with other nations to obtain needed data.

John Christy of the University of Alabama in Huntsville and Roy Spencer of NASA are collecting satellite data that give a global picture of temperatures in the lower atmosphere. This data goes back only to 1979. It's valuable because it includes the vast stretches of ocean and third-world countries where no regular measurements are available. As more years of data are added, separating true climate trends from brief weather changes will become easier.

Volunteers across the USA fill out the nation's weather picture

To figure out what the greenhouse effect might do to the Earth's climate, scientists use the world's fastest supercomputers to make increasingly elaborate computer models.

Some of the most important data for those models will come from people like John Maddox of Rome, Ga. He's following a family tradition of volunteer weather observations begun by his great-great-grandfather, Reuben Norton, in 1855.

Maddox is one of more than 11,500 volunteers across the United States who keep meticulous weather observations to send once a month to the National Climatic Data Center in Asheville, N.C.

There, says Richard Heim of the Asheville center, he and other scientists use the unique, long record of observations to track what the nation's weather has been doing. Long-term volunteer reports from a single location are the only way to separate real climate change from apparent changes.

"In the 1940s and '50s many [weather] stations moved from downtown to airports," Heim says. "Also as cities have grown, we have had to adjust for the urban heat island effect" — the higher temperatures in built-up areas. "The best way to do this is by comparing reports to nearby stations, especially rural stations that haven't moved. We can do that in the United States. We've got a nice, extensive network" of volunteer stations.

Maddox, who is director of development, alumni, and public relations for the

John Maddox checks high and low temperatures from the instrument shelter in his yard.

Darlington School in Rome, Ga., took over the observations from a cousin in 1974. "He sent letters to all of his cousins saying 'I've been doing this for 20 years and I can't go anymore, it's time for some of you guys to take over.' I hadn't thought about doing it too much before that. But I certainly didn't want 136 years to go down the drain," Maddox says.

The family's observations continued even during the Civil War and both world wars. The Rome station is one of only 588 volunteer stations with more than 100 years of records. The National Weather Service honored Maddox in 1991 with its Thomas Jefferson Award for volunteer observers. Jefferson was a dedicated weather observer and record keeper.

"Way back in the old days," Heim says, "the volunteers took their readings, wrote them down on a form and mailed them in once a month. In this modern age of high technology, they get their piece of paper and pen out each day and write down the readings and at the end of the month they send the form in. It would be nice if we could have the observers tied in with computers and satellites, but the cost would be prohibitive." Once the reports arrive in Asheville they are made part of the computer data base available to researchers.

What makes someone willing to record weather observations each day of the year or arrange for someone else to do when he or she is away?

"It kind of gets in your blood," Maddox says. "I remember as a teenager, really as an elementary [school] child, when I'd go and spend the night with my grandmother on a Friday night, and I'd go with my aunt on Saturday mornings, we'd measure the temperatures and rainfalls, and we'd walk on down to the Fifth Avenue Bridge where we'd do our river readings, and we'd walk on down to Western Union. At that time she'd telegraph those river readings into Mobile to the Corps of Engineers.

"And probably, I guess all of the grandchildren and cousins had done that at one time or another. It's just been a family tradition for 136 years."

The historical picture

Where detailed day-to-day records aren't available, such as for much of the 19th century, accounts of big storms, major freezes and unusually hot summers help scientists reconstruct climates of the past. For example, old maps showing the extent of glaciers in the Alps help show that in the 17th century, Europe was colder than it is today. Detailed records of wine production in France can help climatologists reconstruct year-to-year weather changes.

Nature also has left us millions of clues. Fossil plants, pollen and animals are good climate indicators, since certain plants and animals thrive in particular climates. Growth rings of trees tell the story of good and poor growing seasons far into the past. Bubbles of air in ice caps show what the atmosphere was like when falling snow trapped the air. Rocks deposited by glaciers continue to help researchers map the extent and timing of glaciers around the world.

Using such indirect evidence, scientists have a good

idea of the ups and downs of Earth's climate in the past. Such knowledge supplies good tests for the computer models being used to try to forecast future climates.

"By studying the past, by documenting what's happened in the past, we have a ground truth," or a picture of what has happened, says John Kutzbach of the University of Wisconsin at Madison. "If we can come up with physical explanations, put them into our models and check to see whether the model calculates a climate very different from the present, but one that agrees with what actually happened, then we've increased our confidence that we're on the right track."

Astronomical, geological effects on climate

We'll divide the things that affect our climate into two groups. Those in the first group are like one-way streets; the effects travel one way. The forces affect the climate without climate affecting them.

The second, far more complicated, set of forces are like two-way streets; the effects travel both ways. These forces affect the climate and are, in turn, themselves affected by the climate.

First we'll look at the one-way forces. The major "one-way" force is the amount of solar energy reaching the Earth. Other "one-way" forces are collisions of large objects (such as asteroids) with the Earth, geological factors (such as movements of the continents or the growth of mountains) and volcanoes.

As we saw in Chapter 2, the sun is the source of energy that drives the weather. It's natural to assume that changes in the amount of solar energy reaching the Earth would affect climate. Earlier we saw how solar astronomer Jack Eddy gathered evidence linking a reduction in sunspots with the Little Ice Age of the 15th to 18th centuries.

Theories about how changes in the sun's energy output could affect weather or climate are controversial. Not all scientists agree with Eddy about the sunspot-Little Ice Age link, but most have come to accept theories of how changes in the Earth's orbit can affect climate.

In the 19th century, scientists speculated that orbital changes could account for the series of ice ages that have marked the Earth's geological history. But neither a detailed theory nor convincing evidence was available. Milutin Milankovitch, a Serbian mathematician, worked out a theory of Earth orbit in the 1920s and 1930s and gave a timetable of changes that could trigger the beginnings and ends of ice ages. Finally, by the 1980s

geologists had collected enough evidence to show the timing of ice ages agreed with Milankovitch's theory.

Milankovitch found three factors that determine how much solar energy reaches different parts of the Earth: the shape of the orbit, the time of year the Earth is nearest the sun, and the Earth's tilt. All three factors are on different schedules, ranging from around 22,000 to 100,000 years. At times the cycles work together to start or end ice ages.

While the sun is by far the most important astronomical force affecting climate, evidence is growing that large objects, such as asteroids more than five miles wide, have collided with the Earth in the past, throwing up clouds of dust and starting huge fires that could prevent sunlight from reaching the surface, thus cooling the Earth for years. It's likely that such a collision around 65 million years ago caused the extinction of large animals, including the dinosaurs, along with plant and animal life in the oceans and on land. Since such collisions have happened in the past, they could happen again. Some scientists are urging increased astronomical observations that would detect any such objects in the future, with the goal of one day being able to destroy or deflect them.

Geological effects on climate include the changing size and arrangement of the continents and oceans, and the growth or erosion of mountains over millions of years. How much of the Earth is covered by water and the locations of land, especially mountains, make big differences in climate.

While the movements of continents take millions of years to affect the climate, a volcanic eruption can spew enough material into the air within hours to cause many months of weather changes around the globe.

A case that's often cited is the tremendous explosion of Mount Tambora, in what is now Indonesia, in April 1815. It ejected as much as 48 cubic miles of material, killing an estimated 10,000 people immediately. Dust from the volcano spread around the Earth and created spectacular sunsets for a year. Though not a completely proven, scientific cause and effect, the dust is often linked to the so-called 1816 "year without a summer" in western Europe and eastern North America, which was unusually chilly with widespread crop failures.

The connection between Tambora and the "year without a summer" is based on indirect evidence. Only a few people were taking regular weather measurements anywhere in the world in 1815 and 1816. And, of course, they were unable to collect upper-atmosphere

Some of the most important clues to past climates come from tiny, ocean creatures known as foraminifera.

These creatures — some only about one-fifth as wide as a human hair — have shells that have become part of the sediment on the bottom of the oceans.

The shapes and makeup of the shells depend on climatic conditions at the time they form.

Scientists, using foraminifera found in core samples drilled from the bottoms of oceans all over the world, have worked out the Earth's major temperature swings for about 150 million years.

data that we routinely gather now.

When the Mount Pinatubo volcano in the Philippines began a series of "major" and "cataclysmic" eruptions on June 9, 1991, scientists were able to track in great detail the effects of the 22 million tons of sulfur dioxide gas it spewed. It shot debris and gas up about 100,000 feet in the two cataclysmic eruptions on June 15 and 16, and as high as 20,000 feet in several of its weaker eruptions. Most of the debris fell on the Philippines and ash turned day into night in Manila 55 miles away.

The sulfur dioxide gas, however, stayed in the air, mixing with the small amounts of water in the stratosphere to form tiny drops of sulfuric acid. This created a thin haze that spread around the Earth. Normal stratospheric winds carried the haze over most of the Earth by early 1992.

NASA's Earth Radiation Budget satellites showed that during 1992 about 4.7 percent more solar energy than normal was reflected away from the Earth. Other satellite measurements found that Earth's lower atmosphere began cooling in late 1991, dipping about 1°F and not returning to average or above-average readings until 1993, after the haze was gone.

Meteorologists weren't able to say positively that Pinatubo caused the unusually cool summer of 1992 and cold winter of 1992-93 in parts of the USA, but it could have helped trigger the weather patterns responsible.

Not all large eruptions affect the globe as Tambora and Pinatubo did. For example, the 1980 eruption of Mount St. Helens in Washington had no global effect because the main explosion was more to the side of the volcano than straight up into the stratosphere. Since the gases and dust stayed in the lower atmosphere, rain washed them from the air in a few days.

Interactions inside the climate system

Now that we've looked at the "one-way" forces affecting climate, let's turn to those on a two-way street, those that are affected by the climate even as they affect it. These parts of the climate system are:

- The atmosphere.
- The hydrosphere — Earth's water, especially the oceans.
- The cryosphere — ice caps, sea ice and glaciers.
- The biosphere — all living things.

Understanding how the Earth's climate works and making sense of the debate over climate change requires knowing how the atmosphere, hydrosphere, cyrosphere

and biosphere work together, with changes in each part pushing and pulling the other parts. The web of connections is extremely complex and many details are not completely understood, although scientists do have a good general picture of how the various parts interact.

A good deal of the climate system's complexity arises from "feedbacks" that can aid or inhibit changes.

Positive and negative feedback

Positive feedback pushes the climate further in one direction. Negative feedback tends to slow the climate's movement in a particular direction.

Snow and ice are examples of positive feedback. They reflect more sunlight back to space than bare ground or water, helping cool the region. Therefore, the snow and ice are even less likely to melt and precipitation is more likely to be snow than rain.

Positive feedback can work the other way, too. It can encourage melting, once started, to continue. People in places covered by winter snow see this every spring. As warm days begin, the snow melts slowly at first. But once a few bare spots appear, the sun begins to warm the snow-free ground. As the ground warms up, it warms the air above it. This helps melt more snow. Those first, few spots of bare ground start positive feedback that melts snow.

Clouds offer an example of negative feedback for surface temperatures. A day that starts out with clear skies will warm up. But sometimes clouds form, stopping the warming by blocking the sun. The growth of clouds is a negative feedback to the day's warming. In fact, if the clouds grow into thunderstorms, downdrafts can bring down cool air, causing the temperature to drop.

While these examples are easy to understand, the ultimate effect of clouds on climate is not obvious.

Clouds cast a shadow on climate understanding

"Clouds are a substantial problem in the climate business," says Kevin E. Trenberth of the National Center for Atmospheric Research. Trying to understand how clouds affect the climate and how the addition of heat-trapping gases into the atmosphere could change their effects is sending researchers to all parts of the Earth for elaborate, long-running studies. The graphic on Page 204 shows what was involved in one such experiment. It's typical of the ongoing projects.

One of the larger efforts is the U.S. Department of Energy's Atmospheric Radiation Measurement Program. In 1992, it established sites in north-central Oklahoma

Mary Shelley's novel *Frankenstein* is a prime example of how climate can affect literature.

Mary, her husband, Percy, and others including the poet Lord Byron, spent the summer of 1816 on Lake Geneva in Switzerland.

There as elsewhere in Europe and North America, it was "a wet, ungenial summer," in Shelley's words. "Incessant rain often confined us for days to the house."

To pass the time, the group decided to write ghost stories. The story Shelley began that summer of how Dr. Frankenstein created his monster was published in 1818 as *Frankenstein; or, the Modern Prometheus*.

COMPLEX INTERPLAY DETERMINES CLIMATE

Climate — the average weather, including normal temperature and precipitation ups and downs over long time periods — makes parts of the Earth fertile farmland, arctic tundra, desert, tropical rain forest or arid plains. Climate is determined by complex relations among scores of factors ranging from the Earth's orbit around the sun, to ocean and air movements, down to microbes that release gases into the air. These interactions among the air, oceans and land vary on time scales from minutes to millions of years.

Comets or asteroids can hit, darkening the sky with debris that reduces the sunlight reaching the ground.

Changes in the sun's energy output and in Earth's orbit reduce or increase sunlight reaching Earth.

WHERE THE CONTINENTS WERE ABOUT 100 MILLION YEARS AGO

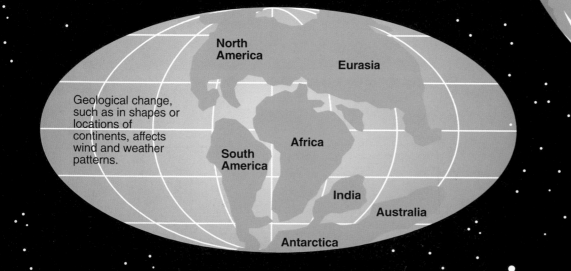

North America

Eurasia

Geological change, such as in shapes or locations of continents, affects wind and weather patterns.

Africa

South America

India

Australia

Antarctica

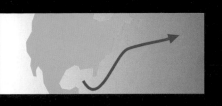

Currents, such as the Gulf Stream, help create weather. Winds warmed by the Gulf Stream give Western Europe a pleasant climate. If the Gulf Stream weakens, Europe will turn colder.

Polar ice helps create stronger storms by increasing temperature contrasts ...

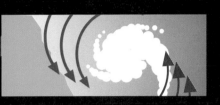

... but the stronger storms mix warm and cold air, weakening the contrasts.

Clouds reflect sunlight, helping cool the Earth.

Clouds also can help warm Earth by radiating infrared energy downward.

If forests burn, less water evaporates, reducing rainfall. Also, carbon dioxide is added to the air.

Water vapor, carbon dioxide and other gases in the air help warm Earth, the "greenhouse effect."

Oceans absorb and give up carbon dioxide. Variations in amounts absorbed or given up change the amounts in the air.

Volcanoes hurl material into the stratosphere, which reduces the amount of sunlight reaching the surface.

Scientists look to clouds for answers

Since the mid-1980s, nations with weather satellites have been building a data base of cloud information to be used in computer models of the climate. As part of this effort, scientists also are trying to learn more about the detailed structure of certain kinds of clouds and how these details relate to large-scale weather patterns. These drawings show how scientists studied high clouds during a research project in southeastern Kansas in the fall of 1991. Similar intense studies of all kinds of clouds in different parts of the world are continuing into the 21st century.

1 Data from eight satellites are compared to data gathered in and around the clouds.

2 A variety of airplanes fly above, in and below clouds making measurements and collecting ice crystals. Some — such as the ER-2, formerly the U-2 spy plane — fly as high as 65,000 feet.

The information sought includes . . .

Sizes and shapes of ice crystals making up clouds.

Amounts of sunlight reflected by various cloud particles.

Amounts of infrared energy emitted by various particles.

Air motions and turbulence inside clouds.

The "water budget" or amounts of vapor, liquid and ice and how they change.

How clouds are linked to large-scale weather patterns such as jet streams.

Ground-based instruments included ...

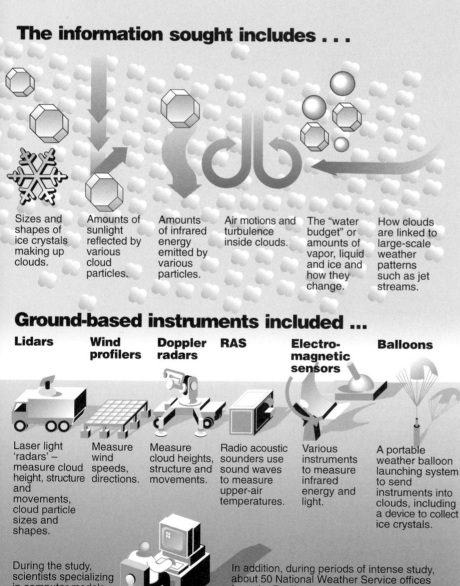

Lidars
Laser light 'radars' – measure cloud height, structure and movements, cloud particle sizes and shapes.

Wind profilers
Measure wind speeds, directions.

Doppler radars
Measure cloud heights, structure and movements.

RAS
Radio acoustic sounders use sound waves to measure upper-air temperatures.

Electro-magnetic sensors
Various instruments to measure infrared energy and light.

Balloons
A portable weather balloon launching system to send instruments into clouds, including a device to collect ice crystals.

During the study, scientists specializing in computer models worked closely with those making the observations.

In addition, during periods of intense study, about 50 National Weather Service offices from the Pacific to the Appalachians launched four, instead of the normal two, weather balloons each day for a more complete picture of the upper atmosphere.

---

and south-central Kansas and scattered instruments over approximately 55,000 square miles of the two states. A tropical Pacific Ocean site, centered around Manus Island in Papua New Guinea, was added in 1996. An Arctic site was planned near Barrow, Alaska. Scientists expect that continuing detailed measurements over long periods of time will help them better understand the interactions among solar energy, infrared energy, clouds, water vapor and other substances in the air.

How much sunlight a cloud reflects and how much infrared energy it radiates downward depend on the cloud's altitude, its thickness, and the nature of its water droplets or ice crystals.

As we saw in Chapter 3, the uneven heating of the Earth's surface causes winds. Since clouds help heat or cool the Earth's surface, they have an effect on the winds. These winds, in turn, can move clouds and bring in dry air to evaporate them or humid air to thicken them.

Clouds, then, help create complicated positive or negative feedback that is difficult to represent in computer models of the climate. Yet, computer models that accurately represent the effects of clouds are needed to answer questions such as: How is warming likely to affect weather patterns? What kinds of clouds will the new patterns create and where will they be? And finally, will the most likely kinds of clouds enhance or hinder global warming?

How ice plays a role

Large amounts of frozen water in the Greenland and Antarctic ice caps — as sea ice around the polar regions and mountain glaciers elsewhere — help create our climate by increasing the temperature contrast between polar regions and the middle latitudes, including the United States.

As the drawing on the next page shows, air over ice-covered polar regions grows colder, than it would otherwise be, increasing the contrast between it and air farther from the poles. As we saw in Chapter 3, strong cold-warm air contrasts supply more energy for storms than weak contrasts. But stronger winds caused by larger contrasts push more arctic air south and bring more warm air north, warming the polar regions. The system creates a negative feedback.

Warren Washington of the National Center for Atmospheric Research in Boulder, Colo., notes another effect of sea ice: Locking up water as ice reduces the amount of sea water that evaporates into the air. If polar regions warmed up enough in the future to melt some of

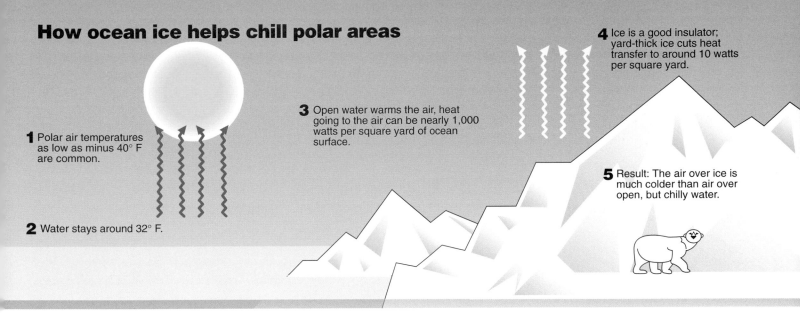

How ocean ice helps chill polar areas

1 Polar air temperatures as low as minus 40° F are common.

2 Water stays around 32° F.

3 Open water warms the air, heat going to the air can be nearly 1,000 watts per square yard of ocean surface.

4 Ice is a good insulator; yard-thick ice cuts heat transfer to around 10 watts per square yard.

5 Result: The air over ice is much colder than air over open, but chilly water.

---

the ice, more water would evaporate into the air, which could increase snowfall not only over the ocean, but also over nearby land.

Water's role in the climate

The term "hydrosphere" refers to all of the water on Earth. Water affects climate by taking up and releasing heat as it changes among its phases — vapor, liquid and ice. Both the amount of water that falls as precipitation and the amount available to be evaporated help determine a region's climate.

Oceans are a major player in climate. They hold large amounts of heat and move it around via global-scale currents.

Since the oceans cool or warm slowly, they slow the temperature changes in the air above them. This is why places near the ocean don't have temperature contrasts as large as places far inland. Summer winds are cool and winter winds are warm.

Scientists refer to this slowing of temperature changes by the oceans as "thermal inertia."

Ocean currents, such as the Gulf Stream in the Atlantic, play a key role in climate by moving warm or cool water from one place to another. Ocean temperatures also help determine the locations of large high- and low-pressure areas such as the Bermuda High and the Aleutian Low.

The atmosphere also affects the oceans. Winds help create currents and the amounts and locations of clouds help determine the ocean's temperature. More than the winds, however, move the large ocean currents such as the Gulf Stream.

The Gulf Stream is really the most prominent part of a global "conveyor belt" of ocean water. The conveyor brings warm water northward in the Atlantic; the faster the belt moves, the warmer the North Atlantic becomes. The belt's movement is known as the "thermohaline circulation" from the Greek words for heat and salt. Water moving north at the Atlantic's surface loses heat to the winds and becomes more salty because of evaporation. In the Labrador Sea between Canada and Greenland and around Iceland, the now-heavier water sinks and more warm water flows north to replace it. This thermohaline sinking is the engine that drives the conveyor belt.

After sinking, the cold, salty water flows south deep in the Atlantic to the Southern Ocean around Antarctica. It eventually rises in the Indian and Pacific oceans and some begins flowing northward again on the surface.

Large speed changes in the conveyor belt, even stops and starts, explain wild swings in Northern Hemisphere temperatures during the ice ages. But the belt has been running relatively steadily the last 8,000 or so years.

Many scientists are convinced that smaller speed changes, which occur every few decades, affect our current weather. Scientists need to learn more about the conveyor belt to distinguish changes caused by its normal speedups and slowdowns from weather changes that greenhouse warming might be causing. As shown in the graphic at the bottom of Page 206, a warmer climate could slow the conveyor belt.

How living things affect climate

The word "biosphere" refers to all of the Earth's living things — from microbes to the largest plants and ani-

How a warmer climate could add ice to Antarctica

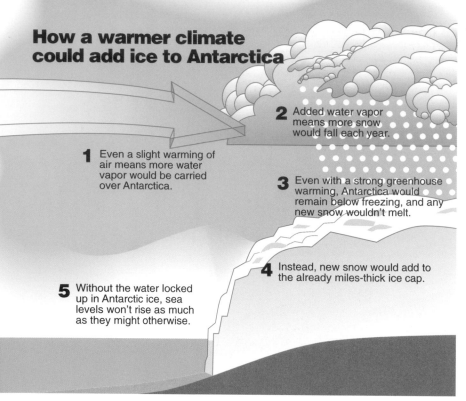

1 Even a slight warming of air means more water vapor would be carried over Antarctica.

2 Added water vapor means more snow would fall each year.

3 Even with a strong greenhouse warming, Antarctica would remain below freezing, and any new snow wouldn't melt.

4 Instead, new snow would add to the already miles-thick ice cap.

5 Without the water locked up in Antarctic ice, sea levels won't rise as much as they might otherwise.

mals. All have the potential to affect the climate.

Carbon dioxide has an elaborate natural cycle. Plants take it from the air to make plant material, which becomes part of animals that either eat plants or eat other animals that eat plants. Marine creatures end up carrying large amounts of carbon to the bottom of the ocean when they die and sink, locking up carbon for thousands of years.

Large amounts of carbon are locked up in living things, the oceans, fossil fuels, the soil and rocks. Carbon dioxide is released into the air when vegetation decomposes or is burned. Burning fossil fuels, which were once living things, adds carbon dioxide to the air.

Methane is another greenhouse gas that is in the air naturally, but human activities increase it. It is created by reactions involving microbes. The biggest source is natural wetlands — methane is often called "swamp gas" — but rice paddies add almost as much to the air. The world's third largest source is fermentation in the stomachs of animals such as cattle and sheep.

Nitrous oxide is also a greenhouse gas being increased by human activity. Nitrate and ammonium fertilizers are thought to add nitrous oxide to the air as they break down.

CFCs are the only greenhouse gases produced solely by humans.

Warmer climate could weaken the Gulf Stream

A National Center for Atmospheric Research computer model that combines realistic ocean temperatures and currents with the atmosphere shows how a greenhouse warming could cause part of northern Europe to turn cooler.

Similar computer models without the ocean currents do not produce the European cooling.

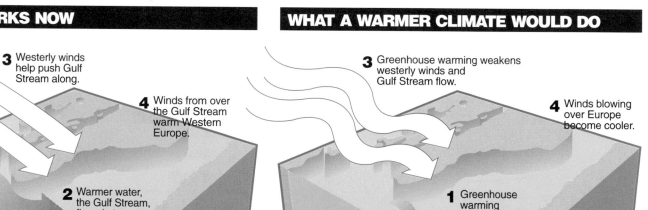

HOW IT WORKS NOW

3 Westerly winds help push Gulf Stream along.

4 Winds from over the Gulf Stream warm Western Europe.

2 Warmer water, the Gulf Stream, flows in to replace sinking water.

1 Cold, salty water sinks, flows southward deep in the Atlantic.

WHAT A WARMER CLIMATE WOULD DO

3 Greenhouse warming weakens westerly winds and Gulf Stream flow.

4 Winds blowing over Europe become cooler.

1 Greenhouse warming increases Atlantic rain, making water less salty.

2 Fresher water doesn't sink as fast, and the Gulf Stream slows down.

Greenland

Canada

Europe

U.S.

Gulf Stream

Africa

South America

Changes in the tropical Pacific affect faraway places

Patterns of sea-surface temperatures, atmospheric pressure, location of the largest thunderstorms and trade-wind speeds along the equator in the Pacific are linked and change from year to year on an irregular schedule. These changes affect weather patterns around the world. The system is known as the El Niño-Southern Oscillation.

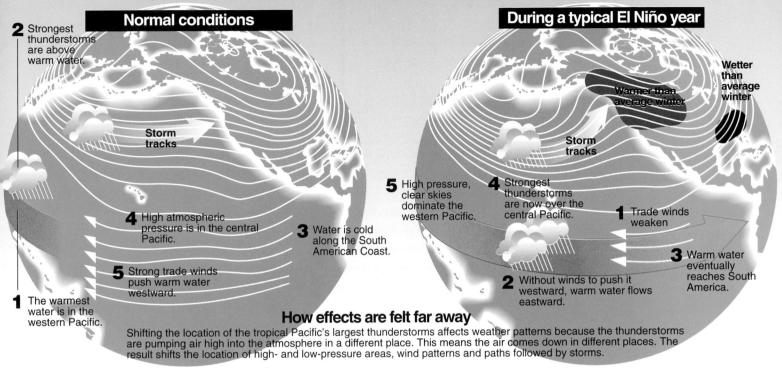

Normal conditions

2 Strongest thunderstorms are above warm water.

Storm tracks

4 High atmospheric pressure is in the central Pacific.

3 Water is cold along the South American Coast.

5 Strong trade winds push warm water westward.

1 The warmest water is in the western Pacific.

During a typical El Niño year

Warmer than average winter

Wetter than average winter

5 High pressure, clear skies dominate the western Pacific.

4 Strongest thunderstorms are now over the central Pacific.

1 Trade winds weaken

2 Without winds to push it westward, warm water flows eastward.

3 Warm water eventually reaches South America.

How effects are felt far away

Shifting the location of the tropical Pacific's largest thunderstorms affects weather patterns because the thunderstorms are pumping air high into the atmosphere in a different place. This means the air comes down in different places. The result shifts the location of high- and low-pressure areas, wind patterns and paths followed by storms.

The biosphere is a two-way street. The climate's effects on living things are as obvious as the difference between the vegetation of a desert and that of a rain forest. Climate variations change plant and animal life.

How El Niño works

The best-known example of how the ocean and atmosphere are interconnected is the set of patterns known as El Niño, or more properly, the El Niño-Southern Oscillation (ENSO).

While the land near the coast of Peru is a desert, the nearby ocean is one of the world's most productive fishing grounds. Sea life is abundant because cold water, rich in nutrients, is coming up — called upwelling — along the coast as winds push surface water away.

The name "El Niño" comes from the Spanish-speaking people who live along the Pacific Coast of Peru. Each year in late December, a southward moving current warms the water. The Peruvians started calling the warm current *El Niño* — boychild, for the Infant Jesus — because it comes around Christmas. Every few years the ocean warming is greater than normal and leads to disruption of the usually abundant fish and other marine life. It also can bring flooding rains to the desert of north-

western Peru. Until 1957, El Niño was thought to affect only South America's west coast. Now we know it's part of a global chain of ocean and atmospheric events.

In 1904, British mathematician Gilbert Walker became director of scientific observatories in India. He wanted to find a way to predict the Indian monsoon. His hope was to forecast dry growing seasons in time to avoid famine like the one that followed the 1899-1900 monsoon failure. In 1924, Walker described a seesaw pattern in tropical Pacific air pressure. He found when air pressure is low around Australia, it's high to the east, around Tahiti. When the pressure is high around Australia, it's low in Tahiti. This is the Southern Oscillation. It was interesting, but at the time didn't seem important.

Then in 1957 and 1958, a strong El Niño occurred off the South American Coast. This happened during the International Geophysical Year when more measurements than usual were being made. These showed that water was warming all the way to the west of the International Date Line, a quarter of the way around the globe. The extensive Geophysical Year measurements also caught unusual air-pressure and wind patterns in the northern Pacific and over North America.

By the 1970s, scientists realized that El Niño and the

Southern Oscillation are part of a huge ocean-atmosphere system that changes storm tracks. The effects are felt far away from its tropical home. The drawings on Page 207 show how the system works.

An extremely strong El Niño-Southern Oscillation in 1982 and 1983 focused new international attention on the phenomenon.

Records going back to the 19th century show that some El Niño years are wet along the U.S. West Coast, some dry, others average. During a 1976-77 El Niño, northern California suffered record drought. But the 1982-83 event brought a series of storms from the fall into the spring that did millions of dollars in damage along the West Coast, including an estimated $200 mil-

lion damage in California alone.

French Polynesia normally goes years between hurricanes, but the 1982-83 El Niño brought five, including one that left 25,000 people homeless on Tahiti. The El Niño was linked to floods in Louisiana, Florida, Cuba, Ecuador, Peru and Bolivia and to droughts in Hawaii, Mexico, southern Africa, the Philippines, Indonesia, and Australia. In Australia, the drought was the worst on record.

Most El Niños last about two years and occur at three-to-seven year intervals. But a strong El Niño in 1991-93 was followed by a weaker one in 1994-95. The 1991-93 El Niño brought severe droughts to Indonesia and other areas in the western, tropical Pacific and was

A leading weather educator does his teaching outside the classroom

Like most scientists, Warren Washington is a teacher as well as a researcher. But as a senior scientist at the National Center for Atmospheric Research (NCAR), he doesn't have to face lecture halls full of undergraduates or grade final exams.

Instead he travels several times a year from his Boulder, Colo., home to Washington, D. C., to help government officials understand knotty issues involving the atmosphere. By the mid-1990s, he had served every administration from Carter through Clinton.

"One thing I've found working with people in Washington," he says, is "they like to talk with scientists in a small group so they can ask questions and not feel they are coming off as being uninformed." These briefings are part of a scientist's duty to educate, he says.

Washington served on the President's National Advisory Committee on Oceans and Atmosphere from 1978 to 1984. In May 1995 Clinton named him to a six-year term on the National Science Board, which advises the president and Congress on U.S. policies in science, engineering, and education. In 1994 he was president of the 11,000-member American Meteorological Society.

Washington grew up in Portland, Ore.

Warren Washington, climate modeler

Though his father had a college degree as a teacher, he worked as a Union Pacific Railroad dining car waiter on the Portland-to-Chicago run because he couldn't get a teaching job in Oregon. The state didn't hire its first black teacher until the 1940s.

Warren Washington says a high school chemistry teacher sparked his interest in science. "She wouldn't answer questions in the normal way teachers do. She'd say, 'Gee, let's see if we can find out.' Rather than just give out an answer, she'd put you in the research mode, encourage you to find the answers for yourself." He earned a bachelor's degree in physics and a master's degree in meteorology at Oregon State University and a doctorate in meteorology from Pennsylvania State Uni-

versity.

As a researcher at NCAR since 1963, Washington has seen his specialty grow from computer modeling of the atmosphere to more complex models that include the oceans and other influences on climate such as sea ice.

He says he wants to be a role model for all minority students, not only blacks, who are interested in science. "I want to show that people such as myself have been successful." Washington says he also serves as a mentor to students who come to NCAR for summer study. This includes talking with them about their interests, helping them select the best schools for advanced studies and giving them names of people to call. "I consider this one of my responsibilities. People helped me along at important points in my career; this is my pay back."

He also works to convince younger students that "science should be viewed as a viable career option." What would he tell a student to encourage interest in science? "Science is fascinating. It gives you an opportunity to understand the environment and to explain how nature is behaving. The climate and weather systems need to be understood further. There's a lot of exciting challenges in the future."

blamed for southern Africa's worst drought in a century. In the United States, however, the two El Niños may have helped end severe drought that had plagued California since 1986. Heavy precipitation fell on California in 1992-93 and again in 1994-95.

El Niño refers to all of the conditions associated with warmer-than-normal water in the tropical Pacific from the International Date Line to South America. When the water in this area is colder than normal, it's called La Niña, the "little girl." La Niña also has global effects, but not as striking as El Niño's. Often La Niña brings dry weather to the Southwestern and South-Central U.S. This occurred from the fall of 1995 into the summer of 1996.

In response to the 1982-83 El Niño, the World Meteorological Organization began the Tropical Ocean and Global Atmosphere (TOGA) program. From 1985 to 1994, scientists tried to learn if some of El Niño's worst effects could be predicted.

As part of the experiment, approximately 70 weather buoys were moored across the tropical Pacific. These radioed back air and ocean temperatures via satellite to weather centers. Comparing that data to weather events around the world let TOGA scientists build computer models to predict El Niño's effects. In a 1996 report, the U.S. National Research Council said the project "largely fulfilled, and in some ways exceeded, its objectives. The buoys are still sending back the data and the computer models are showing how El Niño and La Niña directly affect weather in the tropics."

By 1996, Peru, Brazil, and Australia were using forecasts to predict El Niño drought or rain three months ahead. The forecasts let farmers plant crops most suitable for expected conditions. For example, in Brazil's state of Ceara normal yearly grain production is 650,000 metric tons. In 1987, an El Niño-related drought cut production to 100,000 tons. In 1991, forecasters told Brazilian authorities that another El Niño drought was likely. Authorities made extraordinary efforts to inform farmers, who planted varieties of grain more appropriate to dry conditions. Even though the 1992 El Niño drought was almost as severe as the one in 1987, production fell only to 530,000 tons.

While El Niño helps give a good idea of what seasons will be like in some tropical locations, its effects are not as clear-cut across North America. The U.S. National Weather Service uses El Niño to help make seasonal forecasts for parts of the USA, but it has to consider many other factors as well.

Other ocean-atmosphere patterns

Although researches have been looking into other ocean effects on weather for decades, the success of El Niño predictions helped spark more interest in looking for ways to use ocean conditions to help predict the weather, especially to forecast what kind of average conditions to expect months or even a year in the future.

"Oceans help to load the dice. They bias the probabilities of certain [weather] outcomes. We're trying to find out precisely how they do that," says Randall Dole, director of the National Oceanic and Atmospheric Administration's (NOAA) Climate Diagnostics Center in Boulder, Colo. If forecasters know how the ocean has loaded the dice, or is going to load the dice, they potentially could do a better job of saying things such as, "the odds favor next winter being cold in the Eastern USA and northern Europe," or "the next 20 years should bring as many hurricanes as the 1940s and 1950s."

Atlantic patterns are not as clear-cut as the tropical Pacific's El Niño swings. But there is a general west-to-east flow of wind and storms. So Atlantic patterns could be a key to European, north African and central Asian weather. Even though North America is "upstream" from Atlantic weather systems, they influence the USA's weather, says Gerald Bell of the National Weather Service's Climate Prediction Center. In fact, he says, an Atlantic pattern was a major player in making the 1995-96 winter snowy in the East. This is a pattern of air pressures and jet stream winds known as the "North Atlantic Oscillation."

From 1980 to 1994, the oscillation sent winter winds from the Atlantic over Europe. These relatively mild, humid ocean winds made the winters, on average, wet and warm across Europe, far into Russia. This pattern also made southern Europe and northern Africa dry and allowed mild winter weather in the Eastern U.S. Then, Bell says, in late 1995, the pattern shifted dramatically. Cities in the Eastern U.S. set snowfall records, central and northern Europe had a cold, dry winter and storms ended a long-lasting drought across Portugal, Spain and parts of northern Africa.

For the U.S., the North Atlantic Oscillation helps set the stage for a hard winter like 1995-96. But unlike Western Europe, the Atlantic is only one of several weather makers for the USA.

Computers and climate

The computer models used to study climate have

In 1957, Jacob Bjerknes (1897-1975) worked out the connection between El Niño and the atmospheric anomalies in the northern Pacific and North America.

He was the son of Vilhelm Bjerknes, the leader of the Norwegian meteorologists whose World War I era work began making meteorology a real science.

While working with his father's group in 1919, Jacob introduced the idea of warm, cold and occluded fronts and explained how they are related to extratropical cyclones.

How the greenhouse effect works

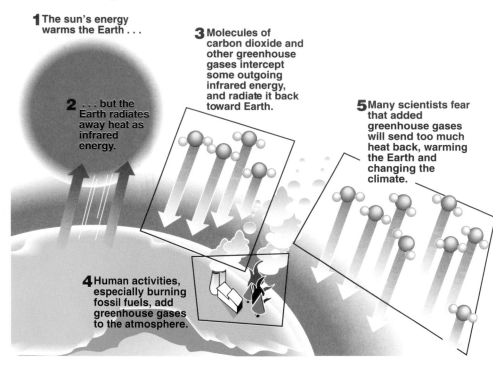

1 The sun's energy warms the Earth . . .

2 . . . but the Earth radiates away heat as infrared energy.

3 Molecules of carbon dioxide and other greenhouse gases intercept some outgoing infrared energy, and radiate it back toward Earth.

4 Human activities, especially burning fossil fuels, add greenhouse gases to the atmosphere.

5 Many scientists fear that added greenhouse gases will send too much heat back, warming the Earth and changing the climate.

some things in common with those used to forecast weather that we describe in Chapter 11. Like the weather forecasting models, they use laws of physics that describe the atmosphere's changes. And, like weather forecasting models, supercomputers are needed to run today's climate models.

But there are differences. Climate models have to take into account things weather forecasting models can ignore, such as ocean currents, and the increase or decrease of certain kinds of plants or animals.

"All climate models are imperfect, but some of these models are useful," says Kevin Trenberth of the National Center for Atmospheric Research. "We can test models on the annual cycle, the change from winter to summer. We can test using the eruption of Mount Pinatubo. We can test them on simulating climates of the distant past. We can test them on some aspects of year-to-year climate such as El Niño."

During the 1990s, many models, especially those that took ocean currents into account, were becoming more accurate. Many were generating El Niños about as often as nature was. After Mount Pinatubo erupted in 1991, some models did a good job of predicting the actual global temperature dip and rise caused by the coming and going of the volcano's global haze.

To continue improving models, scientists need to learn more about how the climate works. They are focus-ing on the effects of clouds, and links between atmosphere and ocean and atmosphere and land.

Effects of greenhouse gases, pollution haze

Usually the term greenhouse effect has a negative connotation. It's usually associated with global warming and as one of the dire consequences of industrial growth. But without the greenhouse effect, life as we know it would not exist on Earth. Our planet's average temperature would be closer to 0°F than the 59°F average it now enjoys.

To clear up the confusion, we could call the greenhouse effect that keeps our planet warm the "natural greenhouse effect." As we saw in Chapter 2, when the sun's energy reaches Earth's surface some of it is reflected back and some absorbed. The absorbed energy warms the Earth, which in turn radiates heat back toward space as infrared energy. Clouds, water vapor, carbon dioxide and other gases in the atmosphere absorb some of the outgoing infrared energy, which heats them. The clouds and gases then radiate the energy in all directions, with some coming back to Earth. In effect, some of the energy remains trapped in our atmosphere, warming the planet.

We could call the greenhouse effect that could cause global warming the "enhanced greenhouse effect." It works the same way as the natural greenhouse effect, but the extra carbon dioxide and other gases that we release into the atmosphere increase the amount of energy that becomes trapped. The Earth gets warmer.

How do greenhouse gases work?

Greenhouse gases absorb and radiate some frequencies of the infrared energy the Earth gives off. Water vapor and carbon dioxide are the two most important greenhouse gases, but the focus is on carbon dioxide because it's the most important one that humans are adding to the atmosphere.

In the early 1990s, scientists began calculating how some other pollutants in the air tend to cool the Earth. Power plant and some factory smokestacks emit sulfur dioxide, which changes to tiny droplets — an aerosol — in the air. This is the same kind of aerosol the Pinatubo volcano spread around the Earth, high in the stratosphere. The aerosols from smokestacks stay in the lower

How ozone is destroyed over Antarctica

Ozone concentrated from around 10 to 20 miles up in the stratosphere is a vital shield against much of the sun's dangerous ultraviolet radiation. Until recent years destruction and creation of ozone by natural processes were generally in balance, keeping the shield intact. Now, however, various man-made substances threaten the ozone shield.

POLAR VORTEX SETS THE STAGE

While CFCs and other substances are destroying stratospheric ozone all over the world, the most dramatic destruction occurs over Antarctica during the Southern Hemisphere spring. Unique weather conditions during the winter set the stage for the spring losses.

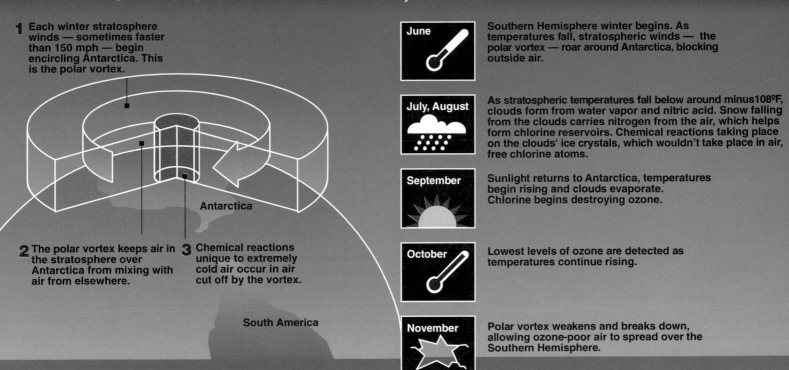

1 Each winter stratosphere winds — sometimes faster than 150 mph — begin encircling Antarctica. This is the polar vortex.

Antarctica

2 The polar vortex keeps air in the stratosphere over Antarctica from mixing with air from elsewhere.

3 Chemical reactions unique to extremely cold air occur in air cut off by the vortex.

South America

HOW OZONE IS DESTROYED

To see what happens, we'll follow a June through November cycle in Antarctica

June — Southern Hemisphere winter begins. As temperatures fall, stratospheric winds — the polar vortex — roar around Antarctica, blocking outside air.

July, August — As stratospheric temperatures fall below around minus108°F, clouds form from water vapor and nitric acid. Snow falling from the clouds carries nitrogen from the air, which helps form chlorine reservoirs. Chemical reactions taking place on the clouds' ice crystals, which wouldn't take place in air, free chlorine atoms.

September — Sunlight returns to Antarctica, temperatures begin rising and clouds evaporate. Chlorine begins destroying ozone.

October — Lowest levels of ozone are detected as temperatures continue rising.

November — Polar vortex weakens and breaks down, allowing ozone-poor air to spread over the Southern Hemisphere.

---

atmosphere for only a few days before being washed out by rain. But more are always being added. As you would expect, industrialized parts of the world have a thicker pollution haze than undeveloped regions. But other kinds of aerosols, such as particles from the burning of forests, are common in some tropical areas.

Some aerosols not only reflect sunlight that would otherwise reach the ground, they also can serve as condensation nuclei that help water vapor condense to form clouds. These clouds, in turn, can reflect more sunlight away from the Earth.

In 1995, Thomas Karl and Richard Knight of the National Climatic Data Center, and George Kukla and Joyce Gavin of Columbia University conducted a detailed study of the relationship between aerosols and temperature. They found that while the Earth as a whole had warmed since the 1950s, summer temperatures in eastern North America, Europe, and eastern Asia had cooled.

These are areas where the most aerosols are emitted. The effect shows up strongest in summer because there's more sunlight to be reflected.

The researchers also found that summer temperatures in much of the U.S. were lower from the 1950s to 1970 when aerosol emissions were increasing. Then, after 1970, when the Clean Air Act began lowering emissions, the temperatures began rising. The researchers used elaborate statistical procedures to almost completely rule out other causes for the temperature increase after 1970. While the cooling is localized, it's large enough to affect global average temperatures.

By the mid-1990s scientists were including the cooling effects of such aerosols in their global climate models, and the models began producing more accurate pictures of the climate. The cooling effect of aerosols is one of the reasons why scientists have lowered their projections of how much the Earth should warm by the end of the 21st

CFCs were invented in 1928 by Thomas Midgley, an industrial chemist who worked for General Motors. Midgley was looking for a fluid to use in home refrigerators that wouldn't be toxic or flammable. CFCs replaced dangerous ammonia.

Going literally 'to the ends of the Earth' to measure lost ozone

In June 1990, 70 nations met in London and agreed to stop all production of chlorofluorocarbons (CFCs) by the end of the century. The meeting opened a new chapter in the history of pollution control.

It was a response based on abstract scientific evidence of invisible chemical reactions that are destroying ozone more than 10 miles above the Earth. Those urging the ban couldn't point to anything obvious; there wasn't anything like evil-smelling waste flowing into a river.

Even more amazing, until British scientists discovered the Antarctic ozone hole only five years before, scientists hadn't suspected these chemical reactions were going on.

Susan Solomon, who specializes in upper atmospheric chemistry at the NOAA Aeronomy Laboratory in Boulder, Colo., was one of the leaders of the project that first theorized the CFC connection and then helped collect the data to prove it.

She first became interested in atmospheric chemistry when she was a senior majoring in chemistry at the Illinois Institute of Technology in Chicago because "it's not chemistry that takes place in a test tube, it's chemistry that takes place on the whole planet."

She's been at the NOAA lab after earning her Ph.D. degree at the University of California at Berkeley in 1981.

"It has been incredibly satisfying to be able to first start thinking about what could have caused the ozone hole … and to be among the first, if not the first, to point out that it was so intense in the Antarctic because of the reactions that might happen on the surface of polar stratospheric clouds."

Before 1985, scientists suspected CFCs were destroying ozone, but at the predicted rate only five percent would be destroyed in the 21st century. The unexpected discovery of the ozone hole (a 40 percent loss each spring) didn't fit any of the existing theories. New theories were quickly devised. One held that winds were bringing in ozone-poor air from elsewhere. Others speculated that solar activity could be destroying ozone.

Solomon was among those who suspected that chlorine from CFCs was responsible. But, how could enough chlorine and ozone be coming together in the stratosphere's extremely thin air to account for such a huge loss? Solomon theorized that polar stratospheric clouds could be responsible.

In August 1986, Solomon led a group of researchers to Antarctica. They used a highly sensitive spectrometer that detected minuscule amounts of various gases in the stratosphere by measuring wavelengths of sunlight that reached the earth. Since it was the polar winter, they had to use the sunlight reflected off the moon.

Solomon found herself standing on a roof in the Antarctic winter cold, holding a mirror to focus moonlight into the instrument. For her, this was a new side of science. "Everything I had done prior to that was with a keyboard, the only instrument I knew. It was the most remarkable experience of my life, one that I feel very privileged to have had the opportunity to do." By 1987, Solomon's measurements, combined with others, including from high-flying airplanes, had left no doubt that stratospheric clouds were enabling chlorine, which comes mainly from CFCs, to destroy huge amounts of ozone.

In January 1997, Solomon made her fourth trip to the Antarctic, going to the South Pole for the first time, with a group looking at the need to rebuild the South Pole station. "It was amazing for me, standing in place where (British explorer Robert) Scott stood almost 85 years to the day, knowing that he was going to die. [All five in Scott's party died on the way back.] It was also almost exactly 10 years to day from when I made my first measurements in the Antarctic. It was really quite moving."

In the 1990s, Solomon helped figure out the effects of the 1991 Mount Pinatubo volcanic eruption on ozone, began looking into the effects of CFC substitutes, and turned more to studying how atmospheric chemistry affects climate.

"I don't worry about finding things to do because I figure in the 21st century we're going to have more people than we've ever had on this planet. And they are going to be putting stuff into the atmosphere. We're going to need to understand all that chemistry."

Susan Solomon, ozone researcher

century. The best estimate of how much the Earth's average climate would warm between 1990 and 2100 has been lowered from about 5.4°F to about 3.6°F.

How it all works

Global warming doesn't mean merely adding a few degrees to today's average high and low temperatures. We would still have frigid outbreaks, but heat waves would become more common. Some places would become drier, others wetter. Some could become cloudier, others clearer. The biggest climate change questions revolve around the possible regional effects. Trenberth says calling it "global heating" instead of "global warming" would be less confusing.

Trenberth points out that while carbon dioxide and other greenhouse gases absorb heat that would otherwise escape into space from the Earth, only a part of the heat goes into warming the air. The rest goes into evaporating water into the air.

This "highlights the important role of water in air-conditioning our system," Trenberth says. "When you run out of water, as in a drought, you're likely to get a heat wave. The onset of the drought will occur quicker, because the wilting of plants occurs quicker. The drought could be a little worse."

Paradoxically, he notes, the fact that warmer air has more water vapor also means that if a storm occurs, there's more water to get into that storm, whether it's a thunderstorm or a major snowstorm. Someone who understands this isn't likely to use a blizzard to "prove" that the whole idea of global warming is false. This argument can't be pushed very far the other way either: A particular blizzard, flood or drought doesn't "prove" that we're seeing the beginning of climate change. Floods and blizzards were occurring long before people started adding carbon dioxide to the air.

Is our weather changing?

When you look at monthly averages for the entire USA, one feature that does stand out is that since about 1970, the USA's precipitation has tended to be above normal. While the nationwide yearly averages have been wet, some areas have been dry, including California, some other parts of the West, northern New England and parts of the Southeast.

Temperatures have averaged on the high side since the mid-1980s, but not quite as warm as during the 1930s. Unlike the warmth of the 1980s and 1990s, the

warm years in the '30s were dry.

Averages over large areas and long periods of time tend to dampen out extremes. A close look at the USA's day-to-day weather shows that weather did seem to become more extreme toward the end of the 20th century, says Thomas Karl of the Climatic Center.

He's not talking about events such as blizzards or hurricanes. He is talking about more subtle measurements such as an increase in the number of days rain falls or the number of "extremely heavy" rain storms, which are defined as storms with more than two inches of rain.

Karl says that since 1970 the USA on average has had about 2 percent more days a year with rain than it had in the previous two decades. This works out to about six more rainy days a year. The U.S. also has become about one percent more cloudy since 1970s compared with the 1950s and 1960s.

Karl says these increases fit with the overall ability of a warmer atmosphere to hold more water vapor, and weather balloon observations are showing more water in the air. He says extreme events become more common because water isn't added evenly to warmer air. A large share of the water vapor in the atmosphere is added in the tropics. Satellite photos of the tropical regions show many relatively cloud-free areas with a few huge thunderstorms or clusters of storms. These storms "are where all of the pumping [of moisture into the air] is going on," Karl says. As air moves north from the tropics to feed storms elsewhere, it isn't uniformly humid, but has extremely humid areas and drier areas. A storm that latches onto humid air will produce heavier precipitation.

The slight rise of the USA's temperatures — 0.6°F to 0.8°F — over the 20th century mainly comes from increases during the winter and spring, Karl says. "Temperatures during summer and autumn have changed little after dropping from conditions of the warm 1930s." The increased precipitation in the late 20th century has tended to be mostly in the summer and fall. This would tend to hold down temperatures in those seasons. Also, Karl says, the West has warmed more than the East.

Temperature increases since the 1980s have shown up more in daily low temperatures than in high temperatures. In other words, nights have warmed while daytime temperatures have changed little. This could be a reflection of the increase in clouds, which tend to keep nights from cooling off .

The evidence shows the USA's weather has changed during the 20th century but in subtle ways that few peo-

Large amounts of chlorofluorocarbons (CFCs) are in the air because they are so useful:

Refrigerants: coolant fluid in refrigerators, freezers and air conditioners.

Solvents: cleaners for electronic circuit boards and other assemblies.

Aerosols: propellants in aerosol cans. The U.S., Canada, Norway and Sweden banned most such uses in 1979.

Foam production: soft foam, packaging material and insulating fiber.

Fire fighting: halons used in some fire extinguishers.

Some highlights of the 1995 Intergovernmental Panel on Climate Change report:

- An increase in the globe's average temperature of about 3.6°F is likely from 1990 to 2100.
- The global average ground-level temperature has increased between 0.5°F and 1°F since the late 19th century.
- The 20th century's global temperatures are probably at least as warm as any other century since at least 1400 AD.
- There isn't enough data to determine whether the weather has become more extreme.
- Global sea level is expected to rise by about 20 inches by 2100.

ple would notice. The question is: Would these climate changes be occurring even if people weren't adding more carbon dioxide and other gases to the air? Or are they the early signs that humans are changing the Earth's climate? Karl says that by comparing the changes with climate models, odds are strong that the changes do reflect climate change being caused by humans. But they are also within ranges that could be caused by natural climatic fluctuations.

The greenhouse debate

Uncertainties and disagreements among climate models combined with the complexities of the climate system fuel the debate about what, if anything, should be done to head off or reduce possible greenhouse warming.

Daniel Albritton, director of NOAA's Aeronomy Laboratory in Boulder, Colo., says there's no doubt that the amounts of greenhouse gases in the atmosphere have increased and that these gases will decrease the amount of infrared radiation escaping from the Earth. "The response [of the atmosphere] is where the questions are legion," he says. "How is the climate — this loose-jointed, physical, biological system that we have such poor understanding of — going to respond?"

Will clouds turn out to provide a positive or negative feedback to warming? What is the role of the oceans with their thermal inertia? How much carbon dioxide are the oceans absorbing and how much will they absorb as they warm up? These are some of the questions at the heart of the debate. And scientists are only beginning to explore how the biosphere and atmosphere interact.

If the models are correct, the greenhouse warming in the 21st century would be the fastest climate change in the Earth's history, certainly the biggest humans have ever faced. It could also make the Earth warmer than ever before in human history.

These factors, among others, mean scientists can't offer any clear-cut answer about the greenhouse effect to policy makers.

Since greenhouse gases, especially carbon dioxide, are the main agents of a potential global warming, most scientists agree the best response is to reduce the amount of carbon dioxide being added to the atmosphere. This becomes an economic and political question.

The key to reducing carbon dioxide emissions is to burn less fossil fuel. But that could slow the economies of the wealthy countries. Poor countries could suffer even more. China, for instance, has huge deposits of coal, but burning coal adds more carbon dioxide to the atmosphere than any other fuel.

One side argues that the incomplete scientific knowledge doesn't justify the economic risk of cutting back on fossil fuel. The other side argues that the potential consequences are too great not to act.

Those in favor of reducing the burning of fossil fuels also argue that even if global warming wasn't a concern, we would reap other benefits: cutting air pollution, saving money, and perfecting new "climate friendly" technologies.

Scientists who study the climate expect to have more clear-cut answers during the first decade of the 21st century, but even then they expect uncertainties to remain. Decisions — which include decisions to do nothing — will continue to be based more on convictions about the right balance between potential environmental harm and potential economic harm than on hard and fast scientific fact.

A clear danger: Ozone destruction

The potential danger to the Earth's ozone layer wasn't discovered until 1974, but the story goes back to 1928 when chlorofluorocarbons (CFCs) were invented to replace dangerous compounds, such as ammonia, then used in home refrigerators. Until the mid-1970s, CFCs seemed like a miracle chemical. Even though they contain chlorine, which can be dangerous, they didn't break down in the atmosphere or react with other chemicals. As long as it's locked up in CFCs, chlorine is harmless. CFCs were a perfect fluid for refrigerators and during the 1950s they were also found to work as spray-can propellants and as solvents.

More CFCs were escaping into the air, but since they were considered harmless, it was of little concern. No need to feel guilty about the Freon — the brand name of one of the widely used CFCs — that escaped from your car's air conditioner or from the old refrigerator you junked.

During the 1960s, as the U.S., the former Soviet Union, Britain and France were talking about building supersonic airliners, concerns arose that nitrogen oxides from aircraft engine exhaust would be injected high in the stratosphere and would eat away at the ozone layer. Paul Crutzen of the Max-Planck Institute for Chemistry in Mainz, Germany, showed in 1970 that nitrogen oxides react with ozone without being consumed, which would speed up the loss of ozone. His work also added to the general knowledge of upper-atmosphere chemistry. But the U.S. and the former Soviet Union dropped plans to build supersonic airliners, leaving only a few British-French Concords, not the fleets of high-flying jets that had aroused concern.

Around this time, Sherwood Rowland and Mario Molina, who were then at the University of California at Irvine, began to wonder what was happening to all the CFCs being released into the atmosphere. By 1974 their studies led to the conclusion that the CFCs were rising into the stratosphere where they would eventually break down, freeing their chlorine, allowing it to destroy ozone. This was the beginning of concern about CFCs and the ozone layer.

The Royal Swedish Academy of Sciences awarded the 1995 Nobel Prize in chemistry to Crutzen, Molina, and Rowland. It said they contributed to "our salvation from a global environmental problem that could have catastrophic consequences."

Some wonder how CFCs, which are heavier than air, can rise into the stratosphere. Thunderstorms, especially those that tower high into the tropical sky, carry them. As we saw in Chapter 5, warm air rises to form thunderstorms. This warm air carries CFCs, and other relatively heavy substances aloft, much in the way a hot air balloon takes heavier-than-air humans aloft.

Scientists had known since 1881 that ozone in the stratosphere blocks much of the sun's dangerous ultraviolet radiation from reaching the Earth's surface. Scientists later discovered that an increase in ultraviolet radiation could cause skin cancer and harm other living things, including crops.

Ozone is a form of oxygen. Each molecule of ordinary oxygen consists of two oxygen atoms. Each molecule of ozone consists of three oxygen atoms. Small amounts of ozone occur naturally near the surface, but larger concentrations build up in the stratosphere about 10 to 20 miles above the Earth.

In the stratosphere, ultraviolet radiation from the sun is absorbed as it hits ozone molecules, breaking them apart. The absorbed energy warms the stratosphere. But most important, by absorbing ultraviolet radiation, ozone acts as a shield.

After ultraviolet energy breaks apart stratospheric ozone molecules, nearly all of the freed oxygen atoms eventually recombine into ozone. Until recently, the natural destruction and creation of ozone were generally in balance; the shield was intact.

But CFC molecules reaching the stratosphere tip the balance. Ultraviolet radiation breaks the CFC molecules apart, freeing chlorine atoms. These chlorine atoms destroy ozone. The chemical reactions are complicated, but the end result is that each chlorine atom can destroy thousands of ozone molecules. Now we have chlorine atoms as well as ultraviolet radiation destroying ozone molecules. Natural processes that create ozone no longer can keep up with the destruction.

In 1979, concern about the threat to the stratosphere's ozone layer led the United States, Canada, Norway and Sweden to ban the use of CFCs as propellants in most spray cans. Worldwide use of CFCs leveled off.

But, says Susan Solomon of NOAA's Aeronomy Laboratory, "use started back up in the early 1980s with the revolution in consumer electronics. CFCs are used to clean electronic chips. Your personal computer, your VCR, your telephone answering machine, all those things you didn't have 10 years ago, you now have partly because they were cleaned with CFCs."

As CFC use grew, scientific concern about the danger to stratospheric ozone increased. But all of the chlorine CFCs added to the stratosphere weren't destroying ozone right away. Researchers found that other substances in the stratosphere were "locking up" large numbers of chlorine atoms in chemical "reservoirs." Until 1985, scientists expected that with so much chlorine locked up in these reservoirs, only about five percent of the stratosphere's ozone would be destroyed by the year 2050.

Ozone was being destroyed, but the rate seemed to be slow enough to allow time to find a solution. Makers of CFCs argued that without evidence of massive ozone destruction, banning such useful products didn't make sense.

In 1985, evidence of massive ozone destruction caught scientists by surprise when the British Antarctic Survey reported ozone was disappearing from the sky over its Halley Bay, Antarctica, station during the Southern Hemisphere spring. The British scientists measured losses as great as 40 percent over Antarctica. Measurements from satellites and other ground stations confirmed the British report.

Someone came up with the phrase "ozone hole" to describe this unexpected thinning of the ozone layer over Antarctica. No matter what people called it, scientists now had to find out what was going on. Was something freeing chlorine from its stratospheric reservoirs?

Researchers, including Solomon, found in 1986 and

While it is needed in the stratosphere, ozone is a pollutant in the lower part of the atmosphere.

Lower-atmosphere ozone forms from exhaust emissions and other pollution. It burns eyes and lungs and harms plants.

This ozone breaks down before drifting up into the stratosphere. And, there is no practical way to gather up polluting ozone and transport it to the stratosphere where it is needed.

1987 that the unique winter conditions in the stratosphere over Antarctica — the coldest part of the stratosphere — were freeing chlorine from its reservoirs, enabling it to destroy ozone.

While loss of ozone over Antarctica wasn't an immediate threat to life, scientists know that in the summer, air over the South Pole mixes with the rest of the stratosphere. Ozone-poor air from around the South Pole could eventually dilute ozone elsewhere. Also, scientists wondered, could whatever was happening over Antarctica eventually happen elsewhere?

Researchers who went to the Antarctic in 1986 with the National Ozone Expedition returned pointing to chlorine from CFCs as a main ozone depleter. In September 1987, the United Nations adopted the "Montreal Protocol on Substances that Deplete the Ozone Layer." Among other things it required a 50 percent reduction in CFC production by the year 2000.

Also in 1987, a series of aircraft flights over Antarctica, including flights by NASA's version of the former U-2 spy airplane, the ER-2, added details to the 1986 discoveries. The 1987 flights provided clear evidence of how Antarctica's polar stratospheric clouds were key factors in the chain of chemical reactions that freed chlorine from its reservoirs. Even more important, the 1987 research left no doubt that chlorine from CFCs and bromine from chemicals known as halons — used in some fire extinguishers — were destroying ozone.

In 1987 and 1988, scientists confirmed that reactions similar to those over Antarctica were going on over the Arctic, but not to the same extent as over the colder Antarctic.

In 1990, when the nations that had signed the 1987 Montreal Protocol met in London to review the treaty, scientists presented convincing evidence of how CFCs lead to ozone destruction. This evidence persuaded the policy makers to speed up reductions in CFC emissions and to phase out all CFC production — but not use — by the year 2000.

End in sight to ozone depletion

During the early 1990s, the Antarctic ozone hole grew and scientific evidence became stronger that man-made substances were responsible. The nations that had signed the original treaties met annually from 1992 through 1995 to tighten restrictions on ozone-destroying substances. In mid-1995 scientists found that the amounts of CFCs in the lower atmosphere had dropped by about 1.5 percent as a result of the restrictions. Ozone-depleting gases in the stratosphere, where they do their damage, were expected to peak by 1998 and then begin declining. This delay occurs because air in the lower atmosphere takes several years to reach the stratosphere.

Just as the 1991 eruption of the Mount Pinatubo volcano created a natural, global experiment for climate modelers, it helped ozone scientists, including Susan Solomon, learn more about how volcanoes affect the ozone layer. Their key finding was that the volcano's aerosols provide sites for chemical reactions that free chlorine, which destroys ozone. Solomon stresses that the volcanic particles themselves don't destroy ozone, but only make it easier for man-made substances do the job.

If nations continue to follow the treaties, the ozone layer should begin to heal itself early in the 21st century, but Antarctic ozone holes aren't likely to end until some time after the middle of the century. Other volcanoes like Pinatubo would slow recovery.

Other concerns are rising. The U.S. is starting to look into the possibility of building a second-generation of supersonic transports. Research into building such aircraft includes seeing how they might affect the ozone layer. This time around, unlike in the 1960s, researchers have knowledge and techniques for studying the upper atmosphere that Crutzen, Molina, and Rowland could have only dreamed of when they did their pioneering ozone research.

Linking ozone destruction, greenhouse warming

Ozone exists at the surface of the Earth, not just in the stratosphere. Here there's too much at times. Chemical reactions involving pollution from vehicles and other sources, along with natural substances emitted by trees, form photochemical ozone smog, especially on warm days. (Trees emit natural hydrocarbons. One example: terpenes, which give evergreen forests their unique odor.) This smog can be dangerous to health. Low-level ozone also tends to heat the Earth as it absorbs ultraviolet radiation. This increases the greenhouse effect.

"We've learned an important lesson out of the CFC saga: One of the key things to look at is the lifetimes of chemicals we're going to put into the air," Solomon says. The invention of CFCs solved some important problems, including making home refrigerators safe. Since CFCs didn't react with other substances, no one saw a potential problem. The fact that they would last more than 100 years didn't seem important. Now we know, Solomon says, any time a new substance is developed we should first ask: "Is it going to have any negative impact?" Then "if we're not sure there is going to be a negative impact, we need to ask how long could any possible impact last."

While ozone destruction and greenhouse warming are separate problems, they are related. In fact, Solomon says, ozone depletion has slowed greenhouse warming. When ozone in the stratosphere absorbs ultraviolet radiation it heats up. This warming is why temperatures that steadily drop as you go higher stop dropping and even begin getting warmer as you go into the stratosphere.

With less ozone in the stratosphere, the air isn't heated as much. In turn, less heat is radiated downward to warm the surface of the Earth. When climate modelers began including this effect, their models did a better job of producing results close to the actual climate.

Even though the interactions between man-made substances and stratospheric ozone are complicated, they are not nearly as complicated as the climate system. Also, Solomon says, she and others who discovered what was happening to ozone "had things we could measure in a straightforward way over a limited time period."

Definite amounts of ozone are being destroyed. Scientists have found firm evidence that forms of chlorine and bromine that destroy ozone are in the air at the very time and place the ozone is being destroyed.

"There's just not any simple thing you can measure for the climate problem," she says. Even the biggest change in average temperature expected as a result of greenhouse warming is smaller than normal, daily temperature changes. There's no unambiguous measurement to be made, such as the destruction of 40 percent of the stratosphere's ozone over Antarctica each spring.

Benefits of understanding climate

"Climate is not a constant thing. It has always changed and it will always change," says Richard Heim of the Climate Center in Asheville. "One thing that we would like to do as a people is know what exactly has happened in the past and get some handle on what we can expect in the future, so we can plan for it. … Maybe we can affect it for better or worse. Hopefully we'll be smart enough to affect it for better."

Understanding climate is an extension of understanding the weather. Someone who is weatherwise is less likely to become a victim of violent weather than a person who doesn't realize what's going on when the sky darkens. Nations, or a world, that understand climate are less likely to become victims of climatic change.

There's another benefit, a more personal one. Climate, like weather, is an intellectual challenge. Exploring climate can draw you into history, geology, astronomy, chemistry, physics and all of the branches of biology as well as meteorology, mathematics and computer science. Weather and climate are a good entry into the insights and pleasures of science — and the secrets of nature itself.

USA record breakers, weather makers

From 100°F scorchers in Alaska to 12°F chillers in Hawaii, the USA's weather is full of surprises. Here is a state-by-state guide to weather records and events.

Rainfall amounts are records for 24-hour periods. Tornado numbers are based on data from 1953-1995, lightning deaths from 1959-1995, and are from the National Climatic Data Center or the National Weather Service's Warning and Forecast Branch. State weather statistics are based on National Weather Service reports. Many weather events are from *The American Weather Book* by David M. Ludlum (Houghton Mifflin Company).

ALABAMA
Hottest: 112°F, Sept. 5, 1925
Coldest: -27°F, Jan. 30, 1966
Rainfall: 20.33", Axis, 4/13/55
Tornadoes (avg. each year): 21
Tornado deaths (avg. each year) 6
Lightning deaths (1959-95): 88
March 21, 1932: Tornadoes sweep across state, killing 269 people, injuring another 1,874. Damage: estimated $5 million. Outbreak among worst in U.S. history with 10 violent (F4-F5) tornadoes.
March 12-15, 1993: "Storm of the Century" buries Alabama to Maine under 1 to 5 feet of snow, closing every major airport and interstate on the East Coast. Storm equates to Category 3 hurricane, swamps Florida with 12-foot storm surge, destroying 600 homes. 100 mph winds whip snow into 14-foot drifts in N.C. Record cold follows: 2° in Birmingham, -11° in Syracuse, N.Y. 270 die. Damage: $3-6 billion.

ALASKA
Hottest: 100°F, June 27, 1915
Coldest: -80°F, Jan. 23, 1971
Rainfall: 15.20", Angoon, 10/12/82
Tornadoes (avg. each year): 0
Tornado deaths (avg. each year): 0
Lightning deaths (1959-95): 0
April 3, 1898: Snowslide near Chilkoot Pass, Yukon Territory, hits during the Gold Rush, buries 142 and kills 43.
Aug. 8-20, 1967: Great Fairbanks Flood. More than 10 inches of rain falls over a two-week period — average for the month is three inches. $100 million in damages.

ARIZONA
Hottest: 128°F, June 29, 1994
Coldest: -40°F, Jan. 7, 1971
Rainfall: 11.40", Workman Creek, 9/4-5/70
Tornadoes (avg. each year): 3
Tornado deaths (avg. each year): 0
Lightning deaths (1959-95): 60
July 10, 1959: Yuma, the hottest urban area in the U.S., caps a season-long heat wave with a peak reading of 118°F. Temperatures rise above 100°F every day for a month.
Sept. 4-6, 1970: Great Labor Day Storm causes severe flooding in central Arizona. Rivers rise 5 to 10 feet per hour. Near Sunflower, waters crest 36 feet above creek bed; 23 die.

ARKANSAS
Hottest: 120°F, Aug. 10, 1936
Coldest: -29°F, Feb. 13, 1905
Rainfall: 14.06", Big Fork, 12/3/82
Tornadoes (avg. each year): 19
Tornado deaths (avg. each year): 3
Lightning deaths (1959-95): 110
Nov. 25, 1926: Tornado hits Belleville and Portland, kills 53, $630,000 in damages.
Nov. 15, 1988: One of the most active severe-weather days of the season, tornado kills 6 south of Little Rock; power lines and electricity out over much of the state.

CALIFORNIA
Hottest: 134°F, July 10, 1913
Coldest: -45°F, Jan. 20, 1937
Rainfall: 26.12", Hoegees Camp, 1/22-23/43
Tornadoes (avg. each year): 5
Tornado deaths (avg. each year): 0
Lightning deaths (1959-95): 21
Aug. 1, 1917: After a severely dry spring and summer, an electrical storm over Trinity County ignites 80 forest fires. Lightning strikes 150 times in one five-square-mile area.
Jan. 11-16, 1952: Snowstorm sets state records: 44 inches at Marlette Lake; 149 inches at Tahoe.

COLORADO
Hottest: 118°F, July 11, 1888
Coldest: -61°F, Feb. 1, 1985
Rainfall: 11.08", Holly, 6/17/65
Tornadoes (avg. each year): 26
Tornado deaths (avg. each year): 0
Lightning deaths (1959-95): 99
March 10, 1884: Two snowslides join to form an avalanche that buries railroad camp at Woodstock in southwestern Colorado; 14 of the 17 at the camp are killed.
Oct. 5-6, 1911: Extremely unusual downpours in the high San Juan Mountains cause floods on the Upper Rio Grande, Doloes and San Miguel rivers, flooding 30 blocks of Alamosa and washing out roads across the region. Remnants of a tropical storm bring rain instead of normal fall snow.

CONNECTICUT
Hottest: 106°F, July 15, 1995
Coldest: -32°F, Feb. 16, 1943
Rainfall: 12.77", Burlington, 8/19/55
Tornadoes (avg. each year): 1
Tornado deaths (avg. each year): 0

Lightning deaths (1959-95): 13
Jan. 5, 1835: One of 19th century's worst cold waves dropped temperatures to -23°F in Hartford.
March 9-11, 1936: All-New England Flood of 1936. Two torrential rainstorms combine with melting snow to cause over $100 million in property damage to six states. On March 21 the Connecticut River at Hartford reaches its highest elevation: 37 feet above flood stage; 24 people die.

DELAWARE
Hottest: 110°F, July 21, 1930
Coldest: -17°F, Jan. 17, 1893
Rainfall: 8.50", Dover, 7/13/75
Tornadoes (avg. each year): 1
Tornado deaths (avg. each year): 0
Lightning deaths (1959-95): 15
Jan. 19-20, 1961: Snowstorm moving up the Atlantic coast dumps up to 11 inches across the state, one death.
Jan. 4, 1992: Resorts hit with 70 mph winds. Storm hits at perigean spring tide, bringing 12-14 foot waves. Sand dunes and boardwalks at Rehoboth Beach and Bethany Beach sustain heaviest damage.

FLORIDA
Hottest: 109°F, June, 29, 1931
Coldest: -2°F, Feb. 13, 1899
Rainfall: 38.70", Yankeetown, 9/5/50
Tornadoes (avg. each year): 48
Tornado deaths (avg. each year): 1
Lightning deaths (1959-95): 357
Feb. 12-13, 1899: The state's most widespread snowstorm brings snow as far south as Tampa on the West Coast and Daytona Beach on the East Coast.
Sept. 2, 1935: "Labor Day" Hurricane — strongest U.S. hurricane on record — sends a 15-foot storm surge across parts of the Florida Keys, killing more than 400, many of them at a Civilian Conservation Corps camp. Barometric pressure of 26.35 inches at Long Key is the lowest surface pressure ever recorded in the U.S.

GEORGIA
Hottest: 112°F, July 24, 1952
Coldest: -17°F, Jan. 27, 1940
Rainfall: 18.00", St. George, 8/28/11
Tornadoes (avg. each year): 20
Tornado deaths (avg. each year): 2

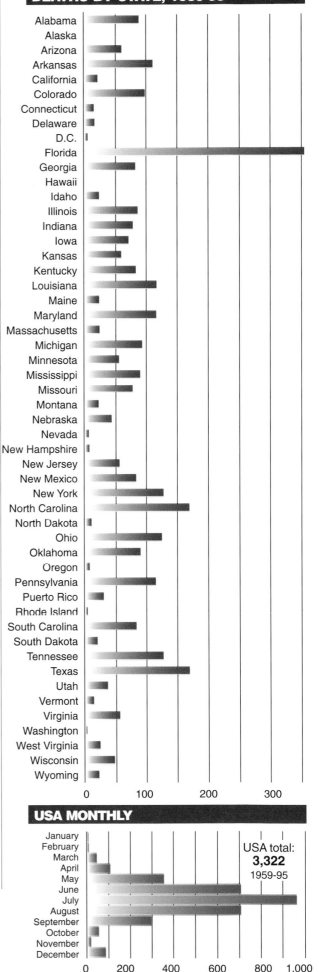

DEATHS BY STATE, 1959-95

State	Deaths
Alabama	
Alaska	
Arizona	
Arkansas	
California	
Colorado	
Connecticut	
Delaware	
D.C.	
Florida	
Georgia	
Hawaii	
Idaho	
Illinois	
Indiana	
Iowa	
Kansas	
Kentucky	
Louisiana	
Maine	
Maryland	
Massachusetts	
Michigan	
Minnesota	
Mississippi	
Missouri	
Montana	
Nebraska	
Nevada	
New Hampshire	
New Jersey	
New Mexico	
New York	
North Carolina	
North Dakota	
Ohio	
Oklahoma	
Oregon	
Pennsylvania	
Puerto Rico	
Rhode Island	
South Carolina	
South Dakota	
Tennessee	
Texas	
Utah	
Vermont	
Virginia	
Washington	
West Virginia	
Wisconsin	
Wyoming	

(axis: 0, 100, 200, 300)

USA MONTHLY

January	
February	
March	
April	
May	
June	
July	
August	
September	
October	
November	
December	

USA total: **3,322** 1959-95

(axis: 0, 200, 400, 600, 800, 1,000)

Normal annual temperature in °F (48 states)

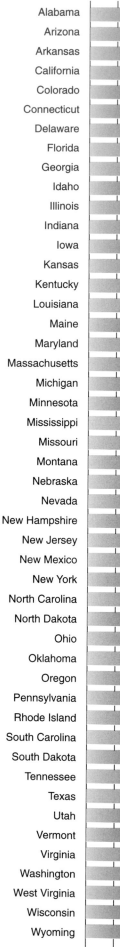

Alabama
Arizona
Arkansas
California
Colorado
Connecticut
Delaware
Florida
Georgia
Idaho
Illinois
Indiana
Iowa
Kansas
Kentucky
Louisiana
Maine
Maryland
Massachusetts
Michigan
Minnesota
Mississippi
Missouri
Montana
Nebraska
Nevada
New Hampshire
New Jersey
New Mexico
New York
North Carolina
North Dakota
Ohio
Oklahoma
Oregon
Pennsylvania
Rhode Island
South Carolina
South Dakota
Tennessee
Texas
Utah
Vermont
Virginia
Washington
West Virginia
Wisconsin
Wyoming

0   10   20   30   40   50   60   70   80

Lightning deaths (1959-95): 82
April 6, 1936: Tornado hits Gainesville, kills 203 and injures 934; $123 million in damages.
Jan. 7-9, 1973: Four inches of freezing rain hits Atlanta, stopping most traffic and closing businesses and schools for several days. Damage estimated at $25 million.

HAWAII
Hottest: 100°F, April 27, 1931
Coldest: 12°F, May 17, 1979
Rainfall: 38.00", Kilauea Plantation, 1/24-25/56
Tornadoes (avg. each year): 0
Tornado deaths (avg. each year): 0
Lightning deaths (1959-95): 0
Dec. 31, 1987: New Year's Eve storm ravages Hawaii Kai, Honolulu; floods cause millions in damage.
Sept. 11, 1992: Hurricane Iniki, Hawaii's most costly natural disaster, hits with winds 130-160 mph. 90 percent of Kauai destroyed, 7 people dead. $1.8 billion damage.

IDAHO
Hottest: 118°F, July 28, 1934
Coldest: -60°F, Jan. 18 , 1943
Rainfall: 7.17", Rattlesnake Creek, 11/23/09
Tornadoes (avg. each year): 2
Tornado deaths (avg. each year): 0
Lightning deaths (1959-95): 23
Aug. 12, 1910: After a dry summer, lightning storm ignites forest fire in Bitterroot Mountain Range on Montana border. Several small towns are obliterated, 16 rail bridges burn, 163 die, including 78 firefighters.
Feb. 11, 1959: Sun Valley gets 38 inches of snow in 24 hours, the state's 24-hour snowfall record.

ILLINOIS
Hottest: 117°F, July 14, 1954
Coldest: -35°F, Feb. 3, 1996
Rainfall: 16.94", Aurora, 7/17-18/96
Tornadoes (avg. each year): 27
Tornado deaths (avg. each year): 4
Lightning deaths (1959-95): 86
Dec. 20, 1836: Arctic air rushes into central part of the state, dropping temperatures from 40°F to 0°F in minutes. There are tales of men frozen to saddles and chickens frozen on the spot.
Jan. 26, 1967: Chicago's greatest snowstorm dumps 23 inches in 29 hours, a single-snowstorm record. Wind gusts of 60 mph send drifts 12 feet high. Total losses: $150 million.

INDIANA
Hottest: 116°F, July 14, 1936
Coldest: -36°F, Jan. 19, 1994
Rainfall: 10.50", Princeton, 8/6/05
Tornadoes (avg. each year): 20
Tornado deaths (avg. each year): 5
Lightning deaths (1959-95): 79
Jan. 13-14, 1937: About 13,000 people flee as Ohio River floodwaters cover about 90 percent of Jeffersonville. In the Evansville area, about 90,000 people evacuate.
April 11, 1965: Worst tornadoes in state history devastate Midwest. Nearly 300 die.

IOWA
Hottest: 118°F, July 20, 1934
Coldest: -47°F, Feb. 3, 1996
Rainfall: 16.70", Decatur County, 8/5-6/59
Tornadoes (avg. each year): 32
Tornado deaths (avg. each year): 1
Lightning deaths (1959-95): 67
March 14, 1870: Severe snowstorm strikes Iowa and Minnesota. Estherville, Iowa, newspaper *North-*

ern Vindicator coins the word "blizzard."
June-Aug. 1993: Iowa, Kansas and Missouri receive year's worth of rain in four months resulting in disastrous flooding on Mississippi and Missouri rivers in The Great Flood of 1993. Des Moines water treatment plant floods, cutting water supply to 250,000 for 12 days. New record river level of 46.9 feet at St. Louis, Mo., July 21, broken 11 days later at 49.6 feet, 19.6 feet above flood stage. Total losses: $15-20 billion. 48 dead.

KANSAS
Hottest: 121°F, July 24, 1936
Coldest: -40°F, Feb. 13, 1905
Rainfall: 12.59", Burlington, 5/31-6/1/41
Tornadoes (avg. each year): 47
Tornado deaths (avg. each year): 4
Lightning deaths (1959-95): 56
May 25, 1955: Tornado destroys the small town of Udall, killing 80 people.
Sept. 3, 1970: Hailstone weighing 1.67 pounds, 17.5 inches in circumference — biggest on record — falls on Coffeyville.

KENTUCKY
Hottest: 114°F, July 28, 1930
Coldest: -34°F, Jan. 28, 1963
Rainfall: 10.40", Dunmor, 6/28/60
Tornadoes (avg. each year): 9
Tornado deaths (avg. each year): 2
Lightning deaths (1959-95): 82
March 27, 1890: Tornado outbreak throughout Ohio Valley, 78 people die in Louisville; $4 million in damages.
Jan. 13-24, 1937: Great Ohio River Flood puts nearly 70 percent of Louisville under water; 175,000 people evacuate, including all 35,000 people in Paducah.

LOUISIANA
Hottest: 114°F, Aug. 10, 1936
Coldest: -16°F, Feb. 13, 1899
Rainfall: 22.00", Hackberry, 8/28-29/62
Tornadoes (avg. each year): 25
Tornado deaths (avg. each year): 2
Lightning deaths (1959-95): 117
Feb. 1784 and Feb. 1899: Ice floes block the Mississippi River at New Orleans, only two times on record.
June 25-28, 1957: Hurricane Audrey kills 390. Gulf waters surge 25 miles inland; heavy damage to offshore oil installations.

MAINE
Hottest: 105°F, July 10, 1911
Coldest: -48°F, Jan. 19, 1925
Rainfall: 8.05", Brunswick, 9/11/54
Tornadoes (avg. each year): 2
Tornado deaths (avg. each year): 0
Lightning deaths (1959-95): 22
Oct. 3-4, 1869: Severe hurricane called "Saxby's Gale" (predicted 11 months earlier by British officer S.M. Saxby) hits Maine and Bay of Fundy. Pours more than 6 inches of rain on New England, causes greatest 6-state flood of the 19th century.
Jan. 17-18, 1979: Portland battered by snowstorm, sets single-storm and greatest 24-hour snowstorm records: 27.1 inches in two days.

MARYLAND
Hottest: 109°F, July 10, 1936
Coldest: -40°F, Jan. 13, 1912
Rainfall: 14.75", Jewell, 7/26-27/1897
Tornadoes (avg. each year): 3
Tornado deaths (avg. each year): 0
Lightning deaths (1959-95): 116

Nov. 10, 1915: Tornado hits southern Maryland. Charles County particularly hard hit; 17 die.
Aug. 23, 1933: "Chesapeake-Potomac" Hurricane moves up Chesapeake Bay with major flooding in Virginia, Maryland and the District of Columbia, does more than $17 million damage in Maryland.

MASSACHUSETTS
Hottest: 107°F, August 2, 1975
Coldest: -35°F, Jan. 12, 1981
Rainfall: 18.15", Westfield, 8/18-19/55
Tornadoes (avg. each year): 3
Tornado deaths (avg. each year): 2
Lightning deaths (1959-95): 24
Winter 1856: Thoreau calls it the "Long Snowy Winter" in Walden, uncommon cold lasts through March. Late season storm dumps 12 inches on Boston.
June 9, 1953: The Worcester County Tornado. Tornado travels 46-mile path. Winds estimated at 250 mph at Holden. Kills 94, injures 1,288, leaves 10,000 homeless.

MICHIGAN
Hottest: 112°F, July 13, 1936
Coldest: -51°F, Feb. 9, 1934
Rainfall: 9.78", Bloomingdale, 8/31-9/1/14
Tornadoes (avg. each year): 16
Tornado deaths (avg. each year): 5
Lightning deaths (1959-95): 92
Nov. 9, 1913: Freshwater Fury, disastrous windstorm over the Great Lakes. At Port Huron gale-force winds blow over 24 hours; 10 ore carriers lost, 270 die.
Nov. 11-12, 1940: Armistice Day Blizzard, cyclonic center over Michigan, and Northern Plains. 69 ships lost on Lake Michigan, 73 deaths in Michigan.

MINNESOTA
Hottest: 114°F, July 6, 1936
Coldest: -60°F, Feb. 2, 1996
Rainfall: 10.84", Fort Ripley, 7/21-22/72
Tornadoes (avg. each year): 19
Tornado deaths (avg. each year): 1
Lightning deaths (1959-95): 55
March 4, 1966: North Dakota and Minnesota ravaged by severe blizzard for over 100 hours. Wind gusts to 100 mph; snowfall 35 inches. All transportation stops for three days.
Jan. 10, 1975: Minnesota's Storm of the Century, severe blizzard with barometric low at 28.55 inches; wind chill -50°F to -80°F at Duluth; 35 storm-related deaths.

MISSISSIPPI
Hottest: 115°F, July 29, 1930
Coldest: -19°F, Jan. 30, 1966
Rainfall: 15.68", Columbus, 7/9/68
Tornadoes (avg. each year): 24
Tornado deaths (avg. each year): 8
Lightning deaths (1959-95): 90
Aug. 17-18, 1969: Hurricane Camille hits with winds 170-200 mph. Death toll reaches 143 along Gulf Coast, 27 missing. Noise from storm reaches 120 decibels, equivalent to noise from a rocket engine.
Jan. 17, 1982: Huge cold outbreak brings below-zero temperatures. Jackson has a low of -5°F.

MISSOURI
Hottest: 118°F, July 14, 1954
Coldest: -40°F, Feb. 13, 1905
Rainfall: 18.18", Edgerton, 7/20/65
Tornadoes (avg. each year): 27
Tornado deaths (avg. each year): 3

Lightning deaths (1959-95): 80
March 18, 1925: Tri-State Tornado careens through Missouri, Illinois and Indiana. Most destructive in U.S. history. Kills 689, injures 1,980. Damage: $16 million.
Sept. 12-13, 1977: Two storms pass over Kansas City; rainfall 16 inches. Flash flood conditions kill 25; 17 in automobiles.

MONTANA
Hottest: 117°F, July 5, 1937
Coldest: -69.7°F, Jan. 20, 1954
Rainfall: 11.50", Circle, 6/20/21
Tornadoes (avg. each year): 5
Tornado deaths (avg. each year): 0
Lightning deaths (1959-95): 21
Jan. 11-13, 1888: Blizzard reported at time as most disastrous ever known in Montana, the Dakotas, and Minnesota. Gale winds, drifting snow, and below-zero temperatures kill thousands of cattle.
Jan. 20, 1954: Rogers Pass, in Lewis and Clark County, records lowest temperature ever in 48 contiguous states: -69.7° F.

NEBRASKA
Hottest: 118°F, July 24, 1936
Coldest: -47°F, Feb. 12, 1899
Rainfall: 13.15", York, 7/8-9/50
Tornadoes (avg. each year): 38
Tornado deaths (avg. each year): 1
Lightning deaths (1959-95): 41
March 23, 1913: Easter Sunday Tornado cuts 5 mile path through Omaha, kills 94.
July 6, 1928: Famous Potter Hailstorm. Biggest hailstone: 1.5 pounds, 17 inches in circumference.

NEVADA
Hottest: 125°F, June 29, 1994
Coldest: -50°F, Jan. 8, 1937
Rainfall: 7.40", Lewer's Ranch, 3/18/07
Tornadoes (avg. each year): 1
Tornado deaths (avg. each year): 0
Lightning deaths (1959-95): 6
Nov. 1889 - March 1890: The White Winter. Unusually cold temperatures and large amounts of snow devastate the state's livestock industry.
Summer 1934: Height of the worst drought in Nevada history, occurred during the Great Depression. Livestock industry wiped out; banks fail. Lake Tahoe at its lowest level. Washoe Lake totally dry.

NEW HAMPSHIRE
Hottest: 106°F, July 4, 1911
Coldest: -46°F, Jan. 28, 1925
Rainfall: 10.38", Mount Washington, 2/10-11/70
Tornadoes (avg. each year): 1
Tornado deaths (avg. each year): 0
Lightning deaths (1959-95): 8
Sept. 3-9, 1821: The Windy Week. On Sept. 3, William Redfield's hurricane hits, followed by The Great New Hampshire Whirlwind, one of the most violent in New England history. 5 die.
March 22, 1936: Great floods on rivers from Maine to Ohio. In New Hampshire hundreds are homeless, mills and factories are destroyed.

NEW JERSEY
Hottest: 110°F, July 10, 1936
Coldest: -34°F, Jan. 5, 1904
Rainfall: 14.81", Tuckerton, 8/19/39
Tornadoes (avg. each year): 2
Tornado deaths (avg. each year): 0
Lightning deaths (1959-95): 55
July 5, 1900: Lightning hits Standard Oil refinery at Bayonne, start-

ing a three-day fire; kills 1, injures 3; smoke rises 13,000 feet; damage: $2 million.
Jan. 6-8, 1996: Major "noreaster" dubbed the "Blizzard of '96" paralyzes the East from Washington to Boston; 19 to 48 inches of snow falls, drifts reach 8 feet; sets 24-hour snowfall record in New Brunswick with 22.6 inches, "snowiest in a century." Severe flooding from heavy rain and rapidly melting snow hits same area 10 days later. 187 dead, $3 billion damage.

NEW MEXICO
Hottest: 122°F, June 27, 1994
Coldest: -50°F, Feb. 1, 1951
Rainfall: 11.28", Lake Maloya, 5/18-19/55
Tornadoes (avg. each year): 9
Tornado deaths (avg. each year): 0
Lightning deaths (1959-95): 82
Dec. 14-16, 1959: Record snowfall of 40 inches falls on Corona.
Dec. 12-20, 1967: Snow falls for 11 days. Snowmobiles and airlifts bring in food; isolated Native American reservations hard hit.

NEW YORK
Hottest: 108°F, July 22, 1926
Coldest: -52°F, Feb. 18, 1979
Rainfall: 11.17", New York City, Central Park, 10/9/03
Tornadoes (avg. each year): 5
Tornado deaths (avg. each year): 0
Lightning deaths (1959-95): 128
Dec. 26, 1947: Snowstorm in New York City. In 24 hours, a record 26 inches falls in Central Park; 32 inches in suburbs; 27 deaths.
Jan. 28 - Feb. 1, 1977: The Blizzard of '77, worst in the history of Buffalo. Wind gusts hit 69 mph. Wind chill was minus 50°F. This storm plus snow earlier in January brings a record 68.3 inches of snow for one month. It is part of 43 consecutive days of snowfall.

NORTH CAROLINA
Hottest: 110°F, Aug. 21, 1983
Coldest: -34°F, Jan. 21, 1985
Rainfall: 22.22", Altapass, 7/15-16/16
Tornadoes (avg. each year): 13
Tornado deaths (avg. each year): 1
Lightning deaths (1959-95): 168
Aug. 17, 1899: San Cirriaco Hurricane. 7 vessels founder on the beach; 6 others disappear. 50 deaths.
Sept. 11-13, 1984: Hurricane Diana hits with winds over 135 mph at Cape Fear; 2 die.

NORTH DAKOTA
Hottest: 121°F, July 6, 1936
Coldest: -60°F, Feb. 15, 1936
Rainfall: 8.10", Litchville, 6/29/75
Tornadoes (avg. each year): 18
Tornado deaths (avg. each year): 0
Lightning deaths (1959-95): 11
Jan. 21, 1918: Chinook winds push temperature at Granville from minus 33°F in the morning to 50°F by afternoon, a rise of 83°F.
April 8, 1952: Large ice dam on the Missouri River breaks, sending water rushing downstream. River's flow increases from 75,000 cubic feet a second to 500,000 cubic feet a second. River rises 8 feet at Bismarck between 10 a.m. and 6 p.m.

OHIO
Hottest: 113°F, July 21, 1934
Coldest: -39°F, Feb. 10, 1899
Rainfall: 10.51", Sandusky, 7/12/66
Tornadoes (avg. each year): 15
Tornado deaths (avg. each year): 3

Lightning deaths (1959-95): 124
Jan. 27, 1937: Ohio River crests at 74.1 feet at the 60-foot flood wall at Portsmouth.
April 3, 1974: Super Tornado Outbreak across 11 states. Ohio hit hardest. In all 315 fatalities — 42 in Ohio. Xenia nearly demolished.

OKLAHOMA
Hottest: 120°F, June 27, 1994
Coldest: -27°F, Jan. 18, 1930
Rainfall: 15.68", Enid, 10/10-11/73
Tornadoes (avg. each year): 53
Tornado deaths (avg. each year): 4
Lightning deaths (1959-95): 90
April 27, 1942: Tornado hits Rogers and Mayes Counties. Worst damage in Pryor. 52 die, damage: $2 million.
April 9, 1947: Southern Plains Tri-State Tornado runs 221 miles through Texas, Oklahoma and Kansas; injures 980; kills 169.

OREGON
Hottest: 119°F, Aug. 10, 1898
Coldest: -54°F, Feb. 10, 1933
Rainfall: 10.17", Glenora, 12/21/15
Tornadoes (avg. each year): 1
Tornado deaths (avg. each year): 0
Lightning deaths (1959-95): 7
Jan. 28, 1868: Cold weather freezes the Columbia River, horse-drawn sleighs and pedestrian traffic cross from Vancouver, Wash., (then a territory) to Portland.
June 14, 1903: Heppner Disaster at Willow Creek. Flash flood from torrential rainstorm sweeps away one-third of the town, 236 die.

PENNSYLVANIA
Hottest: 111°F, July 10, 1936
Coldest: -42°F, Jan. 5, 1904
Rainfall: 34.50", Smethport, 7/17/42
Tornadoes (avg. each year): 10
Tornado deaths (avg. each year): 1
Lightning deaths (1959-95): 114
May 31, 1889: Great Johnstown Flood kills more than 2,200 when the South Fork Dam breaks, about 24 miles upstream from the city, sending a wall of water rushing down the valley.
June 22, 1972: Hurricane Agnes Floods, most costly weather disaster to date, ravages East Coast, Wilkes-Barre hard hit, 122 killed, $2.1 billion in damages.

RHODE ISLAND
Hottest: 104°F, Aug. 2, 1975
Coldest: -25F, Feb. 5, 1996
Rainfall: 12.13", Westerly, 9/16-17/32
Tornadoes (avg. each year): 0
Tornado deaths (avg. each year): 0
Lightning deaths (1959-95): 4
Sept. 23, 1815: Great September Gale. Legendary hurricane hits near Saybrook, Conn., travels up New England coast, creates floods and winds of 100 mph. Floods inundate Providence. Storm wave hits Narragansett Bay; 12 deaths.
Sept. 21, 1938: New England Hurricane of '38, also called the Long Island Express, brings winds of 150 mph to New York, Connecticut, Massachusetts, New Hampshire and Canada. Rhode Island is hardest hit, especially Providence; storm surge kills 41 at Misquamicut beach alone. In all, 600 die, including 250 in Rhode Island; 60,000 left homeless.

SOUTH CAROLINA
Hottest: 111°F, June 28, 1954
Coldest: -19°F, Jan. 21, 1985
Rainfall: 13.25", Effingham, 7/14-15/16

221

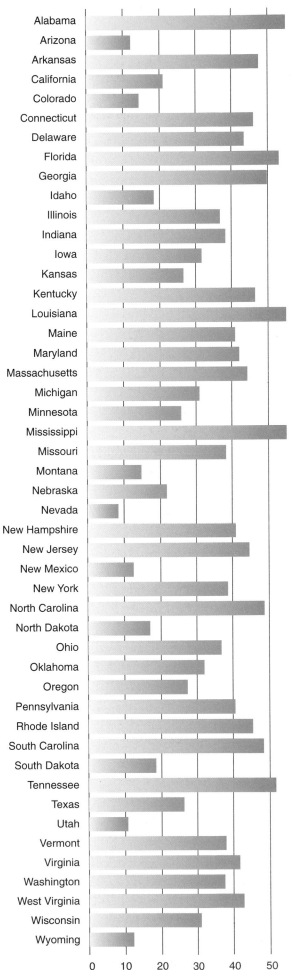

Normal annual precipitation in inches (48 states)

Alabama
Arizona
Arkansas
California
Colorado
Connecticut
Delaware
Florida
Georgia
Idaho
Illinois
Indiana
Iowa
Kansas
Kentucky
Louisiana
Maine
Maryland
Massachusetts
Michigan
Minnesota
Mississippi
Missouri
Montana
Nebraska
Nevada
New Hampshire
New Jersey
New Mexico
New York
North Carolina
North Dakota
Ohio
Oklahoma
Oregon
Pennsylvania
Rhode Island
South Carolina
South Dakota
Tennessee
Texas
Utah
Vermont
Virginia
Washington
West Virginia
Wisconsin
Wyoming

0   10   20   30   40   50

Tornadoes (avg. each year): 11
Tornado deaths (avg. each year): 1
Lightning deaths (1959-95): 82
Feb. 19, 1884: Great Tornado Outbreak of the South; 60 tornadoes, 800 killed, 2,500 injured.
Sept. 21-22, 1989: Hurricane Hugo hits near Charleston with 135 mph winds. Strongest hurricane to hit the U.S. since Camille in 1969. Kills 21 in continental U.S., 28 in Caribbean. Damage: more than $10 billion.

SOUTH DAKOTA

Hottest: 120°F, July 5, 1936
Coldest: -58°F, Feb. 17, 1936
Rainfall: 8.00", Elk Point, 9/10/1900
Tornadoes (avg. each year): 26
Tornado deaths (avg. each year): 0
Lightning deaths (1959-95): 20
April 14, 1873: Great Easter Blizzard strikes without warning. Winds gust at 40 mph for three days; 80 percent of livestock perish.
June 9-10, 1972: Torrential rain causes devastating flash floods in the Black Hills and floods Rapid City, 242 die. Damage: $163 million.

TENNESSEE

Hottest: 113°F, Aug. 9, 1930
Coldest: -32°F, Dec. 30, 1917
Rainfall: 11.00", McMinnville, 3/28/02
Tornadoes (avg. each year): 11
Tornado deaths (avg. each year): 2
Lightning deaths (1959-95): 126
Nov. 20, 1900: Tornado outbreak in Tennessee, Arkansas, and Mississippi kills 73 — 19 in Tennessee.
Feb. 2, 1951: Severe ice storm from Texas to Pennsylvania. Tennessee is hardest hit. Telephone and power out for 7 to 10 days. $100 million loss.

TEXAS

Hottest: 120°F, Aug. 12, 1936
Coldest: -23°F, Feb. 8, 1933
Rainfall: 43.00", Alvin, 7/25-26/79
Tornadoes (avg. each year): 131
Tornado deaths (avg. each year): 10
Lightning deaths (1959-95): 169
Sept. 8, 1900: Galveston Hurricane. USA's worst weather disaster. Wind estimated at more than 120 mph. Weather instruments were torn away when gusts hit 100 mph. 20-foot storm surge rips apart city. More than 7,200 die.
May 11, 1953: Tornado hits near Waco, along 23-mile path, demolishes 850 homes, 114 killed.

UTAH

Hottest: 117°F, July 5, 1985
Coldest: -69°F, Feb. 1, 1985
Rainfall: 6.00", Bug Point, 9/5/70
Tornadoes (avg. each year): 1
Tornado deaths (avg. each year): 0
Lightning deaths (1959-95): 36
Feb. 17, 1926: Second heaviest snowfall in Utah history sends avalanche of snow, rocks and timber down Sap Gulch onto Bingham; engulfs houses in the mining town; buries 75, kills 40.
May 16, 1952: Winds blowing out of canyons of the Wasatch Mountains blast into the Salt Lake City-Ogden area at more than 80 mph, doing an estimated $1 million damage.

VERMONT

Hottest: 105°F, July 4, 1911
Coldest: -50°F, Dec. 30, 1933
Rainfall: 8.77", Somerset, 11/3-4/27
Tornadoes (avg. each year): 0
Tornado deaths (avg. each year): 0
Lightning deaths (1959-95): 13
Feb. 1, 1920: Sea level barometer reading sets new high for continental USA, 31.14 inches at Northfield. Following behind is a four-day snow and sleet storm Feb. 4-7 that paralyzes the region.
Dec. 25-28, 1969: Post Christmas Snowstorm dumps 39 inches in Montpelier, sets single-storm record in Burlington with 29.8 inches.

VIRGINIA

Hottest: 110°F, July 15, 1954
Coldest: -30°F, Jan. 22, 1985
Rainfall: 27.00", Nelson County, 8/20/69
Tornadoes (avg. each year): 6
Tornado deaths (avg. each year): 0
Lightning deaths (1959-95): 53
Jan. 21, 1863: Heavy rain turns Union General Ambrose Burnside's campaign into the infamous "Mud March," after his troops lose the Battle of Fredericksburg.
Aug. 17, 1969: Hurricane Camille, after moving up the Appalachians from Mississippi, unleashes torrential rain that causes disastrous flooding of the James River. In 8 hours, over 27 inches of rain falls in Nelson County. Damage: $116 million. In Camille's path 256 die, 113 in Virginia and West Virginia.

WASHINGTON

Hottest: 118°F, Aug. 5, 1961
Coldest: -48°F, Dec. 30, 1968
Rainfall: 14.26", Mt. Mitchell, 11/23-24/86
Tornadoes (avg. each year): 1
Tornado deaths (avg. each year): 0
Lightning deaths (1959-95): 3
March 1, 1910: U.S.'s most disastrous avalanche, at Wellington Train Station, when two Great Northern trains, stalled at Stevens Pass in the Cascade Mountains, are swept over a ledge and into the canyon below. 96 die.
Feb. 1-2, 1916: Seattle's greatest 24-hour snowfall, 21.5 inches.

WEST VIRGINIA

Hottest: 112°F, July 10, 1936
Coldest: -37°F, Dec. 30, 1917
Rainfall: 19.00", Rockport, 7/18/1889
Tornadoes (avg. each year): 1
Tornado deaths (avg. each year): 0
Lightning deaths (1959-95): 24
April 28, 1928: Late season heavy snow in central Appalachians; Bayard gets 35 inches.
Feb. 26, 1972: Torrential rain causes collapse of Coal Slag Dam in Logan County, 17-mile valley floods, 118 die, $10 million in damages.

WISCONSIN

Hottest: 114°F, July 13, 1936
Coldest: -54°F, Jan. 24, 1922
Rainfall: 11.72", Mellen, 6/24/46
Tornadoes (avg. each year): 19
Tornado deaths (avg. each year): 1
Lightning deaths (1959-95): 47
Feb. 21-22, 1922: Most destructive Midwest ice storm. Ice on power lines up to 4 inches in diameter; $10 million in property losses; 1 death.
April 11-12, 1965: Palm Sunday Tornado Outbreak. 37 tornadoes rage for 9 hours over Wisconsin, Illinois, Indiana, Ohio and Michigan. 271 die, 5,000 are injured. Damage: $200 million.

WYOMING

Hottest: 114°F, July 12, 1900
Coldest: -66°F, Feb. 9, 1933
Rainfall: 6.06", Cheyenne, 8/1/85
Tornadoes (avg. each year): 10
Tornado deaths (avg. each year): 0
Lightning deaths (1959-95): 21
Oct. 29, 1917: Lowest October temperature for the contiguous 48 states recorded at Soda Butte: minus 33°F.
Aug. 1, 1985: Cheyenne Floods. In three hours a storm dumps 6.06 inches of rain on Cheyenne. This follows 10 days of rain. Some city streets are covered with 6 feet of water. 12 die.

Glossary

Absolute humidity: The mass of water vapor in a given volume of air.
Acre-foot: The amount of water needed to cover one acre under a foot of water.
Adiabatic process: A change of temperature without heat being added or taken away. In meteorology, the change in air temperature caused by pressure changes as the air rises or sinks.
Advection: Horizontal movement of any meteorological property, such as warmth or humidity.
Advection fog: Fog formed by warm, humid air flowing over cool ground or water.
Aeronomy: The science of the physics and chemistry of upper atmospheres of planets, including Earth.
Air mass: A large body of air with relatively uniform characteristics, such as temperature, humidity.
Altimeter: A special type of aneroid barometer used in airplanes to measure altitude.
Aneroid barometer: A device to measure air pressure that uses an aneroid, which is a sealed, flexible metal bellows with some air removed that expands and contracts with air-pressure changes.
Antarctic Circle: Latitude 66 degrees, 32 minutes south. Area to the south is the Antarctic.
Apparent temperature: A measure of the danger to human health of various combinations of high temperature and high humidity.
Arctic Circle: Latitude 66 degrees, 32 minutes north. Area to the north is the Arctic.
Atmosphere: The air surrounding the Earth.
Back door cold front: A cold front that moves from the northeast, instead of the more usual north-west or north, in the eastern United States.
Barograph: A device for recording air pressure.
Barometer: A device used to measure air pressure.
Beaufort Wind Scale: Scale used to classify wind speed, devised in 1805 by British Admiral Francis Beaufort to classify winds at sea.
Biosphere: The Earth's living things.
Blizzard: Snow falling with winds faster than 35 mph and visibility of one-quarter mile or less over an extended time period.
Calcium magnesium acetate (CMA): A chemical compound used to melt ice.
Chaos theory: The theory that some systems, such as weather, are ultimately unpredictable because of the effects of small-scale events that can't be included in the prediction equations.
Chlorofluorocarbons (CFCs): Man-made substances used as coolant and computer-chip cleaner, which have been shown to destroy stratospheric ozone when they break down.
Chromosphere: The thin layer of gas on the sun's surface.
Climate: Average weather over a long time period, usually 30 years.
Climate model: Mathematical model containing equations that describe climatic interactions.
Cloud seeding: The use of silver iodide, dry ice or other substances to enhance precipitation.
Cold front: A warm-cold air boundary with the cold air advancing.
Condensation: The change of a vapor to liquid.
Condensation nuclei: Small particles in the air that attract water, and encourage condensation.
Conduction: Transfer of heat within a substance or from one substance to another by molecular action.
Continental air mass: An air mass that forms over land, making it generally dry. It may be warm or cold.
Convection: Transfer of heat by the movement of the heated material. In meteorology, the up and down air motions caused by heat.
Convective storms: Storms created by rising warm air, thunderstorms.
Coriolis Effect: The apparent curving motion of anything, such as wind, caused by Earth's rotation, first described in 1835 by French scientist Gustave-Gaspard Coriolis.
Corona: Outer layer of the sun's atmosphere. Also, a circle of light around the sun or moon caused by clouds.
Cut off low: An area of upper-atmosphere low pressure that is cut off from the general west-to-east wind flow.
Cyclone: An area of low-atmospheric pressure with winds blowing around it, counterclockwise in the Northern Hemisphere, clockwise in the Southern Hemisphere.
Cryosphere: The Earth's ice.
Derecho: Wind storms created by thunderstorms during which winds blow in straight lines.
Dew: Water droplets formed by condensation of water vapor.
Dew point: Measure of humidity given in terms of temperature at which dew will start to form.
Diamond dust: Tiny ice crystals that float in the air creating pillars of light.
Doppler radar: Radar that measures speed and direction of a moving object, such as wind.
Downburst: Wind blasting down from a thunderstorm or shower.
Drizzle: Falling water drops with a diameter less than .02 inch.
Drought: Abnormal dryness for a particular region.
Dry adiabatic lapse rate: Rate at which rising air cools or sinking air warms when no water phase changes are occurring; in both cases 5.4ºF per 1,000 feet.
Dryline: A boundary between warm, dry air and warm, humid air along which thunderstorms form, often found on the southern Plains.
Electromagnetic radiation: Energy that moves in the form of disturbances in electrical and magnetic fields. Light and radio waves are examples.
El Niño: Linked ocean and atmospheric events, which have world-wide effects, characterized by warming of water in the tropical Pacific from around the International Date Line to the coast of Peru.
Equinox: Times when the sun crosses the equator. The spring or vernal equinox occurs around March 21, the autumnal equinox around Sept. 21.
Extratropical cyclone: A large-scale weather system that forms outside the tropics with a low-pressure center.
Flash flood: Flooding with a rapid water rise.
Fog: A cloud with its base on the ground.
Freezing: The phase change of water from liquid to solid.
Freezing nuclei: Small particles in the air that encourage the formation of ice.
Freezing rain: Supercooled raindrops that turn to ice when they come in contact with something.
Front: Boundary between air masses of different densities, and usually different temperatures.
Frost: Water vapor that has turned to ice on an object.
Fujita Scale: Wind damage scale created by Theodore Fujita.
Funnel cloud: A rotating column of air extending from a cloud, but not reaching the ground.
Glaze: A coat of smooth ice created when supercooled drops of water spread out before freezing.
Glory: Colored rings around an object's shadow.
GOES: Geostationary Operational Environmental Satellite, a U.S. weather satellite in an orbit that keeps it above the same place on the equator.
Graupel: Form of ice created when supercooled water droplets coat a falling ice crystal.
Greenhouse effect: Warming of a planet caused by the absorption and re-emission of infrared energy by molecules in the atmosphere.
Ground fog: A layer of fog, often less than 200 feet high, that forms when the ground cools.
Gulf Stream: A warm ocean current that flows from the Gulf of Mexico across the Atlantic to the European Coast. It helps warm Western Europe.
Gust front: Wind flowing out from a thunderstorm.
Hail: Balls of ice that grow in thunderstorm updrafts.
Halo: Any of the rings or arcs of light around the sun or moon caused by ice crystal clouds.
Heat lightning: Glowing flash in clouds. No thunder is heard because heat lightning is too far away.
High: An area of high-atmospheric pressure, also called an anticyclone.
Hurricane: A tropical cyclone with winds of 74 mph or more.
Hydrosphere: The Earth's water.
Ice Age: A geological time period during which sheets of ice cover extensive parts of the Earth.
Ice pellets: Falling drops of frozen water, also called sleet.
Intertropical Convergence Zone: The area near the equator, called "The Doldrums" by sailors, where the trade winds converge.
Inversion: Stable air condition in which air near the ground is cooler than air at a higher altitude.
Jet stream: A narrow band of upper-atmosphere wind with speeds greater than 57 mph.
Kilopascal: A metric unit of air pressure. Millibars divided by 10.

Latent heat: Energy stored when water evaporates into vapor or ice melts into liquid. It's released as heat when vapor condenses or water freezes.
Lightning: A visible discharge of electricity produced by a thunderstorm.
Little Ice Age: Period from the mid-16th century to the mid-19th century during which average global temperatures were lower than during previous and subsequent periods.
Long wave: A south-to-north wave that appears in the normal west-to-east flow of upper-atmosphere winds.
Low: An area of low-atmospheric pressure.
Maritime air mass: An air mass that forms over an ocean, making it humid. It may be warm or cold.
Meridional flow: A north to south to north flow of high-altitude winds.
Mesocyclone: A rotating, upward-moving column of air in a thunderstorm that can spawn tornadoes.
Mesoscale: In meteorology, weather systems and events up to about 250 miles across.
Meteor: Any natural phenomena in the atmosphere.
Meteorological bomb: An extratropical cyclone in which the center pressure drops an average of one millibar an hour for 24 hours. Usually refers to storms off the U.S. East Coast.
Microburst: A downburst less than 2.5 miles in diameter.
Mid latitudes: Region of the Earth outside the polar and tropical regions, between latitudes 23.5 degrees and 66.5 degrees.
Millibar: A metric unit of air pressure measurement.
Moist adiabatic lapse rate: The rate at which rising air cools or sinking air warms when water is changing phases in the air. The rate varies.
Monsoon: Persistent, widespread, seasonal winds. Usually summer winds from the ocean bring rain, while winter winds from the land are dry.
Mountain waves: Up and down air motions created as wind flows over mountains.
Mountain winds: Winds that blow either up or down mountains, caused by different rates of heating and cooling of mountaintops and valleys.
Multicell storms: Thunderstorms consisting of clusters of single-cell thunderstorms.
National Centers for Environmental Predications (NCEP): National Weather Service center in Camp Springs, Md., prepares worldwide computer forecasts.
National Climatic Data Center (NCDC): The National Oceanic and Atmospheric Administration office in Asheville, N.C., keeps climate records.
National Hurricane Center (NHC): National Weather Service office in Coral Gables, Fla., tracks and forecasts hurricanes and other weather in the Atlantic, Gulf of Mexico, Caribbean Sea, and parts of the Pacific.
National Severe Storms Laboratory (NSSL): National Oceanic and Atmospheric Administration Laboratory in Norman, Okla. Studies severe thunderstorms.
National Weather Service: Federal agency observes and forecasts weather. Formerly the U.S. Weather Bureau. It's part of the National Oceanic and Atmospheric Administration, which is part of the Department of Commerce.
Neap tides: Tides at the moon's first and third quarters, characterized by a small rise and fall.
NEXRAD: Next Generation Weather Radar system installed in the 1990s by the National Weather Service, the Defense Department and the Federal Aviation Administration.
Numerical forecasting or prediction: Use of computers to solve mathematical equations and produce weather forecasts.
Occluded front: A boundary between cool, cold, and warm air masses.
100-year-floods: Water levels that, on average, should occur once a century. This is the same as a water level with a 100 to 1 chance of occurring in any single year.
Ozone: Form of oxygen with molecules that consist of three oxygen atoms compared to two atoms for ordinary oxygen molecules.
Ozone hole: The destruction of about 40 percent of the ozone in the stratosphere over Antarctica each spring.
Perigean spring tides: Unusually high tides caused by the Earth and moon being at their nearest approach to each other at the same time the Earth, sun and moon are in a nearly straight line.
Phase changes: Changes of a substance among solid, liquid and gas forms.
Polar regions: Regions of the Earth north of 66.5 degrees north latitude around the North Pole, and south of 66.5 degrees south latitude around the South Pole.
Polar vortex: Strong, winter, upper-atmosphere winds around the polar regions.
Precipitation fog: Fog that forms when precipitation falls into cold air.
Prefrontal squall lines: Lines of thunderstorms ahead of an advancing cold front.
Pressure gradient force: Force acting on air caused by air pressure differences.
Rain: Falling water drops with a diameter greater than 0.02 inch.
Rainbow: Arc or circle of colored light caused by the refraction and reflection of light by water droplets.
Relative humidity: The ratio of the amount of water vapor actually in the air compared to the amount the air can hold at its temperature and pressure. This is expressed as a percentage.
Retrograding: Westward movement of mid latitude weather systems instead of the usual eastward movement.
Ridge: An elongated area of high-atmospheric pressure, running generally north-south, at the surface or aloft.
Rime: Ice formed when tiny drops of water freeze on contact, creating tiny balls of ice with air between them.
River Forecast Centers: National Weather Service offices responsible for flood forecasting and, in some cases, water-supply forecasting.
Saffir-Simpson Hurricane Damage Potential Scale: A 1-5 scale, developed by Robert Simpson and Herbert Saffir that measures hurricane intensity.
Saturation: Point at which the amount of water vapor in the air is greatest for the air's temperature and pressure.
Sea breeze: Winds blowing inland from any body of water.
Sea smoke: Fog that forms when cold air flows over warm water. Also called steam fog.
Severe thunderstorm: A thunderstorm with winds faster than 57 mph or hailstones three-quarters of an inch or larger in diameter.
Short wave: A bend, or wave of wind, only tens of miles long that moves along in the flow of upper-atmosphere winds.
Shower: Intermittent precipitation, either rain or snow, of short duration, which may be light or heavy.
Sleet: In the U.S., generally refers to frozen raindrops.
Solar constant: The amount of solar energy that reaches the top of the atmosphere.
Solar energy: The energy produced by the sun.
Solar flare: Explosions at the sun's surface that send energy outward.
Solitary wave: Up and down air motion that moves across the Earth as a wave.
Solstice: The two points in the Earth's orbit at which the sun is the farthest north or farthest south of the equator.
Snow: Precipitation composed of ice crystals.
Spring tides: Tides that occur at full and new moons and that generally bring the month's highest high tides and lowest low tides.
Squall line: A line of thunderstorms.
Stable air: Air in which temperature and moisture discourage formation of updrafts and downdrafts. Clouds will be low and flat. Any precipitation will be steady.
Stationary front: A warm-cold air boundary with neither cold nor warm air advancing.
Storm Prediction Center: National Weather Service center in Norman, Okla., issues watches for severe thunderstorms and tornadoes across the nation.
Storm surge: Quickly rising ocean water levels associated with hurricanes that can cause widespread flooding.
Storm tracks: Paths that storms generally follow.
Stratosphere: The layer of the atmosphere from about 7 to 30 miles up.
Sublimation: Phase changes of water directly either from vapor into ice or from ice to vapor.
Sun dogs: Splotches of light on one or both sides of the sun caused by ice crystal clouds.
Sunspots: Large cooler regions on the sun's surface that appear dark.
Supercell: A fierce thunderstorm that usually lasts several hours, often spinning out a series of strong tornadoes.
Supercooled water: Water that has cooled below 32°F without forming ice crystals.
Supersaturation: Point at which relative humidity is above 100 percent in the air.
Synoptic scale: Large-scale weather events and systems, generally more than 200 miles across.
Syzygy: The lining up of the sun, Earth, and moon in a nearly straight line.
Temperate climate: Climate that is neither polar nor tropical.
Temperate zones: The areas from roughly 30 degrees to 60 degrees latitude in both hemispheres.
Terminal Doppler Weather Radar (TDWR): Doppler radars being installed at major airports to detect microbursts.
Thunder: Sound produced by a lightning discharge.
Thunderstorms: Localized storms that produce lightning and therefore thunder.
Tides: Movements of water, the Earth's crust and the atmosphere caused by gravitational pull of the moon and sun.
TIROS: Television Infrared Observational Satellite; U.S. satellites in polar orbit about 530 miles up.
Tornado: A strong, rotating column of air extending from the base of a cumulonimbus cloud to the ground.
Trade winds: Global-scale winds in the tropics that blow generally toward the west in both hemispheres.
Transpiration: Release of water vapor into the air by plants.
Tropical cyclone: A low-pressure system in which the central core is warmer than the surrounding atmosphere.
Tropical depression: A tropical cyclone with maximum sustained winds near the surface of less than 39 mph.
Tropical storm: A tropical cyclone with 39 to 74 mph winds.
Tropics: Region of the Earth from latitude 23.5 degrees north — the Tropic of Cancer — southward across the equator to latitude 23.5 degrees south — the Tropic of Capricorn.
Tropopause: The boundary between the troposphere and the stratosphere.
Troposphere: The lower layer of the atmosphere, up to 7 or 8 miles above the Earth.
Trough: An elongated area of low-atmospheric pressure, running generally north-south, either at the surface or aloft.
Typhoon: A tropical cyclone with winds 75 mph or more in the north Pacific, west of the International Date Line.
Unstable air: Air in which temperature and moisture are favorable for the creation of updrafts and downdrafts that can create clouds that sometimes grow into thunderstorms. Precipitation will be showery.
Upslope fog: Fog that forms in humid air flowing uphill.
Vapor channel: Satellite sensor that detects invisible water vapor in the atmosphere.
Warm front: A warm-cold air boundary with the warm air advancing.
Water vapor: The invisible gaseous form of water.
Waterspout: A tornado or weaker vortex from the bottom of a cloud to the surface of a body of water.
Wind chill factor: Effect of wind blowing away the warmed air near the body.
Wind shear: Any sudden change in wind speed or direction.
Zonal flow: A west-to-east flow of winds.

Acknowledgments

Our thanks to the following people and institutions without whose cooperation this book would not have been possible:

American Airlines: Tim Miner
American Meteorological Society: Ira Geer, Richard Hallgren
American Red Cross: Susan Pyle, Norma Schoenhoeft
Atmospherics Inc.: Thomas J. Henderson
California Department of Transportation: Pat Miller, John Qualls
Chevron Chemical Company: Karl Hoenke
City College of New York: Stanley D. Gedzelman
Colorado State University: Nolan Doesken, William Gray
Federal Aviation Administration: David Sankey
Federal Highway Administration: Brian Chollar
Herbert Saffir Consulting Engineers: Herbert Saffir
Illinois Water Survey: Kenneth Kunkel
Keystone Gliderport: Tom Knauff
KHOU-TV, Houston: Neil Frank
Massachusetts Institute of Technology: Edward Lorenz, Frederick Sanders
Monroe Elementary School, Enid, Okla.: Lori Painter
National Aeronautics and Space Administration: Wayne Darnell, Edwin F. Harrison, Keith Henry, Dave McDougal, Joanne Simpson
National Center for Atmospheric Research: Ben Bernstein, Joan Frisch, Charles Knight, Margaret Lemone, John McCarthy, Marcia Politovich, Wayne Sand, Kevin Trenberth, Warren Washington
National Oceanic and Atmospheric Administration (not including National Weather Service): Ron Alberty, Daniel Albritton, Peter Black, Bill Brennan, William Brown, Randall Dole, Chris Ennis, Stanley Goldenberg, Richard Heim, David Jorgensen, JoAnn Joselyn, Thomas Karl, Neal Lott, Sam McCown, Barbara McGehan, Rich Murnan, Jack Parrish, Thomas N. Pyke, Jr., Erik Rasmussen, Tom Ross, David Rust, Susan Solomon, Hugh Willoughby, Fergus Wood
NOAA National Weather Service: Tom Adams, Gerald Bell, Jim Belville, Ray Brady, Kimberly Comba, Mark DeMaria, Randee Exler, Dan Fread, Elbert W. Friday Jr., Ronald Gird, Melody Hall, Charles H. Hoffeditz, James Hoke, Michael D. Hudlow, Brian Jarvinen, Robert H. Johns, J.T. Johnson, Eugenia Kalnay, Stephanie Kenitzer, Paul Kocin, Linda Kremkau, Frank Lapore, Miles Lawrence, Dale G. Lillie,

Bud Littin, Ronald McPherson, Frederick Ostby, Bill Read, Barry Reichenbaugh, Bob Rieck, Gody Rivera, David Rodenhuis, Chester Ropelewski, Lans Rothfusz, Joe Schaefer, Ben Scott, Rob Shedd, Robert Sheets, Robert Simpson, Cindy Smith, George Smith, John Sokich, Huug Vanden Dool, Sondra Young

National Research Council, Strategic Highway Research Program: David Minsk

Natural Hazards Research and Applications Information Center: William Riebsame

New York Department of Environmental Conservation: Michael Birmingham, Mark Levanway

Ohio Wesleyan University: Dave Hickcox

Oklahoma Climatological Survey: Kenneth Crawford

Soaring Society of America: Mark Palmer

Southern Pacific Railroad: John Signor, Bob Hoppy

State University of New York at Albany: Duncan Blanchard, Bernard Vonnegut

Tabler and Associates: Ron Tabler

University Corporation for Atmospheric Research: William Bonner, Jack Eddy

University of Chicago: T. Theodore Fujita

University of Delaware: Laurence Kalkstein

University of Illinois at Champaign-Urbana: Mohan Ramamurthy

University of Oklahoma: Howard Bluestein

University of Wisconsin at Madison: John Kutzbach

Weather Services Corp.: Wayne Barnes, Gary Best, Paul Clark, Todd Glickman, Paul Head, Michael Henry, Peter Leavitt, Bill Limmer, Ken McKinley, John Murphy, Mark Nichols, Mike Palmerino, Bob Rice, Jim Serna, Scott Yuknis

Weatherwise magazine: David Ludlum

WUSA-TV, Washington, D.C.: Doug Hill

Graphics and photography credits

viii	**H. Darr Beiser,** USA TODAY
ix	**Patty Wood**
1	**W. Balzer,** Weatherstock
2-3	**Jeff Dionise,** USA TODAY
4	The Galveston Historical Foundation and Rosenberg Library
6	**The Bettmann Archive**
7	**The Bettmann Archive**
11	**Tim Dillon,** USA TODAY
13	**Jeff Dionise,** USA TODAY
14-15	**Jeff Dionise,** USA TODAY
16	**Jeff Dionise,** USA TODAY
17	**Jeff Dionise,** USA TODAY
18	**Robert Hanashiro,** USA TODAY
19 (2)	**Jeff Dionise,** USA TODAY
22	**Jeff Dionise,** USA TODAY
23 (2)	**Jeff Dionise,** USA TODAY
24	**Jeff Dionise,** USA TODAY
26 (top)	**Bob Laird,** USA TODAY
26 (bottom)	**Jeff Dionise,** USA TODAY
27 (2)	**Jeff Dionise,** USA TODAY
28	**Ernie Leyba**
29	**Jeff Dionise,** USA TODAY
31	**Ken Biggs,** Tony Stone Worldwide
32	**Julie Stacey,** USA TODAY
33	**Elys McLean,** USA TODAY
34	**Jeff Dionise,** USA TODAY
35	**Jeff Dionise,** USA TODAY
36 (top)	**Sam Ward,** USA TODAY
36 (bottom)	**Jeff Dionise,** USA TODAY
37	**Jeff Dionise,** USA TODAY
38-39	**Jeff Dionise,** USA TODAY
40	**Jeff Dionise,** USA TODAY
41	**Stephen Lefkovits,** USA TODAY
42 (top)	**Elys McLean,** USA TODAY
42 (bottom)	**Eileen Blass,** USA TODAY
43	**Julie Stacey,** USA TODAY
44 (top)	**Jeff Dionise,** USA TODAY
44 (bottom)	**Robert Hanashiro,** USA TODAY
45	**Jeff Dionise,** USA TODAY
47	**R. Lewis,** Weatherstock
48	**Jeff Dionise,** USA TODAY
49	**Jeff Dionise,** USA TODAY
50	**Julie Stacey,** USA TODAY
51	**Stephen Lefkovits,** USA TODAY
52-53	**Jeff Dionise,** USA TODAY
54-55	**Jeff Dionise,** USA TODAY
56 (2)	**Sam Ward,** USA TODAY
57 (top)	**Jeff Dionise,** USA TODAY
57 (bottom)	**Elys McLean,** USA TODAY
58	**AP/Wide World Photos**
59	**Robert Hanashiro,** USA TODAY
60	**Jeff Dionise,** USA TODAY
62	**Jeff Dionise,** USA TODAY
63	**AP/Wide World Photos**
64 (2)	**Jeff Dionise,** USA TODAY
65	**Donald W. Elliott**
66 (right)	**Jeff Dionise,** USA TODAY
66 (left)	**Sam Ward,** USA TODAY
67 (2)	**Jeff Dionise,** USA TODAY
68	**Sam Ward,** USA TODAY
69	**Jeff Dionise,** USA TODAY
70	**Jeff Dionise,** USA TODAY
71	**Sam Ward,** USA TODAY
72	**Sam Ward,** USA TODAY
73	**Jeff Dionise,** USA TODAY
74	**Sam Ward,** USA TODAY
75 (2)	**Jeff Dionise,** USA TODAY
76-77	**Jeff Dionise,** USA TODAY
78	**Jeff Dionise,** USA TODAY
79	**Robert Hanashiro,** USA TODAY
81	**Donald W. Elliott**
83	*The Coloradoan,* Ft. Collins, Colo.
84	**Sam Ward,** USA TODAY
85	**Bob Laird,** USA TODAY
86 (left)	**Langill & Darling** from the Library of Congress Collection

86 (right)	**AP/Wide World Photos**
87	**Julie Stacey,** USA TODAY
88-89	**Jeff Dionise,** USA TODAY
90 (top)	**Dorothea Lange** from the Library of Congress Collection
90 (left)	**Barbara Ries**
90 (right)	**H. Darr Beiser,** USA TODAY
91	**Sam Ward,** USA TODAY
92	**Lee Shively,** *The Sheveport Times*
93	**Sam Ward,** USA TODAY
95	**Jeff Dionise,** USA TODAY
97	Mt. Washington Observatory
98	**Sam Ward,** USA TODAY
99	**Elys McLean,** USA TODAY
100 (top)	**Keith Carter,** USA TODAY
100 (bottom)	**Doug Nicotera**
101	**Sam Ward,** USA TODAY
102-103	**Jeff Dionise,** USA TODAY
105	**Ernie Leyba**
106	**Sam Ward,** USA TODAY
107 (2)	**Ron Coddington,** USA TODAY
109	**Jeff Dionise,** USA TODAY
110	**Tim Dunn,** *Reno Gazette-Journal*
108	**Jeff Dionise,** USA TODAY
108	**Jeff Dionise,** USA TODAY
112	**Robert Hanashiro,** USA TODAY
113	**Ron Tabler**
115	**Warren Faidley,** Weatherstock
116	**Keith Carter,** USA TODAY
117	**NOAA,** Weatherstock
118	**Sam Ward,** USA TODAY
119 (2)	**Keith Carter,** USA TODAY
120	**Robert Hanashiro,** USA TODAY
121 (top)	**Elys McLean,** USA TODAY
121 (bottom)	**Howard Bluestein**
122	**Jeff Dionise,** USA TODAY
123	**Elys McLean,** USA TODAY
124-125	**Jeff Dionise,** USA TODAY
126 (top)	**Joseph H. Golden,** NOAA
126 (bottom)	**John Zich**
127 (top)	**Julie Stacey,** USA TODAY
127 (bottom)	**Keith Carter,** USA TODAY
128 (top)	**Marty Baumann,** USA TODAY
128 (bottom)	**Rod Little,** USA TODAY
128	**Jeff Dionise,** USA TODAY
133	**Ernie Leyba**
134	**Jeff Dionise,** USA TODAY
137	**Ernie Leyba,** USA TODAY
126-127	**Jeff Dionise,** USA TODAY
132	**Keith Carter,** USA TODAY
139	**NASA**
140 (top)	**Jeff Dionise,** USA TODAY
140 (bottom)	**Keith Carter,** USA TODAY
141	**Keith Carter,** USA TODAY
142 (top)	**Keith Carter,** USA TODAY
142 (bottom)	**Bob Laird,** USA TODAY
143 (top)	**Julie Stacey,** USA TODAY
143 (bottom)	**Sam Ward,** USA TODAY
144	**Keith Carter,** USA TODAY
145	**Keith Carter,** USA TODAY
147	**Patty Wood**
148-149	**Jeff Dionise,** USA TODAY
150	**Acey Harper**
152	**Keith Carter,** USA TODAY
156 (left)	**M. Laca,** Weatherstock
156 (right)	**Eileen Blass,** USA TODAY
157	**H. Darr Beiser,** USA TODAY
158	**Keith Carter,** USA TODAY
159	**Frank Folwell,** USA TODAY
162	**Andrew Itkoff**
163	**H. Darr Beiser,** USA TODAY
164	**Keith Carter,** USA TODAY
165 (top)	**Bob Laird,** USA TODAY
165 (right)	**Suzy Parker,** USA TODAY
166 (top)	**Charles A. Doswell III**
166 (bottom)	**Keith Carter,** USA TODAY
167	(Graphic) **Keith Carter,** USA

	TODAY; (photo) **K. Brewster,** Weatherstock
168	**Marcy Eckroth Mullins,** USA TODAY
169	**Keith Carter,** USA TODAY
170-171	**Jeff Dionise,** USA TODAY
172	**Suzy Parker,** USA TODAY
173	**Robert Hanashiro,** USA TODAY
174-175	(Graphic) **Jeff Dionise,** USA TODAY; (left photo, p. 162) **Barbara Van Cleve,** Tony Stone Worldwide; (right photo, p. 162) **Ronald Holle;** (left photo, p. 163) **Ken McVey,** Tony Stone Worldwide; (right photo, p. 163) **Charles A. Doswell III**
176-177	(Graphic) **Jeff Dionise,** USA TODAY; (left photo, p. 164) **Joe Towers,** Weatherstock; (right photo, p.164) **Ronald Holle;** (photo, p.165) **David Thede,** *Weatherwise* Magazine, Heldref Publications
178	**Stephen Lefkovits,** USA TODAY
181	**Balthazar Korab**
183	**Keith Carter,** USA TODAY
185	**Jeff Dionise,** USA TODAY
186-187	**Rod Coddington,** USA TODAY
188	**Barbara Ries**
190-191	**Keith Carter,** USA TODAY
192	**Robert Hanashiro,** USA TODAY
193	**Marcy Eckroth Mullins,** USA TODAY
194 (2)	**H. Darr Beiser,** USA TODAY
196	**Michael Wyke**
197	**Joseph Sterling,** Tony Stone Worldwide
198	**J.L. Albert,** USA TODAY
199	**Ann States,** SABA
202-203	**Jeff Dionise,** USA TODAY
204	**Marty Baumann,** USA TODAY
205 (top)	**Marcia Staimer,** USA TODAY
206 (top)	**Sam Ward,** USA TODAY
206 (bottom)	**Ron Coddington,** USA TODAY
207	**Sam Ward,** USA TODAY
208	**Robert Hanashiro,** USA TODAY
210	**Julie Stacey,** USA TODAY
211	**Jeff Dionise,** USA TODAY
212	**Robert Hanashiro,** USA TODAY
219	**Jeff Dionise,** USA TODAY
220	**Jeff Dionise,** USA TODAY
221	**Jeff Dionise,** USA TODAY

BIBLIOGRAPHY

If our book has sparked an interest in learning more about weather and climate, you could learn more with these resources:

The amount of weather and climate information on the Internet's World Wide Web is increasing rapidly. The weather section of USA TODAY's Online service offers forecasts, plus stories and graphics on advances in meteorology. It also offers a comprehensive guide to university, government and other relevant web sites. Find it at http://www.usatoday.com/weather/wfront.htm

Weatherwise, a bimonthly magazine, is intended for non-scientists interested in weather. It is published by Heldref Publications, 1319 18th Street N.W., Washington, D.C. 20036.

A college-level textbook, *Meteorology Today* by C. Donald Ahrens offers more details on the topics in this book. It is intended for students who are not majoring in science.

Photographs of clouds and other atmospheric phenomena are featured in two field guides: *The Audubon Society Field Guide to North American Weather* by David Ludlum and *A Field Guide to the Atmosphere* by Vincent J. Schaefer and John A. Day.

A good book for sailors is *Weather at Sea* by David Houghton and Fred Sanders. Ruth Kirk's *Snow* covers wintry topics ranging from avalanches to polar exploration to skiing.

Twister: The Science of Tornadoes and the Making of an Adventure Movie by Keay Davidson has a little about the movie *Twister* and a lot about tornado history and science. It's by far the best introduction to the topic. No similar, up-to-date book about hurricanes exists, but *The Hurricane and Its Impact* by Robert Simpson and Herbert Riehl is a wide-ranging discussion of hurricanes and their effects on land and sea.

Anyone interested in hurricanes should have *Tropical Cyclones of the North Atlantic Ocean,* published by the National Climatic Data Center in Asheville, N.C. In addition to extensive statistical information, it has separate maps for each year since 1871 showing the tracks of all known tropical cyclones in the Atlantic, Caribbean and Gulf of Mexico. A review of each year's hurricanes in the February issue of *Weatherwise* magazine includes a map you can use to keep *Tropical Cyclones* up to date.

The National Climatic Data Center, at the Federal Building, Asheville, N.C. 28801, also publishes other books, brochures and CDs on climate, including data for individual cities.

Some understanding of atmospheric chemistry is needed to follow debates about air pollution, ozone depletion and climate change.

Atmosphere, Climate, and Change by Thomas E. Graedel and Paul J. Crutzen is an introduction to the topic that doesn't require previous knowledge of chemistry, but it is not a book that can be skimmed.

Rainbows, Halos, and Glories by Robert Greenler is a good, detailed introduction to the effects of water and ice on light in the atmosphere.

All About Lightning by Martin A. Uman is a thorough, non-technical introduction to lightning by one of the world's leading lightning experts.

The long chapter "Whither the Weather" in John L. Casti's *Searching for Certainty: What Scientists Can Know About the Future* is a good way to learn more about numerical forecasting and climate modeling. *The Essence of Chaos* by Edward Lorenz, who "discovered" chaos, is based on a series of lectures for the general public.

The Coevolution of Climate & Life by Stephen H. Schneider and Randi Londer is a detailed introduction to climate studies.

John J. Nance's *What Goes Up: the Global Assault on Our Atmosphere* reads almost like a novel as it tells how scientists discovered the way CFCs are destroying ozone.

Two books by Craig F. Bohren, *Clouds in a Glass of Beer* and *What Light Through Yonder Window Breaks,* show how a scientist goes about answering questions about ordinary things, including atmospheric phenomena.

Anyone comfortable with calculus and college physics will find two books useful: *Atmospheric Science, an Introductory Survey* by John Wallace and Peter V. Hobbs is a college-level textbook for science majors. *An Introduction to Three-Dimensional Climate Modeling* by Warren M. Washington and Claire L. Parkinson covers the basics of modeling climate.

Any time a weather-related disaster hits the U.S., the National Oceanic and Atmospheric Administration organizes a survey team to study what happened, how the event was forecast, and how the public and emergency officials responded. Survey team reports are an essential starting point for anyone looking for information on a disaster.

Information about the reports is available from the National Weather Service Public Affairs office at 1325 East-West Highway, Silver Spring, Md. 20910. This office also supplies other information about NWS.

The American Meteorological Society, 1701 K Street N.W., Suite 300, Washington, D.C. 20006, has an extensive educational program to help elementary, middle school and high school teachers use weather to teach science. The American Meteorological Society also publishes several scientific journals and reports of scientific conferences. Its *Glossary of Weather and Climate* is an up-to-date weather "dictionary."